Elements of the Mathematical Theory of Multi-Frequency Oscillations

Mathematics and Its Applications (*Soviet Series*)

Volume 71

Elements of
the Mathematical Theory
of Multi-Frequency
Oscillations

by

A. M. Samoilenko

SPRINGER SCIENCE+BUSINESS MEDIA, B.V.

Library of Congress Cataloging-in-Publication Data

Samoĭlenko, A. M. (Anatoliĭ Mikhaĭlovich)
    [Élementy matematicheskoĭ teorii mnogochastotnykh kolebaniĭ.
  English]
    Elements of the mathematical theory of multi-frequency
  oscillations / by A.M. Samoĭlenko.
        p.   cm. -- (Mathematics and its applications (Soviet series) ;
  v. 71)
    Translation of: Élementy matematicheskoĭ teorii mnogochastotnykh
  kolebaniĭ.
    Includes index.
    ISBN 978-94-010-5557-4   ISBN 978-94-011-3520-7 (eBook)
    DOI 10.1007/978-94-011-3520-7
    1. Nonlinear oscillations.   I. Title.   II. Series: Mathematics
  and its applications (Kluwer Academic Publishers).   Soviet series ;
  71.
  QA867.5.S2613   1991
  515'.352--dc20   91-29110

ISBN 978-94-010-5557-4

Printed on acid-free paper

Translated from the Russian by Yuri Chapovsky

This is the translation of the original work
Mathematical Theory Elements of Multi-Frequency Oscillations
Published by Nauka Publishers, Moscow, © 1987

SERIES EDITOR'S PREFACE

'Et moi, ..., si j'avait su comment en revenir,
je n'y serais point allé.'

Jules Verne

The series is divergent; therefore we may be
able to do something with it.

O. Heaviside

One service mathematics has rendered the
human race. It has put common sense back
where it belongs, on the topmost shelf next
to the dusty canister labelled 'discarded non-
sense'.

Eric T. Bell

Mathematics is a tool for thought. A highly necessary tool in a world where both feedback and non-linearities abound. Similarly, all kinds of parts of mathematics serve as tools for other parts and for other sciences.

Applying a simple rewriting rule to the quote on the right above one finds such statements as: 'One service topology has rendered mathematical physics ...'; 'One service logic has rendered computer science ...'; 'One service category theory has rendered mathematics ...'. All arguably true. And all statements obtainable this way form part of the raison d'être of this series.

This series, *Mathematics and Its Applications*, started in 1977. Now that over one hundred volumes have appeared it seems opportune to reexamine its scope. At the time I wrote

"Growing specialization and diversification have brought a host of monographs and textbooks on increasingly specialized topics. However, the 'tree' of knowledge of mathematics and related fields does not grow only by putting forth new branches. It also happens, quite often in fact, that branches which were thought to be completely disparate are suddenly seen to be related. Further, the kind and level of sophistication of mathematics applied in various sciences has changed drastically in recent years: measure theory is used (non-trivially) in regional and theoretical economics; algebraic geometry interacts with physics; the Minkowsky lemma, coding theory and the structure of water meet one another in packing and covering theory; quantum fields, crystal defects and mathematical programming profit from homotopy theory; Lie algebras are relevant to filtering; and prediction and electrical engineering can use Stein spaces. And in addition to this there are such new emerging subdisciplines as 'experimental mathematics', 'CFD', 'completely integrable systems', 'chaos, synergetics and large-scale order', which are almost impossible to fit into the existing classification schemes. They draw upon widely different sections of mathematics."

By and large, all this still applies today. It is still true that at first sight mathematics seems rather fragmented and that to find, see, and exploit the deeper underlying interrelations more effort is needed and so are books that can help mathematicians and scientists do so. Accordingly MIA will continue to try to make such books available.

If anything, the description I gave in 1977 is now an understatement. To the examples of interaction areas one should add string theory where Riemann surfaces, algebraic geometry, modular functions, knots, quantum field theory, Kac-Moody algebras, monstrous moonshine (and more) all come together. And to the examples of things which can be usefully applied let me add the topic 'finite geometry'; a combination of words which sounds like it might not even exist, let alone be applicable. And yet it is being applied: to statistics via designs, to radar/sonar detection arrays (via finite projective planes), and to bus connections of VLSI chips (via difference sets). There seems to be no part of (so-called pure) mathematics that is not in immediate danger of being applied. And, accordingly, the applied mathematician needs to be aware of much more. Besides analysis and numerics, the traditional workhorses, he may need all kinds of combinatorics, algebra, probability, and so on.

In addition, the applied scientist needs to cope increasingly with the nonlinear world and the extra mathematical sophistication that this requires. For that is where the rewards are. Linear models are honest and a bit sad and depressing: proportional efforts and results. It is in the non-linear world that infinitesimal inputs may result in macroscopic outputs (or vice versa). To appreciate what I am hinting at: if electronics were linear we would have no fun with transistors and computers; we would have no TV; in fact you would not be reading these lines.

There is also no safety in ignoring such outlandish things as nonstandard analysis, superspace and anticommuting integration, p-adic and ultrametric space. All three have applications in both electrical engineering and physics. Once, complex numbers were equally outlandish, but they frequently proved the shortest path between 'real' results. Similarly, the first two topics named have already provided a number of 'wormhole' paths. There is no telling where all this is leading - fortunately.

Thus the original scope of the series, which for various (sound) reasons now comprises five subseries: white (Japan), yellow (China), red (USSR), blue (Eastern Europe), and green (everything else), still applies. It has been enlarged a bit to include books treating of the tools from one subdiscipline which are used in others. Thus the series still aims at books dealing with:

- a central concept which plays an important role in several different mathematical and/or scientific specialization areas;
- new applications of the results and ideas from one area of scientific endeavour into another;
- influences which the results, problems and concepts of one field of enquiry have, and have had, on the development of another.

The world is full of periodic and quasi-periodic motions and oscillations. That is multifrequency oscillations. Mathematically, to a large extent, that means the study of invariant tori of dynamical systems. These constitute the more systematic part of the phase space of a dynamical system as opposed to the chaotic parts and their study is just as (perhaps more) important as the study of chaos.

This monograph constitutes a thorough and up-to-date treatment of invariant tori, including existence, perturbation and stability results. Many of these results involve original work by the author himself. These chapters are preceded by ones on periodic and quasi-periodic functions and on stability theory, thus making the volume selfcontained.

The shortest path between two truths in the real domain passes through the complex domain.

J. Hadamard

La physique ne nous donne pas seulement l'occasion de résoudre des problèmes ... elle nous fait pressentir la solution.

H. Poincaré

Never lend books, for no one ever returns them; the only books I have in my library are books that other folk have lent me.

Anatole France

The function of an expert is not to be more right than other people, but to be wrong for more sophisticated reasons.

David Butler

Amsterdam, August 1991   Michiel Hazewinkel

Contents

Preface

It can be seen from the title that we shall be dealing here with a mathematical theory based on precise notions and definitions. The central object of this theory is an invariant toroidal manifold of a dynamical system considered in a Euclidean space E^n or in its product with an n-dimensional torus, that is, in the space $E^n \times T_m$. Taking into account the fact that a quasi-periodic solution of a dynamical system in E^n "sweeps out" an invariant torus of this sytem, the relation between the theory of invariant tori and the theory of multi-frequency oscillations becomes clear; the existence of such tori is a necessary condition for the existence of multi-frequency oscillations of quasi-periodic solutions of dynamical systems. By defining multi-frequency oscillations as a motion of a dynamical system, describing a recurrent trajectory on an invariant toroidal manifold of the system, we make the invariant toroidal manifold into the main subject of the mathematical theory of multi-frequency oscillations; the existence of such a manifold is sufficient for multi-frequency oscillations of the system to exist.

The original part of the monograph — the last two chapters, is devoted to the following topics: the existence of invariant toroidal manifolds for linear systems in $E^n \times T_m$, perturbation theory of such manifolds for non-linear systems, smoothness and stability properties of these manifolds, the behaviour of trajectories in a small neighbourhood, the study of separatrix manifolds in the case of exponential dichotomy of toroidal manifolds, linear system block decomposition under exponential dichotomy, and substantiation of Galerkin's procedure for finding toroidal manifolds.

The first two chapters of the book contain a theory of quasi-periodic functions, expounded on the basis of the theory of periodic functions of several variables, and some results from the theory of invariant sets and their stability. The material expounded in these chapters is closely related to the theory of multi-frequency oscillations and will be useful to specialists who are not well acquainted with the above-mentioned topics.

Questions concerning asymptotic integration of multi-frequency oscillation systems are excluded from the book. They are considered in a separate work.

The results given in the last two chapters were obtained mainly by the author. The results obtained in conjunction with V.L. Kulik and I.O. Parasyuk form an exception. The content of the book was partially covered in special courses given to graduate students of the department of Mathematics and Mechanics of Kiev State University. The courses have been given regularly since 1974.

The book is designed for specialists working in the area of multi-frequency oscillation analysis, and can be useful to many researchers, students and graduate students who in their work meet oscillatory phenomena.

A.M. Samoilenko.

Introduction

Originally, oscillation theory arose within celestial mechanics. A mathematical idealization of processes and phenomena studied in oscillation theory reduced the problem of describing a given physical system to the study of a linear differential equation, and the oscillation itself to a solution of this equation of "oscillatory" type. The principle of superposition reflected the main regularity of the linear theory and the amplitude-frequency characteristic was the main property of oscillations [54], [155].

The works by Poincaré and Lyapunov [66], [69–71], [101] devoted to the theory of periodic solutions of weakly non-linear systems have given researchers a mathematical tool for studying oscillation processes of "non-linear" nature, which are difficult to interpret in the linear theory. The needs of electrical and electronic engineering contributed to a fast development of this "new direction" and were instrumental in its separation as an independent theory in non-linear oscillation theory in mathematical physics. A fundamental contribution to the solution of problems of the non-linear theory was made by N.M. Krylov and N.N. Bogolyubov [19], [21], [23], [55–57].

Periodic solutions of non-linear systems of differential equations have turned out to be a central object of study. For studying these, a great number of methods have been developed, from numerical to topological; thus periodic solutions have been studied from all sides [44], [51], [52], [66], [71], [80], [87], [92], [97], [99], [122], [133], [136], [137], [140], [141], [144]. Many real processes whose mathematical idealization allow one to regard them as "repeating themselves exactly" in the same periods of time, have been well interpreted within oscillation theory [2], [4], [23], [31], [37], [38], [55–57], [71], [73], [76], [80], [92], [97], [99], [136–138].

Starting in the sixties, a sharp change took place in oscillation theory towards a study of oscillating processes characterized by their "almost exact" repetition in "almost the same" periods of time. Research in this theory has focused on quasi-periodic oscillations, quasi-periodic solutions of the corresponding equations; and the creation of KAM-theory, which is the theory of quasi-periodic solutions of "almost integrable" Hamiltonian systems [6–8], [48], [49], [89–91], [156–160] has been the greatest achievement. Quasi-periodic solutions of "weakly non-linear" systems of general type have been studied by N.N. Bogolyubov [20], [21], [24]. Many problems concerning the analysis of systems which describe quasi-periodic oscillations were solved in papers by Yu.A. Mitropol'skii [24], [77–79], [82], [83]. The papers [103–105] by the author are closed related to the latter.

Quasi-periodic oscillations represent a rather complicated and "sensitive" object of study. The practical effectiveness of finding such oscillations in an individual dynamical system is negligible, for the reason that the frequency basis is "unstable". This shows up well in examples and can be explained theoretically using deep results from the qualitative theory of differential equations: a quasi-periodic oscillation can be easily "destroyed" and turned into a periodic oscillation by a small change in the right-hand side of the system. This fact forces one to look for a "coarser" object of study in the theory of multi-frequency oscillation than a quasi-periodic solution.

In 1967 V.V. Nemytskii in his lectures at the Mathematical School in Ujgorod, analyzed in detail possible answers to the following question: what motion of a dynamical system in E^n should be considered as "oscillatory", and concluded that it is recurrent motion [93]. With such a definition, a minimal set "swept" by the trajectories of a recurrent motion becomes the main object of study in the theory of auto-oscillation.

Until now, among the minimal sets, those "swept" by the trajectories of quasi-periodic motions have been studied best. Following A.A. Andronov and A.A. Witt [2], [3], we can assert that already Poincaré knew that such a set is a torus. Regardless of the fact that the invariant tori of a dynamical sytem in E^n are "carriers" of auto-oscillations of a non-linear system, until recently they were studied only occasionally. Thus, the first deep statements about invariant toroidal manifolds of systems of non-linear mechanics, that is, integral manifolds of toroidal type, were obtained by N.N. Bogolyubov and N.M. Krylov [19], [23], [57], while substantiating the asymptotic methods of non-linear mechanics. Only later were the ideas suggested in these works fully developed by Yu.A. Mitropol'skii, giving rise to the method of integral manifolds used in non-linear mechanics [22], [81]. This is a powerful and practically convenient mathemati-

cal device for studying integral manifolds of a wide class of non-linear systems
of differential equations. They influenced the nature of further developments in
the perturbation theory of toroidal manifolds, and led S.P. Diliberto [145–147],
T. Hale [138], [151], Y. Kurtsveil [61], J. Kyner [153], I. Kupka [154] and others
[11–13], [45], [95] to deep results.

The work of K.O. Friedrichs [149], [150] started a new area in this theory.
His theory of periodic solutions of positive symmetric systems of linear partial differential equations, worked out using methods of functional analysis as
applied in mathematical physics, but used for linear extensions of dynamical
systems on a torus, is an original and distinctive theory of existence of invariant
tori of these systems.

In the sixties, J. Moser developed his rapidly convergent iteration method
[89], [156–159], and applied it, in particular, to non-linear generalizations of
positive symmetric systems. This led to new results on the preservation of
an invariant torus under perturbations. These results were essentially different
from those obtained by the method of integral manifolds. The assumption that
the system of equations considered has a high order of smoothness, needed in
order to apply the method of Moser, were significantly weakened by R. Sacker
[163]. Subsequent studies by J. Moser and R. Sacker [90], [91], [160], [164] have
finished the creation of the perturbation theory of invariant tori of dynamical
systems.

The report given at the Fifth International Conference on non-linear oscillations [107] started a new series of papers on problems of the theory of
perturbation and stability of invariant toroidal manifolds of dynamical systems. The introduction of the notion of a Green's function in the problem of
existence of an invariant torus of a linear extension of a dynamical system on
a torus turned out to be very fruitful and encouraged development in different
areas of this theory. It seemed a return to the ideas of the method of integral
manifolds in the theory of perturbation and stability of invariant tori of systems of non-linear mechanics, and led to new results in this theory [58], [59],
[86], [106], [109], [111–117], [119], [123], [124].

Other work containing deep results on invariant tori of dynamical systems
includes the monographs of M. Hirsch, C. Pugh, M. Shub [152], N.I. Neimark
[92], and D.U. Umbetjanov [132], and papers of Yu.N. Bibikov and V.A. Pliss
[15], [16], I.V. Bronshtein and V.P. Burdijaev [29–30], N. Fenichel [148] and

G. Sell [165–168].

Some aspects of the theory given in this book were developed and applied to different classes of equations [74], [75], [84], [85], [88], [96], [120], [125], [126]. Problems relating to the topic were also considered in [32], [33], [40], [41], [45], [53], [121], [130], [131], [133], [134], [135], [170], [171].

Chapter 1. Periodic and quasi-periodic functions

1.1. The function spaces $C^r(\mathcal{T}_m)$ and $H^r(\mathcal{T}_m)$

Let $f(\phi) = f_1(\phi), \ldots, f_n(\phi))$ be an $\mathbf{R}^n$-valued function of a variable $\phi = (\phi_1, \ldots \ldots, \phi_m)$ that is continuous and periodic with period 2π with respect to each variable ϕ_α $(\alpha = 1, \ldots, m)$. The set of all such functions forms a linear space which we denote by $C(\mathcal{T}_m)$. This space can be turned into a complete normed space by introducing the norm

$$|f|_0 = \max_{\phi \in \mathcal{T}_m} \|f(\phi)\|,$$

where

$$\|f\|^2 = \sum_{i=1}^n |f_i|^2$$

is the Euclidean norm in the space $\mathbf{R}^n$ of the function $f(\phi)$.

In $C(\mathcal{T}_m)$ we distinguish a subspace $C^r(\mathcal{T}_m)$ of functions having continuous partial derivatives up to order r inclusive with respect to the variables ϕ_α $(\alpha = 1, \ldots, m)$. The set $C^r(\mathcal{T}_m)$ can be turned into a complete normed space by introducing the norm

$$|f(\phi)|_r = \max_{0 \leq \rho \leq r} |D^\rho f(\phi)|_0,$$

where D^ρ is the ρth partial derivative with respect to ϕ_α $(\alpha = 1, \ldots, m)$.

Let $P(\phi)$ be a trigonometric polynomial in $C(\mathcal{T}_m)$, that is, $P(\phi)$ is a finite sum of the form

$$P(\phi) = \sum_{\|k\| \leq N} a_k \sin(k, \phi) + b_k \cos(k, \phi) = \sum_{\|k\| \leq N} P_k e^{i(k, \phi)},$$

where $k = (k_1, \ldots, k_m)$ is a vector with integer components, $(k, \phi) = k_1 \phi_1 + \ldots + k_m \phi_m$, a_k and b_k are real and P_k complex coefficients, that is,

$$a_k = -\operatorname{Im} P_k, \quad b_k = \operatorname{Re} P_k, \quad P_{-k} = \overline{P}_k,$$

1

($\overline{P}_k$ is the complex conjugate of P_k), N is an arbitrary non-negative integer. The space of all such polynomials forms a linear space which will be denoted by $\mathcal{P}(\mathcal{T}_m)$.

According to Weierstrass's theorem [10], a function $f(\phi) \in C(\mathcal{T}_m)$ can be uniformly approximated by trigonometric polynomials. This means that for any function $f(\phi) \in C(\mathcal{T}_m)$ there is a sequence of trigonometric polynomials $P_\nu(\phi)$ ($\nu = 1, 2, \ldots$) such that

$$\lim_{\nu \to \infty} |f(\phi) - P_\nu(\phi)|_0 = 0.$$

It follows that the space $C(\mathcal{T}_m)$ is the closure of the space $\mathcal{P}(\mathcal{T}_m)$ of trigonometric polynomials with respect to the norm $|\cdot|_0$. A similar conclusion can also be made for the spaces $C^r(\mathcal{T}_m)$, namely, every such space is the closure of the space of trigonometric polynomials $\mathcal{P}(\mathcal{T}_m)$ with respect to this norm.

Thus, using $\mathcal{P}(\mathcal{T}_m)$ we can form a chain of Banach spaces embedded in each other:

$$C(\mathcal{T}_m) = C^0(\mathcal{T}_m) \supset C^1(\mathcal{T}_m) \supset \ldots \supset C^r(\mathcal{T}_m) \supset C^{r+1}(\mathcal{T}_m) \supset \ldots \supset C^\infty(\mathcal{T}_m),$$
$$(1.1)$$

where $C^\infty(\mathcal{T}_m) = \bigcap_{r=0}^{\infty} C^r(\mathcal{T}_m)$.

Now we can define an inner product $(\cdot, \cdot)_0$ for any two trigonometric polynomials from $\mathcal{P}(\mathcal{T}_m)$

$$P = \sum_{\|k\| \leq N} P_k e^{i(k,\phi)}, \quad Q = \sum_{\|k\| \leq N} Q_k e^{i(k,\phi)} \qquad (1.2)$$

by setting

$$(P,Q)_0 = \frac{1}{(2\pi)^m} \int_0^{2\pi} \ldots \int_0^{2\pi} \langle P,Q \rangle d\phi_1 \ldots d\phi_m = \sum_k \langle P_k, Q_{-k} \rangle. \qquad (1.3)$$

Here $\langle P_k, Q_{-k} \rangle = \sum_{j=1}^n P_k^j Q_{-k}^j$ is the usual inner product of the vectors $P_k = (P_k^1, P_k^2, \ldots, P_k^n)$ and $Q_{-k} = (Q_{-k}^1, Q_{-k}^2, \ldots, Q_{-k}^n)$.

The inner product (1.3) defines a norm $\|\cdot\|_0$ in the space $\mathcal{P}(\mathcal{T}_m)$:

$$\|P\|_0^2 = (P,P)_0 = \frac{1}{(2\pi)^m} \int_0^{2\pi} \ldots \int_0^{2\pi} \|P\|^2 d\phi_1 \ldots d\phi_m = \sum_{\|k\| \leq N} \|P_k\|^2.$$

Taking the closure of $\mathcal{P}(\mathcal{T}_m)$ with respect to this norm we obtain a Hilbert space denoted by $H^0(\mathcal{T}_m)$. The elements of the space $H^0(\mathcal{T}_m)$ are given by series $\sum_k f_k e^{i(k,\phi)}$ for which the sum $\sum_k \|f_k\|^2$ is finite. Here $\sum_k$ denotes $\sum_{k \in \mathbf{Z}^m}$, where $\mathbf{Z}^m$ is the set of all integer-valued vectors $k = (k_1, \ldots, k_m)$.

For the polynomials (1.2) and any non-negative integer r, one can define an inner product $(\cdot, \cdot)_r$ by setting, for example, as in [14], [89]

$$(P, Q) = ((1 - \Delta)^r P, Q)_0 = \sum_{\|k\| \le N} (1 + \|k\|^2)^r \langle P_k, Q_{-k} \rangle, \qquad (1.4)$$

where $\Delta = \sum_{\nu=1}^m \frac{\partial^2}{\partial \phi_\nu^2}$ is the Laplace operator.

The inner product (1.4) induces in $\mathcal{P}(\mathcal{T}_m)$ a norm $\| \cdot \|_r$:

$$\|P\|_r^2 = (P, P)_r = \sum_{\|k\| \le N} (1 + \|k\|^2)^r \|P_k\|^2.$$

Taking the closure of the space $\mathcal{P}(\mathcal{T}_m)$ with respect to this norm we obtain a Hilbert space which we shall denote by $H^r(\mathcal{T}_m)$. The elements of the space $H^r(\mathcal{T}_m)$ are series $\sum_k f_k e^{i(k,\phi)}$ for which the sum $\sum_k (1 + \|k\|^2)^r \|f_k\|^2$ is finite.

Thus, starting with $\mathcal{P}(\mathcal{T}_m)$ we can form together with the chain (1.1) the following chain of Hilbert spaces:

$$H(\mathcal{T}_m) = H^0(\mathcal{T}_m) \supset H^1(\mathcal{T}_m) \supset \ldots \supset H^r(\mathcal{T}_m) \supset H^{r+1}(\mathcal{T}_m) \ldots H^\infty(\mathcal{T}_m),$$

where $H^\infty(\mathcal{T}_m) = \bigcap_{r=0}^\infty H^r(\mathcal{T}_m)$.

We now find out what the structure of the spaces $H^r(\mathcal{T}_m)$ is.

1.2. Structure of the spaces $H^r(\mathcal{T}_m)$. Sobolev theorems

According to the Riesz-Fisher theorem [50], each series $\sum_k f_k e^{i(k,\phi)}$ for which the sum $\sum \|f_k\|^2$ is finite is a Fourier series of a function $f(\phi)$ which is square summable on the cube $\mathcal{T}_m : 0 < \phi_\alpha \le 2\pi$ ($\alpha = 1, \ldots, m$). From this it follows that the space $H(\mathcal{T}_m)$ can be identified with the space of functions periodic in each variable ϕ_α ($\alpha = 1, \ldots, m$) with period 2π, and which are square summable on the cube $\mathcal{T}_m$ of the periods. In other words, $H(\mathcal{T}_m)$ can be identified with the space $\mathcal{L}_2(\mathcal{T}_m)$.

Now let the series $\sum_k f_k e^{i(k,\phi)}$ define an element of $H^r(\mathcal{T}_m)$. For any integer-valued vector $\rho = (\rho_1, \ldots, \rho_m)$, $\rho_\alpha \ge 0$ ($\alpha = 1, \ldots, m$) and $|\rho| = \rho_1 + \rho_2 + \ldots + \rho_m \le r$, the series

$$\sum_k (ik)^\rho f_k e^{i(k,\phi)} = \sum_k i^{|\rho|} k_1^{\rho_1} k_2^{\rho_2} \ldots k_m^{\rho_m} f_k e^{i(k,\phi)}$$

represents an element $f^{(\rho)}$ of $H^{r-|\rho|}(T_m)$. This can be seen from the following estimate:

$$\|f^{(\rho)}\|_{r-|\rho|}^2 = \sum_k (1+\|k\|^2)^{r-|\rho|}\|(ik)^\rho f_k\|^2 =$$

$$= \sum_k (1+\|k\|^2)^{r-|\rho|}(k)^{2\rho}\|f_k\|^2 \le \sum_k (1+\|k\|^2)^r\|f_k\|^2 = \|f\|_r^2$$

We consider the inner product $(f^{(\rho)}, P)_0$ for an arbitrary polynomial $P \in \mathcal{P}(T_m)$. We have

$$(f^{(\rho)}, P)_0 = \sum_{\|k\|\le N} (ik)^\rho \langle f_k, P_{-k}\rangle =$$

$$= (-1)^\rho \sum_{\|k\|\le N} \langle f_k, (-ik)^\rho P_{-k}\rangle = (-1)^{|\rho|}(f, D^\rho P)_0, \qquad (2.1)$$

where $D^\rho = \partial^{|\rho|}/(\partial\phi_1^{\rho_1}\ldots\partial\phi_m^{\rho_m})$ is the $|\rho|$th derivative of the function P with respect to ϕ. We say that a function $f \in \mathcal{L}_2(T_m)$ has the generalized derivative D^ρ if there is a function $f^{(\rho)}$ in $\mathcal{L}_2(T_m)$ satisfying the equality

$$(f^{(\rho)}, P)_0 = (-1)^{|\rho|}(f, D^\rho P)_0 \qquad (2.2)$$

for any polynomial $P \in \mathcal{P}(T_m)$.

Equality (2.1) shows that an element f of $H^r(T_m)$ can be identified with a function that belongs to $\mathcal{L}(T_m)$ and has all generalized derivatives with respect to ϕ up to the order r inclusive. The generalized derivative $D^\rho f = f^{(\rho)}$ has the following Fourier series

$$\sum_k (ik)^\rho f_k e^{i(k,\phi)},$$

where

$$f_k = \frac{1}{(2\pi)^m}\int_0^{2\pi}\ldots\int_0^{2\pi} f(\phi)e^{-i(k,\phi)}d\phi_1\ldots d\phi_m.$$

Indeed, if $\sum_k f_k^{(\rho)}e^{i(k,\phi)}$ is the Fourier series of the function $f^{(\rho)} = D^{(\rho)}f$, then according to equality (2.2) we have

$$\sum_{\|k\|\le N} \langle f_k^{(\rho)} - (ik)^\rho f_k, P_{-k}\rangle = 0$$

for any trigonometric polynomial $P = \sum_{\|k\|\le N} P_k e^{i(k,\phi)} \in \mathcal{P}(T_m)$. By taking $P = \sum_{\|k\|\le N}[f_k^{(\rho)} - (ik)^\rho f_k]e^{-i(k,\phi)}$ we see that $f_k^{(\rho)} = (ik)^\rho f_k$ for any integer

k. Since the series $\sum_k (ik)^\rho f_k e^{i(k,\phi)}$ is the Fourier series of the function $f^{(\rho)}$ belonging to $\mathcal{L}_2(\mathcal{T}_m)$, it satisfies Parseval's equality [67]:

$$\sum_k \|(ik)^\rho f_k\|^2 = \sum_k (k)^{2\rho}\|f_k\|^2 = \|f^{(\rho)}\|_0^2 = \sum_k \|f_k^{(\rho)}\|^2.$$

This leads to the following estimate:

$$\sum_k (1+\|k\|^2)^r\|f_k\|^2 \leq c \sum_{|\rho|\leq r,k} (k)^{2\rho}\|f_k\|^2 = c \sum_{|\rho|\leq r,k} \|f_k^{(\rho)}\|^2 = c \sum_{|\rho|\leq r} \|D^\rho f\|_0^2,$$

which shows that the function f generates an element of the space $H^r(\mathcal{T}_m)$. It shows that the space $H^r(\mathcal{T}_m)$ can be identified with the space of functions periodic with respect to the variables ϕ_α ($\alpha = 1, \ldots, m$) with period 2π that are square summable on the cube of periods $\mathcal{T}_m$ and have generalized derivatives with respect to ϕ up to order r inclusive.

The properties given by the Sobolev theorems [127] form an important characterization of the spaces $H^r(\mathcal{T}_m)$.

Theorem 1. *For $r > m/2 + p$ the inclusion $H^r(\mathcal{T}_m) \subset C^p(\mathcal{T}_m)$ and the estimate*

$$|D^\rho f|_0 \leq c\|f\|_r$$

hold, where $|\rho| \leq p$, c is a positive constant that does not depend on f and f is an arbitrary function belonging to $H^r(\mathcal{T}_m)$.

A proof of Theorem 1 follows from the following sequence of inequalities

$$\sum_k \|(ik)^\rho f_k e^{i(k,\phi)}\| \leq c \sum_k \|k\|^p \|f_k\| \leq$$

$$\leq c \sum_k (1+\|k\|^2)^{(r-p)/2+p/2}\|f_k\|(1+\|k\|^2)^{(p-r)/2} \leq$$

$$\leq c\left(\sum_k (1+\|k\|^2)^r\|f_k\|^2\right)^{1/2}\left(\sum_k (1+\|k\|^2)^{p-r}\right)^{1/2} = c_1\|f\|_r, \qquad (2.3)$$

where, using the condition $r > m/2 + p$, we obtain the following estimate for the constant:

$$c_1^2 = c^2 \sum_k (1+\|k\|^2)^{p-r} \leq c' \sum_{\nu=1}^\infty (1+\nu^2)^{p-r} \sum_{|k|=\nu} 1 \leq$$

$$\leq c_2 \sum_{\nu=1}^\infty \nu^{2p-2r+m-1} = c_2 \sum_{\nu=1}^\infty \nu^{-1-2(r-p-m/2)} = c_2' < \infty.$$

According to (2.3), the series obtained by differentiating the function $f = \sum_k f_k e^{i(k,\phi)}$ term by term not more than p times is uniformly convergent, so that $f \in C^p(T_m)$. Moreover,

$$|D^p f|_0 = \left|\sum_k (ik)^p f_k e^{i(k,\phi)}\right|_0 \leq c_1 \|f\|_r,$$

where c_1 is a constant independent of f.

Theorem 2. *For $s < r$, every set of functions that is bounded in $H^r(T_m)$ is compact in $H^s(T_m)$.*

To prove Theorem 2 we take a bounded sequence of functions $f^{(j)}$ ($j = 1, 2, \ldots$) in $H^r(T_m)$. This means that

$$\|f^{(j)}\|_r < M, \; j = 1, 2, \ldots. \tag{2.4}$$

Let $\sum_k f_k^{(j)} e^{i(k,\phi)}$ be the Fourier series for the function $f^{(j)}$. Then taking (2.4) into account we have

$$\|f_k^{(j)}\| \leq M(1 + \|k\|^2)^{-r/2}$$

and for any k there is a convergent subsequence of the sequence of the coefficients $f_k^{(j)}$ ($j = 1, 2, \ldots$). Without loss of generality we can assume that the sequence $f_k^{(j)}$ itself converges as $j \to \infty$. We set

$$f_N^{(j)} = \sum_{\|k\| \leq N} f_k^{(j)} e^{i(k,\phi)}, \quad R_N f^{(j)} = \sum_{\|k\| > N} f_k e^{i(k,\phi)}$$

and estimate $\|f^{(j)} - f^{(\nu)}\|_s$ for $s < r$ and arbitrary $j, \nu \geq 1$.

For each fixed N the convergence of the coefficients $f_k^{(j)}$, $\|k\| \leq N$, $j \to \infty$ implies that

$$\|f_N^{(j)} - f_N^{(\nu)}\|_s \to 0$$

as $j, \nu \to \infty$.

On the other hand, for fixed j, ν and $s < r$ we have

$$\|R_N f^{(j)} - R_N^{(\nu)}\|_s^2 \leq c(1 + N^2)^{s-r} M^2 \to 0$$

as $N \to \infty$.

But then

$$\|f^{(j)} - f^{(\nu)}\|_s \leq \|f_N^{(j)} - f_N^{(\nu)}\|_s + \|R_N f^{(j)} - R_N f^{(\nu)}\|_s \to 0$$

as $j, \nu \to \infty$.

This proves that the set $f^{(j)}$ ($j = 1, 2, \ldots$) is compact in the space $H^s(T_m)$ for $s < r$.

1.3. Main inequalities in $H^r(\mathcal{T}_m)$

Following [62] it is easy to derive a number of inequalities satisfied by functions belonging to the space $H^r(\mathcal{T}_m)$.

Indeed, the equality

$$(f,g)_r = \sum_k (1 + \|k\|^2)^r \langle f_k, g_{-k}\rangle, \qquad (3.1)$$

which defines an inner product of the elements $f = \sum_k f_k e^{i(k,\phi)}$ and $g = \sum_k g_k e^{i(k,\phi)}$ from $H^r(\mathcal{T}_m)$ makes sense even if $f \in H^{r-s}(\mathcal{T}_m)$, $g \in H^{r+s}(\mathcal{T}_m)$. It is therefore natural to apply formulae (3.1) for elements, one of which is taken from the space $H^{r-s}(\mathcal{T}_m)$ and the other from $H^{r+s}(\mathcal{T}_m)$. It follows from the generalized Schwartz inequality for the inner product $(\cdot, \cdot)_r$ that

$$|(f,g)_r| \le \|f\|_{r-s}\|g\|_{r+s},$$

for elements $f \in H^{r-s}(\mathcal{T}_m)$ and $g \in H^{r+s}(\mathcal{T}_m)$. Thus we obtain the inequality

$$\|f\|_r^2 \le \|f\|_{r-s}\|f\|_{r+s},$$

which gives an estimate of the r-norm of an element f of $H^{r+s}(\mathcal{T}_m)$ in terms of its $(r-s)$-norm and $(r+s)$-norm.

The inequality

$$1 \le a^{r+s} + a^{r_0-s}$$

is obvious for any $a > 0$ and $r > s \ge r_0$. For any $\epsilon > 0$, setting $a = \epsilon^{1/(r-s)}(1+\|k\|^2)$ we obtain the estimate

$$(1 + \|k\|^2)^s \le \epsilon(1 + \|k\|^2)^r + \epsilon^{(r_0-s)/(r-s)}(1 + \|k\|^2)^{r_0}.$$

This proves that when $f \in H^r(\mathcal{T}_m)$, $r > s \ge r_0$ and $\epsilon > 0$,

$$\|f\|_s \le \epsilon\|f\|_r + \epsilon^{-(s-r_0)/(r-s)}\|f\|_{r_0}. \qquad (3.2)$$

Setting in (3.2) $r = 0$, $\epsilon = \|f\|_r^{s/r-1}\|f\|_0^{1-s/r}$, we obtain the inequality

$$\|f\|_s \le 2\|f\|_0^{1-s/r}\|f\|_r^{s/r},$$

which gives an estimate of the s-norm of $f \in H^r(\mathcal{T}_m)$ in terms of its r-norm and 0-norm with $r > s > 0$.

The next two inequalities which relate the norm of a function

$$f = \sum_k f_k e^{i(k,\phi)} \in H^r(\mathcal{T}_m)$$

to the norm of its generalized derivatives $D^\rho f = \sum_k (ik)^\rho f_k e^{i(k,\phi)}$ for $|\rho| \le r$
follow from considerations of the previous section:

$$\|D^\rho f\|_p \le c\|f\|_{|\rho|+p}, \quad \|f\|_r \le c \sum_{|\rho| \le r} \|D^\rho f\|_0$$

for any $0 \le p \le r - |\rho|$ and some constant c which does not depend on f.

The following more complicated inequality which gives an estimate of generalized derivatives of functions is due to Nirenberg [162]:

$$\left\| \, \|D^\rho f\|^{r/|\rho|} \right\|_0^{|\rho|/r} \le c|f|_0^{1-|\rho|/r}\|f\|_r^{|\rho|/r} \tag{3.3}$$

for any $|\rho| \le r$ and some c which is independent of f. Inequality (3.3)
leads to Moser's estimate [89] for the composite of functions $(f \cdot g)(\phi) =$
$f(\phi, g(\phi))$, where $f(\phi, y) \in C^r(\mathcal{T}_m \times T_n)$, T_n is the unit ball in E^n, $g(\phi) =$
$(g_1(\phi), \ldots, g_n(\phi))$ is a function from $H^r(\mathcal{T}_m) \cap C(\mathcal{T}_m)$ taking values in T_n,
$\|g\| \le 1$. The estimate is as follows:

$$\|f(\phi, g(\phi))\|_r \le c|f|_r (1 + \|g\|_r), \tag{3.4}$$

where c is a positive constant independent of f and g,

$$|f|_r = \max_{0 \le \sigma \le r} \max_{(\phi,y) \in \mathcal{T}_m \times T_n} \|D^\sigma f(\phi, y)\|,$$

D^σ is any partial derivative with respect to ϕ, y of order σ.

To prove (3.4) we write down the differentiation formula for the composite
of functions:

$$D^r_\phi f(\phi, g(\phi)) = \sum_{\rho+\sigma \le r} (D^\sigma_\phi D^\rho_y f(\phi, y = g(\phi))) \times$$

$$\times \sum_\alpha c_{\sigma\rho\alpha}(D_\phi g(\phi))^{\alpha_1}(D^2_\phi(g(\phi))^{\alpha_2} \ldots (D^r_\phi g(\phi))^{\alpha_r},$$

where $\alpha_1, \ldots, \alpha_r$, σ and ρ are non-negative positive numbers that satisfy:

$$\alpha_1 + \alpha_2 + \ldots + \alpha_r = \rho, \quad \alpha_1 + 2\alpha_2 + \ldots + r\alpha_r = r - \sigma,$$

D_ϕ^σ and D_y^ρ are any partial derivatives with respect to the variables ϕ and y of order σ and ρ respectively, and the $c_{\sigma\rho\alpha}$ are non-negative constants. By making an estimate of the terms in these formulae using Holder's inequality, we find that

$$\|D_\phi^r f(\phi, g(\phi))\|_0 \le c \sum_{\substack{\rho+\sigma\le r \\ \alpha}} \left\| D_\phi^\sigma D_y^\rho f(\phi, y = g(\phi)) \prod_{\nu=1}^r (D_\phi^\nu g(\phi)^{\alpha_\nu}) \right\|_0 \le$$

$$\le c \sum_{\substack{\rho+\sigma\le r \\ \alpha}} \left\| \|D_\phi^\sigma D_y^\rho f(\phi, y = g(\phi))\|^{r/\sigma} \right\|_0^{\sigma/2} \prod_{\nu=1}^r \left\| \|D_\phi^\nu g(\phi)\|^{r/\nu} \right\|_0.$$

Inequality (3.3) now leads to Moser's estimate (3.4):

$$\|D_\phi^r f(\phi, g(\phi))\|_0 \le$$

$$\le c \sum_{\substack{\rho+\sigma\le r \\ \alpha}} \left\| \|D_\phi^\sigma D_y^\rho f(\phi, y)\|^{r/\sigma} \right\|_0^{\sigma/r} \prod_{\nu=1}^r |g|_0^{(1-\nu/r)\alpha_\nu} \|g\|_r^{(\nu\alpha_\nu)/r} \le$$

$$\le c|f|_r \sum_{\substack{\rho+\sigma\le r \\ \alpha}} \|g\|_r^{1-\sigma/r} \le c_1 |f|_r (1 + \|g\|_r).$$

Note that for a function f that is linear with respect y estimate (3.4) holds when $g \in H^r(\mathcal{T}_m)$.

The next two equalities, which relate the inner product $(\cdot, \cdot)_r$ and the differential operator $1 - \Delta = 1 - \sum_{\nu=1}^m \frac{\partial^2}{\partial\phi_\nu^2} = K$:

$$(f, g)_r = (K^s f, g)_{r-s} = (f, K^s g)_{r-s}, \quad \|K^s f\|_r = \|f\|_{r+2s},$$

easily follow from equality (1.4) which defines the inner product in $H^r(\mathcal{T}_m)$. These equalities hold for elements f and g of $H^\rho(\mathcal{T}_m)$ with the indices being such that the corresponding inner products and norms are defined.

1.4. Quasi-periodic functions. The spaces $C^r(\omega)$

Let $\omega_1, \omega_2, \ldots, \omega_m$ be positive numbers that are incommensurable in the sense that

$$(k, \omega) = k_1\omega_1 + k_2\omega_2 + \ldots + k_m\omega_m \ne 0$$

for any integer vector $k = (k_1, \ldots, k_m)$, $|k| \ne 0$.

A function of a real variable t:

$$F(t) = f(\omega t) = f(\omega_1 t, \ldots, \omega_m t), \tag{4.1}$$

where $f \in C(T_m)$, is called *quasi-periodic*, the set $\omega = (\omega_1, \ldots, \omega_m)$ *its frequency basis*, and the numbers ω_ν $(\nu = 1, \ldots, m)$ the *frequencies* of the function F.

We give a number of properties of quasi-periodic functions. To derive them from the similar properties of periodic functions we use Kronecker's statement [67] on solutions of a system of inequalities. We give it as a separate lemma.

Lemma 1. *Let $\omega_1, \omega_2, \ldots, \omega_m$ be incommensurable real numbers. Then for any set of real numbers $\phi_1, \phi_2, \ldots, \phi_m$ the system of inequalities*

$$|\omega_\nu t - \phi_\nu| < \delta (\mathrm{mod}\, 2\pi), \quad \nu = 1, \ldots, m,$$

has a relatively dense set of solutions for any arbitrary small positive δ.

From Lemma 1 and the definition of a quasi-periodic function one can easily deduce Weil's theorem [172] on the density of the values of a quasi-periodic function F in the set of the values of the function f.

Theorem 1. *The set of values of a quasi-periodic function $F(t) = f(\omega t)$ is dense in the set of values of the function $f(\phi)$.*

Indeed, let $\phi = (\phi_1, \ldots, \phi_m)$ be an arbitrary point of the cube $T_m : 0 \leq \phi_\alpha \leq 2\pi$ $(\alpha = 1, \ldots, m)$, ϵ an arbitrary small number. Since $f(\phi)$ is periodic and continuous, it is uniformly continuous. Therefore we can find a $\delta = \delta(\epsilon)$ such that

$$\|f(\phi) - f(\phi')\| < \epsilon \tag{4.2}$$

for any point $\phi' = (\phi'_1, \phi'_2, \ldots, \phi'_m)$ with $|\phi_\nu - \phi'_\nu| < \delta$, $\nu = 1, \ldots, m$. It follows from Lemma 1 that for $\phi'_1, \ldots, \phi'_m$ and such a δ there exists $\tau \in \mathbf{R}$ such that

$$|\omega_\nu \tau - \phi'_\nu| < \delta \,(\mathrm{mod}\, 2\pi), \quad \nu = 1, \ldots, m.$$

Then from (4.2) we have the estimate

$$\|f(\phi') - F(\tau)\| = \|f(\phi') - f(\omega \tau)\| < \epsilon,$$

which proves Theorem 1.

As a corollary of Theorem 1 we obtain the following equality

$$\sup_{t \in \mathbb{R}} \|F(t)\| = \max_{\phi \in \mathcal{T}_m} \|f(\phi)\| = |f|_0, \tag{4.3}$$

which relates the maximum of a quasi-periodic function $\|F(t)\| = \|f(\omega t)\|$ with the norm $|\cdot|_0$ of a function $f \in C(\mathcal{T}_m)$. Equality (4.3) allows one to deduce a number of properties of quasi-periodic functions. They are known as the approximation theorem, the mean value theorem, and the limit function theorem [64].

Indeed, since the function $f \in C(\mathcal{T}_m)$ can be uniformly approximated by trigonometric polynomials $P_N(\phi) = \sum_{\|k\| \le N} P_k e^{i(k,\phi)}$, it follows from (4.3) that the quasi-periodic function $F(t) = f(\omega t)$ can be uniformly approximated by *quasi-trigonometric* polynomials $P_N(\omega t) = \sum_{\|k\| \le N} P_k e^{i(k,\omega)t}$ such that

$$\sup_{t \in \mathbb{R}} \|F(t) - P_N(\omega t)\| = \max_{\phi \in \mathcal{T}_m} \|f(\phi) - P_N(\phi)\| = |f - P_N|_0 < \epsilon \tag{4.4}$$

for a given $\epsilon > 0$ and some $P_N(\phi) \in \mathcal{P}(\mathcal{T}_m)$. Inequality (4.4) is equivalent to the following approximation theorem.

Theorem 2. *Let $F(t)$ be a quasi-periodic function with the frequency basis $\omega = (\omega_1, \ldots, \omega_m)$. For any $\epsilon > 0$ there exists a quasi-trigonometric polynomial $P_N(t) = \sum_{\|k\| \le N} P_k e^{i(k,\omega)t}$ such that*

$$\sup_{t \in \mathbb{R}} \|F(t) - P_N(t)\| < \epsilon.$$

Furthermore, for any quasi-trigonometric polynomial

$$P(\omega t) = \sum_{\|k\| \le N} P_k e^{i(k,\omega)t}$$

the mean value

$$\lim_{T \to \infty} \frac{1}{T} \int_\tau^{T+\tau} P(\omega t)dt = P_0 = \frac{1}{(2\pi)^m} \int_0^{2\pi} \ldots \int_0^{2\pi} P(\phi)d\phi_1 \ldots d\phi_m$$

exists and does not depend on τ. Thus, there also exists a similar value for a quasi-periodic function F which can be uniformly approximated by these polynomials. This leads to the well-known mean value theorem.

Theorem 3. *Let $F(t) = f(\omega t)$ be a quasi-periodic function. Then the limit*

$$\lim_{T\to\infty} \frac{1}{T} \int_\tau^{T+\tau} F(t)dt = F_0 = \frac{1}{(2\pi)^m} \int_0^{2\pi} \cdots \int_0^{2\pi} f(\phi)d\phi_1 \ldots d\phi_m \qquad (4.5)$$

exists and converges uniformly with respect to $\tau \in \mathbf{R}$.

The limit (4.5) is called the mean value of the function F. As can be seen from (4.5) it does not depend on $\tau \in \mathbf{R}$.

The set of all quasi-periodic functions with a frequency basis ω forms a linear space which we shall denote by $C(\omega)$. By introducing a norm $|\cdot|$ in $C(\omega)$ using

$$|F|_0 = \sup_{t\in\mathbf{R}} \|F(t)\|,$$

we turn $C(\omega)$ into a normed space, which we shall denote by $C^0(\omega)$. Taking into account the fact that the space $C^0(T_m)$ is complete and using equality (4.3) we see that the following theorem holds.

Theorem 4. *The space $C^0(\omega)$ is a Banach space.*

The inner product $(f,g)_0$ of functions $f,g \in C^0(T_m)$ induces an inner product on $C^0(\omega)$ of the functions $F = f(\omega t)$ and $G = g(\omega t) \in C^0(\omega)$. Using (4.5), this inner product is given by the following equality:

$$(F,G)_0 = \lim_{T\to\infty} \frac{1}{T} \int_0^T \langle F(t), G(t)\rangle dt = (f,g)_0. \qquad (4.6)$$

The inner product (4.6) induces on $C^0(\omega)$ the norm

$$\|F\|_0^2 = (F,F)_0.$$

The family of functions $\{e^{i(k,\omega)t}\}$ forms a complete orthonormal set in $C(\omega)$. Any function $F \in C(\omega)$ can be represented as a Fourier series with respect to this set:

$$F(t) \simeq \sum_k F_k e^{i(k,\omega)t}, \quad F_k = \lim_{T\to\infty} \frac{1}{T} \int_0^T F(t)e^{-i(k,\omega)t} dt,$$

and it satisfies Parseval's equality

$$\sum_k \|F_k\|^2 = \|F\|_0^2. \qquad (4.7)$$

In the space $C^0(\omega)$ of quasi-periodic functions we choose the subspace $C^r(\omega)$ of all functions F for which representation (4.1) is defined by a function $f \in C^r(\mathcal{T}_m)$. The function $F \in C^r(\omega)$ has r continuous derivatives with respect to t and each derivative is a quasi-periodic function. The space $C^r(\omega)$ can be turned into a complete space if one defines a norm $|\cdot|_r$ by setting

$$|F|_r = |f|_r$$

whenever $F(t) = f(\omega t)$, where $f \in C^r(\mathcal{T}_m)$.

1.5. The spaces $H^r(\omega)$ and their structure

Suppose that the function $\sum_k f_k e^{i(k,\phi)}$ belongs to the space $H^r(\mathcal{T}_m)$, and let $\omega = (\omega_1, \ldots, \omega_m)$ be positive incommensurable numbers. We form a function $F(t)$ by defining it to be the series $\sum_k f_k e^{i(k,\omega)t}$. The set of functions of this type forms a linear space which we denote by $H^r(\omega)$. For any two functions F and G belonging to $H^r(\omega)$ and given by the series $\sum_k f_k e^{i(k,\omega)t}$ and $\sum_k g_k e^{i(k,\omega)t}$, we can define an inner product and a norm $\|\cdot\|_r$:

$$(F, G)_r = \sum_k (1 + \|k\|^2)^r \langle f_k, g_{-k} \rangle, \quad \|F\|_r^2 = (F, F)_r^2.$$

Since the space $H^r(\omega)$ is the completion with respect to the norm $\|\cdot\|_r$ of the space $\mathcal{P}(\omega)$ generated by polynomials of the form $P(t) = \sum_{\|k\| \leq N} P_k e^{i(k,\omega)t}$, the space $H^r(\omega)$ is a Hilbert space. We set $H^\infty(\omega) = \bigcap_{r=0}^{\infty} H^r(\omega)$.

If a function F is quasi-periodic with the frequency basis $\omega = (\omega_1, \ldots, \omega_m)$, it follows from Parseval's equality (4.7) that its Fourier series $\sum_k F_k e^{i(k,\omega)t}$ belongs to the space $H^0(\omega)$. So, each such series can be identified with a quasi-periodic function for which it is the Fourier series. To find out what other elements of the space $H^0(\omega)$ can be identified with a quasi-periodic function one needs an extension of the above notion of a quasi-periodic function due to P.Bol [25]. This extension leads to the notion of a quasi-periodic function in the sense of Besicovich [142].

Definition. A function $F(t)$ defined on the real line **R** and square summable on any finite line interval is called a *Besicovich quasi-periodic function with* *frequency basis* $\omega = (\omega_1, \ldots, \omega_m)$ if there exists a sequence of quasi-periodic polynomials $P_\nu \in \mathcal{P}(\omega)$ such that

$$\lim_{\nu \to \infty} \overline{\lim_{T \to \infty}} \frac{1}{2T} \int_{-T}^{T} \|F(t) - P_\nu(t)\|^2 dt = 0.$$

The set of all Besicovich quasi-periodic functions with frequency basis ω forms a metric space, where the distance between two functions F and G is defined to be

$$\rho(F, G) = \left\{ \lim_{T \to \infty} \frac{1}{T} \int_{0}^{T} \|F(t) - G(t)\|^2 dt \right\}^{1/2}.$$

We denote this space by $B^2(\omega)$. The following theorem due to Besicovich [142] gives an important property of this space.

Theorem 1. *The space $B^2(\omega)$ is complete.*

The proof of Theorem 1 is constructive and the interested reader can find it in [64].

Since the space $B^2(\omega)$ is complete, the limit of a sequence of quasi-trigonometric polynomials $P_\nu(t) = \sum_{\|k\| \leq N_\nu} f_k^{(\nu)} e^{i(k,\omega)t}$ convergent with respect to the mean square belongs to this space. Because the function $\sum_k f_k e^{i(k,\omega)t}$ in $H^0(\omega)$ is obtained by taking closure of the space $\mathcal{P}(\omega)$ with respect to the norm $\|\cdot\|_0$, it corresponds to a Besicovich quasi-periodic function $F(t)$. Conversely, for any Besicovich quasi-periodic function $F(t) \in B^2(\omega)$ there exists in $B^2(\omega)$ a Cauchy sequence of quasi-trigonometric polynomials $P_\nu(t) \in \mathcal{P}(\omega)$ converging to $F(t)$. The equality

$$\|P_\nu - P_j\|_0 = \left[\lim_{T \to \infty} \frac{1}{2T} \int_{-T}^{T} \|P_\nu(t) - P_j(t)\|^2 dt \right]^{1/2} = \rho(P_\nu, P_j),$$

which holds for the polynomials $P_\nu(t)$ $(\nu = 1, 2, \ldots)$, shows that the sequence P_ν $(\nu = 1, 2, \ldots)$ is a Cauchy sequence in $H^0(\omega)$. Therefore the limit of the sequence P_ν $(\nu = 1, 2, \ldots)$ belongs to the space $H^0(\omega)$. Thus we see that there is a correspondence between the functions of the space $B^2(\omega)$ and the space $H^0(\omega)$. That is the space $H^0(\omega)$ can be identified with the space of Besicovich quasi-periodic functions $B^2(\omega)$. This proves the following theorem.

Theorem 2. *The space $H^0(\omega)$ is the space of Besicovich quasi-periodic functions with frequency basis ω.*

Now the series $\sum_k f_k e^{i(k,\omega)t}$ which represents a function of $H^0(\omega)$ can be obtained from the Fourier series $\sum_k f_k e^{i(k,\phi)}$ of a function $f(\phi) \in H^0(\mathcal{T}_m)$ by formally replacing ϕ by ωt. Since the function $f(\phi)$ is measurable on the cube of periods $\mathcal{T}_m$, it can be identified with every function of $H^0(\mathcal{T}_m)$ that is equal to f almost everywhere on $\mathcal{T}_m$. The set of $\phi = (\omega t)\bmod 2\pi$, $t \in \mathbf{R}$ has zero Lebesgue measure on $\mathcal{T}_m$. Consequently, the function $f(\phi)$ with its Fourier series $\sum_k f_k e^{i(k,\phi)} \in H^0(\mathcal{T}_m)$ can be identified in $H^0(\mathcal{T}_m)$ with a function $f(\phi)$ which on the line $\phi = \omega t$, $t \in \mathbf{R}$, takes values equal to the values of the Besicovich quasi-periodic function $F(t)$ generated by the series $\sum_k f_k e^{i(k,\omega)t} \in H^0(\omega)$. This leads to the following equality

$$F(t) = f(\omega t), \tag{5.1}$$

which associates with the Besicovich quasi-periodic function $F(t)$ a measurable square summable periodic function of many variables $f(\phi)$ on the cube of periods $\mathcal{T}_m$.

Equality (5.1) leads to the following relation

$$f_k = \lim_{T\to\infty} \frac{1}{2T} \int_{-T}^{T} F(t) e^{-i(k,\omega)t}\, dt =$$

$$= \frac{1}{(2\pi)^m} \int_0^{2\pi} \cdots \int_0^{2\pi} f(\phi) e^{-i(k,\phi)}\, d\phi_1 \ldots d\phi_m,$$

which shows that the series $\sum_k f_k e^{i(k,\omega)t}$ is a Fourier series of the Besicovich quasi-periodic function $F(t)$. It also leads to the equality:

$$\sum_k \|f_k\|^2 = \lim_{T\to\infty} \frac{1}{2T} \int_{-T}^{T} \|F(t)\|^2\, dt,$$

which is Parseval's equality for the function $F(t) \in H^0(\omega)$.

We say that a Besicovich quasi-periodic function $F(t)$ has the generalized derivative d^r/dt^r if there exists a function $F^{(r)}(t)$ in $H^0(\omega)$ satisfying the equality

$$(F^{(r)}, P)_0 = (-1)^r (F, d^r P/dt^r)_0 \tag{5.2}$$

for any trigonometric polynomial $P(t) \in \mathcal{P}(\omega)$. The inner product in (5.2) will be understood in the usual sense for quasi-periodic functions:

$$(F, P)_0 = \lim_{T\to\infty} \frac{1}{2T} \int_{-T}^{T} \langle F(t), P(t)\rangle\, dt.$$

Let the series $\sum_k f_k e^{i(k,\omega)t}$ define an element of the space $H^r(\omega)$. The function $F(t)$ with the above Fourier series belongs to the space $B^2(\omega)$. The function $F^{(\rho)}(t)$ with Fourier series $\sum_k [i(k,\omega)]^\rho f_k e^{i(k,\omega)t}$ also belongs to this space for any $0 \le \rho \le r$.

Moreover, since

$$(F^{(\rho)}, P)_0 = \sum_k [i(k,\omega)]^\rho \langle f_k, P_{-k} \rangle =$$

$$= (-1)^\rho \sum_k \langle f_k, [i(-k,\omega)]^\rho P_{-k} \rangle = (-1)^\rho (F, d^\rho P/dt^\rho)_0,$$

$F^{(\rho)}(t)$ is the ρth generalized derivative of the function $F(t)$. This shows that each function $F(t) \in H^r(\omega)$ can be identified with a Besicovich quasi-periodic function having generalized derivatives. Thus the following theorem holds.

Theorem 3. *The space $H^r(\omega)$ contains the Besicovich quasi-periodic functions having generalized derivatives up to order r inclusive.*

However, the space $H^r(\omega)$ does not coincide with the entire space of Besicovich functions having generalized derivatives of order r. This can be seen by considering the following example.

Consider a function

$$F(t) = \sum_{n=1}^{\infty} \frac{\cos(k^n, \omega)t}{\|k^n\|^{m+1}}, \quad m \ge 2, \tag{5.3}$$

where $k^n = (k_1^n, \dots, k_m^n)$ is an integer vector satisfying the conditions:

$$\lim_{n \to \infty} (k^n, \omega) = 0, \quad \lim_{n \to \infty} \|k^n\| = \infty, \quad \|k^n\| \ne 0. \tag{5.4}$$

Since the series

$$\sum_{n=1}^{\infty} \frac{|(k^n, \omega)|^r}{\|k^n\|^{m+1}}$$

converges for $r > 0$ because

$$\sum_{n=1}^{\infty} \frac{|(k^n, \omega)|^r}{\|k^n\|^{m+1}} \le c \sum_{n=1}^{\infty} \frac{1}{\|k^n\|^{m+1}} \le c \sum_{\|k\| \ne 0} \frac{1}{\|k\|^{m+1}} =$$

$$= c \sum_{\nu=1}^{\infty} \nu^{-(m+1)} \sum_{\|k\|=\nu} 1 \le c_1 \sum_{\nu=1}^{\infty} \nu^{-(m+1)+m-1} = c_1 \sum_{\nu=1}^{\infty} \nu^{-2} = c_2 < \infty,$$

the function $F(t)$ is infinitely differentiable. On the other hand, since

$$\sum_{n=1}^{N} \frac{(1 + \|k^n\|^2)^r}{\|k^n\|^{m+1}} \geq \sum_{n=1}^{N} \frac{\|k^n\|^{2r}}{\|k^n\|^{m+1}} = \sum_{n=1}^{N} \|k^n\|^{2r-(m+1)} \geq N$$

for $2r \geq m + 1$, the function $F(t)$ does not belong to space $H^r(\omega)$ when $r \geq (m + 1)/2 > 1$. Thus, the function $F(t)$ has derivatives of any order but does not belong to $H^r(\omega)$ when $r \geq (m + 1)/2 > 1$.

Finally, we also need to see that conditions (5.4) hold for a sequence of integer vectors k^n $(n = 1, 2, \ldots)$. This can be achieved by using the following result from number theory [139].

Lemma. *For any irrational number α, there exists a sequence of integers p_n, q_n $(q_n \geq 2^{(n-1)/2}$ for $n \geq 2)$ such that*

$$\frac{1}{q_n(q_{n+1} + q_n)} < \left| \alpha - \frac{p_n}{q_n} \right| < \frac{1}{q_n q_{n+1}}. \tag{5.5}$$

As a consequence of this lemma we have inequality

$$|k_1^n \omega_1 + k_2^n \omega_2| = k_2^n \omega_1 \left| \frac{\omega_2}{\omega_1} + \frac{k_1^n}{k_2^n} \right| < k_2^n \omega_1 \frac{1}{k_2^n k_2^{n+1}} \leq 2^{-n/2} \omega_1, \tag{5.6}$$

which can be obtained from (5.5) by setting $\alpha = \omega_2/\omega_1$, $k_1^n = -p_n$ and $k_2^n = q_n$ $(\omega_1 > 0)$. Inequality (5.6) leads to conditions (5.4) for $m \geq r$ and $k_j^n = 0$ for $j \geq 3$.

Sobolev theorems similar to those given in §1.2 hold also for the spaces $H^r(\omega)$. The first theorem asserts that $H^r(\omega) \subset C^p(\omega)$ for $r > m/2 + p$, while the second asserts that a bounded set of functions in $H^r(\omega)$ is compact in $H^s(\omega)$ for $s < r$.

Because the Sobolev theorems hold, the space $H^r(\omega)$ has advantages over other spaces of quasi-periodic functions that have derivatives up to order r inclusive. For example, the spaces $\pi^r(\omega)$ which contain all quasi-periodic functions with frequency basis ω and whose rth order derivatives are also quasi-periodic functions. The norm

$$| \cdot |_r = \max_{0 \leq p \leq r} \sup_{t \in \mathbb{R}} \left| \frac{d^p}{dt^p} \right|$$

turns $\pi^r(\omega)$ into a normed space, but the Sobolev theorems do not hold. This can be seen by considering the sequence of functions

$$\sin(k^n, \omega)t, \quad n = 1, 2, \ldots, \, m \geq 2, \tag{5.7}$$

which belong to space $\pi^r(\omega)$ for any $r \geq 0$ only if $k^n = (k_1^n, \ldots, k_m^n)$ are chosen in such a way that

$$|(k^n, \omega)| \leq 1, \quad (k^n, \omega) \to 0 \quad (n \to \infty).$$

Even though the sequence (5.7) is bounded in $\pi^r(\omega)$ it is compact in $\pi^0(\omega)$.

1.6. First integral of a quasi-periodic function

Let

$$F(t) = f(\omega t), \quad f(\phi) \in C(T_m),$$

be a quasi-periodic function with frequency basis $\omega = (\omega_1, \ldots, \omega_m)$ having zero mean value:

$$\overline{F} = \lim_{T \to \infty} \frac{1}{T} \int_0^T F(t) dt = 0.$$

Consider the first integral

$$J(t) = \int_0^t F(t) dt. \tag{6.1}$$

By considering the function

$$F(t) = \sum_{n=1}^{\infty} \frac{\sin(k^n, \omega) t}{n^2 \text{sign}(k^n, \omega)}, \tag{6.2}$$

where $k^n = (k_1^n, \ldots, k_m^n)$ is an integer vector satisfying the conditions

$$|(k^n, \omega)| \leq n^{-2}, \quad \|k^n\| \neq 0, \quad n = 1, 2, \ldots,$$

we see that $J(t)$ is not always a quasi-periodic function.

Indeed, since series (6.2) is uniformly convergent, it can be integrated term by term. Thus,

$$J(t) = \sum_{n=1}^{\infty} \frac{(1 - \cos(k^n, \omega) t)}{n^2 |(k^n, \omega)|} = 2 \sum_{n=1}^{\infty} \frac{\sin^2(k^n, \omega) t/2}{n^2 |(k^n, \omega)|}$$

and therefore

$$\sup_{t \in \mathbb{R}} J(t) \geq 2 \sup_{t \in \mathbb{R}} \sum_{n=1}^{N} \frac{\sin^2(k^n, \omega) t/2}{n^2 |(k^n, \omega)|} \geq$$

$$\geq 2 \lim_{T \to \infty} \frac{1}{T} \int_0^T \sum_{n=1}^{N} \frac{\sin^2(k^n, \omega) t/2}{n^2 |(k^n, \omega)|} dt = \sum_{n=1}^{N} \frac{1}{n^2 |(k^n, \omega)|} \geq N.$$

The last inequality shows that the function $J(t)$ is unbounded on **R**, therefore $J(t)$ is not quasi-periodic.

The following main result which gives the conditions for the function (6.1) to be quasi-periodic is due to P. Bol [25], [43].

Theorem 1. *The first integral $J(t)$ of a quasi-periodic function $F(t)$ is a quasi-periodic function if and only if it is bounded on* **R**.

According to the definition of a quasi-periodic function, it is bounded on **R**. So to prove Theorem 1 we need to show that if the function (6.1) is bounded then it is quasi-periodic. Thus, let

$$|J(t)| \leq M, \ t \in \mathbf{R}.$$

We define the function

$$\Phi_\tau(\phi) = \int_0^\tau f(\omega s + \phi) ds$$

which satisfies the obvious relation

$$\Phi_\tau(\omega t) = J(t + \tau) - J(t), \quad t, \tau \in \mathbf{R}. \tag{6.3}$$

Since the function $J(t)$ is bounded on **R**, it follows that for some sequence of numbers

$$1 \leq N_1 < N_2 < \ldots < N_k < \ldots, \quad \lim_{k \to \infty} N_k = \infty,$$

the sequence

$$\frac{1}{N_k} \int_0^{N_k} J(\tau) d\tau, \quad k = 1, 2, \ldots$$

converges as $k \to \infty$.

We define a sequence of functions in $C(\omega)$ by setting

$$\Psi_k(\omega t) = \frac{1}{N_k} \int_0^{N_k} [\Phi_\tau(\omega t) - J(\tau)] d\tau, \quad k = 1, 2, \ldots.$$

According to equality (6.3) we have

$$\Psi_k(\omega t) = -J(t) + \frac{1}{N_k} \int_0^{N_k} [J(t + \tau) - J(\tau)] d\tau, \quad k = 1, 2, \ldots, \tag{6.4}$$

and after some obvious transformations we obtain

$$\Psi_k(\omega t) + J(t) = \frac{1}{N_k} \int_0^{N_k} \int_0^t f(\omega s + \omega \tau) ds d\tau =$$

$$= \frac{1}{N_k} \int_0^t \int_s^{N_k + s} f(\omega \tau) d\tau \, ds = \int_0^t \frac{J(N_k + s) - J(s)}{N_k} ds, \quad k = 1, 2, \dots. \quad (6.5)$$

Since $J(t)$ is bounded on $\mathbf{R}$ and in view of the relations (6.5), it follows that the limit

$$\lim_{k \to \infty} [\Psi_k(\omega t) + J(t)] = 0 \qquad (6.6)$$

converges uniformly on any bounded interval $I \subset \mathbf{R}$.

  We claim that limit (6.6) is uniform on the whole of $\mathbf{R}$. According to (6.4), this is equivalent to the fact that

$$\lim_{k \to \infty} \frac{1}{N_k} \int_0^{N_k} [J(t + s) - J(s)] ds = 0 \qquad (6.7)$$

is uniform on the whole of $\mathbf{R}$.

  We shall prove (6.7) using ideas of H. Bohr [26], [143], S. Bochner [43]. Following [65], we choose an arbitrary sequence of numbers t_n such that ωt_n converges to ϕ modulo 2π:

$$\lim_{n \to \infty} \omega t_n = \phi \bmod 2\pi, \quad \phi \in \mathcal{T}_m.$$

Consider the sequence $J(t + t_n)$ of shifts of the function $J(t)$. Since

$$J(t + t_n) = J(t_n) + \int_0^t f(\omega s + \omega t_n) ds, \qquad (6.8)$$

this sequence is compact in the sense of local convergence on $\mathbf{R}$, that is, it is uniformly convergent on any bounded interval of $\mathbf{R}$. Let $\hat{J}(t)$ be a limit point of sequence (6.8). Passing to the limit in (6.8) we obtain

$$\hat{J}(t) = \hat{J}(0) + \int_0^t f(\omega s + \phi) ds. \qquad (6.9)$$

Let us show that

$$\sup_{t \in \mathbf{R}} \hat{J}(t) = \sup_{t \in \mathbf{R}} J(t), \quad \inf_{t \in \mathbf{R}} \hat{J}(t) = \inf_{t \in \mathbf{R}} J(t). \qquad (6.10)$$

Indeed, since $\hat{J}(t)$ is a limit function of some subsequence of $\{J(t + t_n)\}$,

$$\inf_{t \in \mathbf{R}} J(t) \le \inf_{t \in \mathbf{R}} \hat{J}(t) \le \sup_{t \in \mathbf{R}} \hat{J}(t) \le \sup_{t \in \mathbf{R}} J(t).$$

Consider the sequence $\hat{J}(t - t_n)$. According to (6.9) we have

$$\hat{J}(t - t_n) = \hat{J}(0) + \int_0^{t - t_n} f(\omega s + \phi)ds = \hat{J}(0) + \int_0^t f(\omega s - \omega t_n + \phi)ds +$$

$$+ \int_0^{-t_n} f(\omega s + \phi)ds = \hat{J}(-t_n) + \int_0^t f(\omega s - \omega t_n + \phi)ds,$$

and hence it is clear that the sequence $\hat{J}(t - t_n)$ is compact in the sense of local convergence on $\mathbf{R}$. Let $\hat{J}_1(t)$ be a limit point of the sequence $\hat{J}(t - t_n)$. Passing to the limit in (6.9) we see that

$$\hat{J}_1(t) = \hat{J}_1(0) + \int_0^t f(\omega s)ds = \hat{J}_1(0) + J(t).$$

Now we have

$$\hat{J}_1(0) + \inf_{t \in \mathbf{R}} J(t) = \inf_{t \in \mathbf{R}}(\hat{J}_1(0) + J(t)) \ge \inf_{t \in \mathbf{R}} \hat{J}(t) \ge \inf_{t \in \mathbf{R}} J(t),$$

$$\hat{J}_1(0) + \sup_{t \in \mathbf{R}} J(t) = \sup_{t \in \mathbf{R}}(\hat{J}_1(0) + J(t)) \le \sup_{t \in \mathbf{R}} \hat{J}(t) \le \sup_{t \in \mathbf{R}} J(t),$$

which is a contradiction if equalities (6.10) do not hold.

We now show that the convergence $J(t + t_n) \to \hat{J}(t)$ is uniform on the whole of $\mathbf{R}$. Suppose the contrary. Then there exists a sequence $\{s_n\} \subset \mathbf{R}$ such that

$$|J(s_n + t_n) - \hat{J}(s_n)| \ge \alpha > 0. \tag{6.11}$$

Without loss of generality we can suppose that the sequence s_n is such that

$$\lim_{n \to \infty} \omega s_n = \phi_0 \bmod 2\pi, \quad \phi_0 \in \mathcal{T}_m.$$

Consider the sequence of functions

$$J(t + s_n + t_n) = J(s_n + t_n) + \int_0^t f(\omega s + \omega s_n + \omega t_n)ds,$$

$$\hat{J}(t + s_n) = \hat{J}(s_n) + \int_0^t f(\omega s + \omega s_n + \phi)ds.$$

Their local limits on **R** are

$$J_1(t) = J_1(0) + \int_0^t f(\omega s + \phi_0 + \phi)ds,$$

$$\hat{J}_2(t) = \hat{J}_2(0) + \int_0^t f(\omega s + \phi_0 + \phi)ds,$$

where $J_1(0)$ and $\hat{J}_2(0)$ are the corresponding limit points of $J(s_n + t_n)$ and $\hat{J}(s_n)$. It follows from equalities (6.10) that

$$\sup_{t \in \mathbf{R}} J_1(t) = J_1(0) + M_1 = \sup_{t \in \mathbf{R}} J(t),$$

$$\sup_{t \in \mathbf{R}} \hat{J}_2(t) = \hat{J}_2(0) + M_1 = \sup_{t \in \mathbf{R}} \hat{J}(t) = \sup_{t \in \mathbf{R}} J(t).$$

But then $J_1(0) = \hat{J}_2(0)$, which contradicts inequality (6.11), according to which the equality $\lim_{\nu \to \infty} J(s_{n_\nu} + t_{n_\nu}) = \lim_{\nu \to \infty} \hat{J}(s_{n_\nu})$ is impossible. This contradiction shows that the convergence

$$J(t + t_n) \rightarrow \hat{J}(t) \tag{6.12}$$

is uniform on the whole of **R**.

We now show that relation (6.12) implies that for any $\epsilon > 0$ there exists a positive number $l = l(\epsilon)$ such that any interval $I = [\tau, \tau + l]$ of **R** contains at least one number $T = T(\epsilon)$ for which the inequality

$$|J(t + T) - J(t)| < \epsilon, \ t \in \mathbf{R} \tag{6.13}$$

is true. For suppose not. Then there exists an $\epsilon_0 > 0$ such that one cannot choose a corresponding $l(\epsilon_0)$. Hence for any $l > 0$ there is an interval $I_\tau = [\tau - l, \tau + l]$ of **R** such that for any point $\theta \in I_\tau$ we have

$$\sup_{t \in \mathbf{R}} |J(t + \theta) - J(t)| \geq \epsilon_0. \tag{6.14}$$

We construct a sequence of intervals $I_{\tau_n} = [\tau_n - l_n, \tau_n + l_n] \ (n = 1, 2, \ldots)$ of this type by taking l_1 to be arbitrary, and l_n, τ_n satisfying the condition

$$l_n > \max_{\nu < n} |\tau_\nu|, \ n = 2, 3, \ldots. \tag{6.15}$$

Condition (6.15) ensures that $\tau_n - \tau_\nu \in I_{\tau_n}$ for $\nu < n$. Consider the sequence $J(t + \tau_p) \ (p = 1, 2, \ldots)$ of shifts of the function $J(t)$. Let $\tau_{p_j} \ (j = 1, 2, \ldots)$ be a subsequence of the numbers τ_n $(n = 1, 2, \ldots)$ such that $\omega \tau_{p_j}$ converges modulo 2π as $j \to \infty$: $\lim_{j \to \infty} \omega \tau_{p_j} = \phi \bmod 2\pi$ for some $\phi \in \mathcal{T}_m$. The subsequence $J(t + \tau_{p_j})$ $(j = 1, 2, \ldots)$ is compact in the sense of uniform convergence on $\mathbf{R}$:

$$J(t + \tau_{p_j}) \to \hat{J}(t) \tag{6.16}$$

uniformly on the whole of $\mathbf{R}$. However, when $p_\nu < p_n$ we have the following estimate for $\sup_{t \in \mathbf{R}} |J(t + \tau_{p_n}) - J(t + \tau_{p_\nu})|$:

$$\sup_{t \in \mathbf{R}} |J(t + \tau_{p_n}) - J(t + \tau_{p_\nu})| = \sup_{t \in \mathbf{R}} |J(t + \tau_{p_n} - \tau_{p_\nu}) - J(t)| \geq \epsilon_0. \tag{6.17}$$

This is true since $\tau_{p_n} - \tau_{p_\nu} \in I_{\tau_{p_n}} = [\tau_{p_n} - l_{p_n}, \tau_{p_n} + l_{p_n}]$.

Inequality (6.17) means that the sequence $J(t + \tau_{p_j})$ is not compact in the sense of uniform convergence on $\mathbf{R}$. The latter statement contradicts relation (6.16). This proves inequality (6.13). Now it is evident that the convergence (6.7) is uniform on $\mathbf{R}$. Indeed, for any $\epsilon > 0$ there exists $l = l(\epsilon)$ such that on any interval $I = [t, t + l(\epsilon)]$ there exists a point $T = T_t(\epsilon)$ such that

$$|J(s + T) - J(s)| < \epsilon, \quad s \in \mathbf{R}.$$

But then, for any $t \in \mathbf{R}$

$$\left| \frac{1}{N_k} \int_0^{N_k} [J(t + s) - J(s)] ds \right| \leq \left| \frac{1}{N_k} \int_t^T J(s) ds \right| + \left| \frac{1}{N_k} \int_{N_k + T}^{N_k + t} J(s) ds \right| +$$

$$+ \left| \frac{1}{N_k} \int_0^{N_k} [J(s + T) - J(s)] ds \right| \leq \frac{2Ml}{N_k} + \epsilon,$$

which proves that the convergence of (6.7) is uniform on $\mathbf{R}$.

Thus we have shown that

$$J(t) = - \lim_{k \to \infty} \Psi_k(\omega t) \tag{6.18}$$

in the sense of uniform convergence on $\mathbf{R}$. It follows that the sequence $\Psi_k(\phi) \in C(\mathcal{T}_m)$ of functions is Cauchy with respect to the norm of the space $C(\mathcal{T}_m)$:

$$\max_{\phi \in \mathcal{T}_m} |\Psi_{k+p}(\phi) - \Psi_k(\phi)| = \sup_{t \in \mathbf{R}} |\Psi_{k+p}(\omega t) - \Psi_k(\omega t)| < \epsilon$$

for arbitrary $\epsilon > 0$, $p = 1, 2, \ldots$, $k \geq N = N(\epsilon)$. Since this space is complete, the sequence $\Psi_k(\phi)$ converges uniformly with respect to $\phi \in \mathcal{T}_m$ to a certain function in the space $C(\mathcal{T}_m)$:

$$\lim_{k \to \infty} \Psi_k(\phi) = \Psi(\phi), \quad \Psi(\phi) \in C(\mathcal{T}_m). \tag{6.19}$$

Relations (6.18) and (6.19) show that the function $J(t)$ belongs to the space $C(\omega)$, as required. From the above proof we obtain the formula

$$J(t) = \bar{J} + \Phi(\omega t),$$

where $\bar{J} = \lim\limits_{T \to \infty} \frac{1}{T} \int_0^T \int_0^\tau f(\omega s) ds \, d\tau$ is the mean value of the function $J(t) = \int_0^t f(\omega s) ds$, $\Phi(\phi)$ is the function in $C(T_m)$ given by the formula

$$\Phi(\phi) = -\lim_{T \to \infty} \int_0^T \int_0^\tau f(\omega s + \phi) ds \, d\tau, \qquad (6.20)$$

where the limit is uniform with respect to $\phi \in T_m$. This formula gives a relation between the functions $f(\phi)$ and $\Phi(\phi)$ which define the corresponding function $F(t)$ and its first integral $J(t)$.

According to (6.20), the Fourier series of the function $\Phi(\phi)$ can be determined from the Fourier series of the function $f(\phi) \simeq \sum_{k \neq 0} f_k e^{i(k,\phi)}$ by using the relation

$$\Phi(\phi) \simeq \sum_{k \neq 0} \frac{f_k}{i(k,\omega)} e^{i(k,\phi)} \qquad (6.21)$$

As is clear from (6.21), the "smoothness" property of the primitive, that is, its membership of the space $C^r(T_m)$ or $H^r(T_m)$ with $r \geq 1$, depends essentially on the arithmetic properties of the frequency basis ω of the function being integrated. In particular, this leads to the following result on the "smoothness" of the primitive.

Theorem 2. *Let $F(t) = f(\omega t) \in H^r(\omega) \cap C(\omega)$, $\overline{F} = 0$ and suppose that the frequency basis $\omega = (\omega_1, \ldots, \omega_m)$ satisfies the inequality*

$$|(k,\omega)| \geq K\|k\|^{-d} \qquad (6.22)$$

for any integer vector $k = (k_1, \ldots, k_m)$, $k \neq 0$, and for constants $K > 0$ and $d > 0$. If the primitive $J(t) = \int_0^t F(t) dt \in C(\omega)$ and $r \geq d$, then $J(t) \in H^{r-d}(\omega)$ and the function $\Phi(\omega t) = J(t) - \overline{J}$ satisfies inequality

$$\|\Phi(\phi)\|_{r-d} \leq K^{-1}\|f(\phi)\|_r.$$

The assertions of the theorem follow from the chain of estimates

$$\|\Phi(\phi)\|_{r-d}^2 = \sum_{k \neq 0} \frac{(1 + \|k\|^2)^{r-d}}{|(k,\omega)|^2}\|f_k\|^2 \leq$$

$$\leq K^{-2} \sum_{k \neq 0} \|k\|^{2d} (1 + \|k\|^2)^{r-d} \|f_k\|^2 \leq$$

$$\leq K^{-2} \sum_{k \neq 0} (1 + \|k\|^2)^r \|f_k\|^2 = K^{-2} \|f(\phi)\|_r^2.$$

From this theorem and taking into account Sobolev's theorem on embedding the spaces $H^r(\mathcal{T}_m)$ into $C^s(\mathcal{T}_m)$ it follows that if the conditions (6.22) hold for the function $F(t) = f(\omega t) \in C^r(\omega)$, then its primitive $J(t) = \int_0^t f(\omega t)dt = \overline{J} + \Phi(\omega t)$ is a member of $C^s(\mathcal{T}_m)$, where s is the largest integer satisfying the inequality

$$r - d - m/2 > s. \tag{6.23}$$

The value $d + m/2$ determines the "loss of smoothness" of a quasi-periodic function in the process of taking its first integral. It should be noted that the set of frequencies $(\omega_1, \ldots, \omega_m)$ which satisfy inequality (6.22) is a rather "powerful" set even for $d = m + 1$. In fact we have the following theorem [9].

Theorem 3. *In the ball $\|\omega\| \leq 1$, the Lebesgue measure of the numbers $\omega = (\omega_1, \ldots, \omega_m)$ that do not satisfy (6.22) approaches to zero for $d = m + 1$ as $K \to 0$.*

A frequency basis satisfying (6.22) for $d = m + 1$ is sometimes called a "strongly incommensurable" basis. As is seen from (6.23), for a strongly incommensurable frequency basis, the loss of smoothness of a function from the space $C^r(\omega)$ due to taking the first integral is determined by the value of $3m/2 + 1$.

1.7. Spherical coordinates of a quasi-periodic vector function

Let the formulas

$$
\begin{aligned}
x_1 &= R \cos \theta_1, \\
x_2 &= R \sin \theta_1 \cos \theta_2, \ldots, \\
x_{n-1} &= R \sin \theta_1 \sin \theta_2 \ldots \sin \theta_{n-2} \cos \theta_{n-1}, \\
x_n &= c + R \sin \theta_1 \sin \theta_2 \ldots \sin \theta_{n-2} \sin \theta_{n-1}
\end{aligned}
\tag{7.1}
$$

define the relations between Cartesian coordinates $x = (x_1, \ldots, x_n)$ and spherical coordinates $\theta = (\theta_1, \ldots, \theta_{n-1})$, $0 \leq \theta_j \leq \pi$ $(j = 1, \ldots, n-2)$, $-\infty < \theta_{n-1} < \infty$ introduced on a sphere $S_R(c)$ of radius R and with centre at the point $x_0 = (0, \ldots, 0, c)$.

Consider a vector function $f(\phi) = (f_1(\phi), \ldots, f_n(\phi))$ with the Cartesian coordinate functions $x_j = f_j(\phi) \in C^r(\mathcal{T}_m)$ for $j = 1, \ldots, n$. For

$$R^2 = R^2(\phi) = f_1^2(\phi) + \ldots + (f_n(\phi) - c)^2 > 0$$

the spherical coordinates

$$\theta = \theta(\phi) = (\theta_1(\phi), \ldots, \theta_{n-1}(\phi)),$$

for this function on the sphere $S_R(c)$ are defined and there arises the question of how the coordinates $\theta(\phi)$ depend on ϕ. The answer is provided by the following theorem.

Theorem. *Let* $f(\phi) \in C^r(\mathcal{T}_m)$ *and*

$$f_{n-1}^2(\phi) + (f_n(\phi) - c)^2 \geq K > 0, \quad \phi \in \mathcal{T}_m. \tag{7.2}$$

Then the spherical coordinates $\theta(\phi) = (\theta_1(\phi), \ldots, \theta_{n-1}(\phi))$ *of the function* $f(\phi)$ *satisfy the conditions*

$$\theta_j(\phi) \in C^r(\mathcal{T}_m) \ \forall j = 1, \ldots, n-2, \quad \theta_{n-1}(\phi) = (p, \phi) + \Phi(\phi), \tag{7.3}$$

where $p = (p_1, \ldots, p_m)$ *is an integer vector,* $\Phi(\phi) \in C^r(\mathcal{T}_m)$.

It should be noted that for $n = 2$ and $\phi = \omega t$ the above theorem is known as Bohr's theorem [26] on the argument of a quasi-periodic function. A proof of the theorem for $n > 2$ can be reduced to the case of $n = 2$. Indeed, suppose that the theorem is true for any pair of functions $f_{n-1}(\phi)$ and $f_n(\phi) - c$ satisfying condition (7.2). Then it follows from the representation

$$f_{n-1}(\phi) = R_1(\phi) \cos \theta_{n-1}(\phi),$$
$$f_n(\phi) = c + R_1(\phi) \sin \theta_{n-1}(\phi)$$

that $R_1(\phi) = \sqrt{f_{n-1}^2(\phi) + (f_n(\phi) - c)^2} \in C^r(\mathcal{T}_m)$ and $\theta_{n-1}(\phi) = (p, \phi) + \Phi(\phi)$, where $p = (p_1, \ldots, p_m)$ is an integer vector, $\Phi(\phi) \in C^r(\mathcal{T}_m)$. Relation (7.2) also holds for the pair of functions $f_{n-2}(\phi)$ and $R_1(\phi)$, which leads to the representation

$$f_{n-2}(\phi) = R_2(\phi) \cos \theta_{n-2}(\phi),$$
$$R_1(\phi) = R_2(\phi) \sin \theta_{n-2}(\phi),$$

where $R_2(\phi) = \sqrt{f_{n-2}^2(\phi) + R_1^2(\phi)} \in C^r(T_m)$ and the function $\theta_{n-2}(\phi)$ has the form of (7.3):

$$\theta_{n-2}(\phi) = (p^{(1)}, \phi) + \Phi_1(\phi), \quad \Phi_1(\phi) \in C^r(T_m). \qquad (7.4)$$

Since the functions R_1 and R_2 are positive, $\sin \theta_{n-2}(\phi) = R_1(\phi)/R_2(\phi) > 0$ for all $\phi \in T_m$ and thus $p^{(1)} = 0$ in (7.4). This proves that the function $\theta_{n-2}(\phi)$ belongs to the space $C^r(T_m)$. By considering further pairs of functions $f_{n-3}(\phi)$ and $R_2(\phi), f_{n-4}(\phi)$ and $R_3(\phi) = \sqrt{f_{n-3}^2(\phi) + R_2^2(\phi)}$, etc., we see that the theorem holds assuming that it is true for $n = 2$. Thus it remains to prove the theorem for $n = 2$. In the proof we shall follow the reasoning of Jensen [65]. We set

$$g_1(\phi) = f_{n-1}(\phi), \quad g_2(\phi) = f_n(\phi) - c$$

and consider the polar coordinates $R(\phi), \theta(\phi)$ for the pair $g_1(\phi), g_2(\phi)$. They are defined by the relations:

$$g_1(\phi) = R(\phi)\cos\theta(\phi), \quad g_2(\phi) = R(\phi)\sin\theta(\phi).$$

In view of inequality (7.2), $R(\phi) \in C^r(T_m)$ and fixing a continuous branch of the function $\tan^{-1}(g_2(\phi)/g_1(\phi))$, for example, by the condition

$$-\frac{\pi}{2} \le \theta(0) = \tan^{-1}\frac{g_2(0)}{g_1(0)} \le \frac{\pi}{2},$$

we can take it to be $\theta(\phi)$.

Since the functions $g_1(\phi), g_2(\phi)$ are periodic and $\theta(\phi)$ is continuous, it follows that for every $1 \le j \le m$ there exists an integer p_j such that

$$\theta(\phi_1, \ldots, \phi_j + 2\pi, \ldots, \phi_m) - \theta(\phi_1, \ldots, \phi_j, \ldots, \phi_m) = 2p_j\pi. \qquad (7.5)$$

We set $p = (p_1, \ldots, p_m)$ and write $\theta(\phi)$ in the form

$$\theta(\phi) = (p, \phi) + \Phi(\phi).$$

It follows from (7.5) that $\Phi(\phi) \in C(T_m)$.

From the relations $\sin\theta(\phi) \in C^r(T_m)$, $\cos\theta(\phi) \in C^r(T_m)$ and $\Phi(\phi) \in C(T_m)$ it becomes evident that $\Phi(\phi) \in C^r(T_m)$.

1.8. The problem on a periodic basis in E^n

Let

$$U(\phi) = (u_1(\phi), \ldots, u_r(\phi)) \tag{8.1}$$

be a $n \times r$-matrix formed by the columns $u_j(\phi)$ $(j = 1, \ldots, r)$ that belong to the space $C^l(\mathcal{T}_m)$ and are linearly independent for all $\phi \in \mathcal{T}_m$ so that

$$\operatorname{rank} U(\phi) = r, \quad \phi \in \mathcal{T}_m.$$

For $r = n$, the vectors $u_j(\phi)$ $(j = 1, \ldots, n)$ form a basis in the Euclidean space E^n for any $\phi \in \mathcal{T}_m$. This basis is called *periodic and orthonormal* if the vectors $u_j(\phi)$ $(j = 1, \ldots, n)$ are orthonormal for every $\phi \in \mathcal{T}_m$:

$$\|u_j(\phi)\| = 1, \quad \langle u_j(\phi), u_i(\phi) \rangle = \delta_{ij}, \quad i, j = 1, \ldots, n,$$

where δ_{ij} is the Kronecker symbol.

For $1 \leq r < n$ there arises the problem whether there exists an $n \times (n - r)$ matrix

$$V(\phi) = (v_1(\phi), \ldots, v_{n-r}(\phi))$$

(the columns $v_j(\phi)$ $(j = 1, \ldots, n - r)$ of which belong to the space $C^l(\mathcal{T}_m)$ and are linearly independent for all $\phi \in \mathcal{T}_m$) such that the system of vectors

$$(U(\phi), V(\phi)) = (u_1(\phi), \ldots, u_r(\phi), v_1(\phi), \ldots, v_{n-r}(\phi))$$

forms a periodic basis in E^n such that

$$\det(U(\phi), V(\phi)) \neq 0, \quad \phi \in \mathcal{T}_m.$$

This problem is called the *problem of extending a periodic r-system of vectors* $U(\phi)$ *to a periodic basis in* E^n, or the *problem on a periodic basis in* E^n.

The problem on a periodic basis in E^n has an elementary solution of

$$n = r + 1.$$

In this case a complementary vector $V(\phi) = (v_1'(\phi), \ldots, v_n'(\phi))$ can be formed using elements of the matrix $U(\phi)$ by setting

$$v_j'(\phi) = (-1)^{j-1} \Delta_j(\phi), \quad j = 1, \ldots, n,$$

where $\Delta_j(\phi)$ is the determinant of the matrix obtained from $U(\phi)$ by taking out the jth row. With such a choice of the vector $V(\phi)$, the matrix $(U(\phi), V(\phi))$ satisfies the condition

$$\det(U(\phi), V(\phi)) = (-1)^{n+1}(\Delta_1^2(\phi) + \cdots + \Delta_n^2(\phi)) \neq 0, \quad \phi \in \mathcal{T}_m,$$

ensuring that the $(n-1)$-system (8.1) can be extended to a periodic basis in E^n by the vector $V(\phi)$.

Generally, the solution of the problem on a periodic basis in E^n depends on relations between r, n and m. We give a solution based on the author's paper [110]. First we prove some auxiliary statements.

Let the column vector

$$e_1(\phi_1, \ldots, \phi_{n-1}) = (e_1^1(\phi_1), \ldots, e_{n-1}^1(\phi_1, \ldots, \phi_{n-1}), e_n^1(\phi_1, \ldots, \phi_{n-1})) \quad (8.2)$$

have the coordinate functions

$$e_j^1(\phi_1, \ldots, \phi_j) = \sin\phi_1 \sin\phi_2 \ldots \sin\phi_{j-1} \cos\phi_j, \quad j = 1, \ldots, n-1,$$
$$e_n^1(\phi_1, \ldots, \phi_{n-1}) = \sin\phi_1 \sin\phi_2 \ldots \sin\phi_{n-2} \sin\phi_{n-1}$$

which give the formulae for transition between Cartesian and spherical coordinates introduced on the sphere $S_1(0)$. For $j = 2, \ldots, n$ we form the column vectors $e_j = e_j(\phi_{j-1}, \phi_j, \ldots, \phi_{n-1})$ by setting

$$e_j = \frac{\partial}{\partial\phi_{j-1}} e_1\left(\frac{\pi}{2}, \ldots, \frac{\pi}{2}, \phi_{j-1}, \phi_j, \ldots, \phi_{n-1}\right). \quad (8.3)$$

Lemma 1. *The system of vectors $(e_1, e_2, \ldots, e_n)$ forms a periodic basis in E^n.*
Indeed, according to formula (8.3), the vector e_j has the form

$$e_j = (0, \ldots, 0, -\sin\phi_{j-1}, \cos\phi_{j-1}\cos\phi_j, \cos\phi_{j-1}\sin\phi_j\cos\phi_{j+1}, \ldots,$$
$$\cos\phi_{j-1}\sin\phi_j \ldots \sin\phi_{n-2}\cos\phi_{n-1}, \cos\phi_{j-1}\sin\phi_j \ldots \sin\phi_{n-1}), \quad (8.4)$$

for $j \geq 2$. This leads to the equality

$$(e_j, e_j) = 1, \quad j = 2, \ldots, n. \quad (8.5)$$

Moreover, if $j \geq 2$ then, as follows from formulae (8.4), we have the chain of inequalities for e_j:

$$e_j = \frac{1}{\sin \phi_1 \sin \phi_2 \ldots \sin \phi_{j-2}} \frac{\partial e_1}{\partial \phi_{j-1}} =$$

$$= \frac{1}{\cos \phi_1 \sin \phi_2 \ldots \sin \phi_{j-2}} \frac{\partial^2 e_1}{\partial \phi_1 \partial \phi_{j-1}} = \ldots$$

$$\ldots = \frac{1}{\cos \phi_1 \cos \phi_2 \ldots \cos \phi_{j-2}} \frac{\partial^{j-1} e_1}{\partial \phi_1 \partial \phi_2 \ldots \partial \phi_{j-1}}.$$

But then

$$e_{j+1} = \frac{1}{\cos \phi_{j-1}} \frac{\partial e_j}{\partial \phi_j},$$

$$e_{j+\nu} = \frac{1}{\cos \phi_{j-1} \sin \phi_j \ldots \sin \phi_{j+\nu-2}} \frac{\partial e_j}{\partial \phi_{j+\nu-1}}, \quad \nu = 2, \ldots,$$

so that the value of $(e_j, e_{j+\nu})$ is proportional to the value of $(e_j, \partial e_j / \partial \phi_{j+\nu-1})$ which is equal to

$$\left(e_j, \frac{\partial e_j}{\partial \phi_{j+\nu-1}} \right) = \tfrac{1}{2} \frac{\partial}{\partial \phi_{j+\nu-1}} (e_j, e_j) = 0.$$

And so the vectors e_j $(j = 1, \ldots, n)$ are orthogonal.

We now return to the matrix (8.1). Schwartz's orthogonalization process enables us to represent it in the form

$$U(\phi) = O(\phi) \cdot T(\phi), \tag{8.6}$$

where $O(\phi)$ is an $n \times r$-matrix with orthonormal columns and $T(\phi)$ is an $r \times r$ non-singular lower triangular matrix. The elements of the two matrices belong to the space $C^l(\mathcal{T}_m)$. We take an arbitrary $n \times (n-r)$-matrix $V(\phi)$ and form the square matrix

$$\Phi(\phi) = (U(\phi), V(\phi)),$$

by extending $U(\phi)$ to an $n \times n$-matrix.

Relation (8.6) shows that

$$(U(\phi), V(\phi)) = (O(\phi)T(\phi), V(\phi)) = (O(\phi), V(\phi)) \begin{pmatrix} T(\phi) & 0 \\ 0 & E \end{pmatrix}, \tag{8.7}$$

where E is the $(n-r)(n-r)$ identity matrix.

Taking into consideration equality (8.7) it is easy to see that the following statement holds.

Lemma 2. *For the r-system of vectors (8.1) to be extendable to a periodic basis in E^n it is necessary and sufficient that the r-system of vectors $O(\phi)$, obtained by orthogonalization of the columns of the matrix $U(\phi)$, be extendable to an orthonormal periodic basis in E^n.*

Indeed, if the matrix $V(\phi) \in C^l(T_m)$ extends $U(\phi)$ to a periodic basis in E^n, then $V(\phi)$ also extends $O(\phi)$ to a periodic basis in E^n because the matrix

$$(O(\phi), V(\phi)) = (U(\phi), V(\phi)) = \begin{bmatrix} T^{-1}(\phi) & 0 \\ 0 & E \end{bmatrix}$$

is periodic and non-singular.

Making the columns of the matrix $(O(\phi), V(\phi))$ orthonormal, we find a periodic matrix $O_1(\phi)$ which extends $O(\phi)$ to a periodic orthonormal basis in E^n.

Lemma 2 relates the problem on a periodic basis in E^n to the problem of finding linearly independent periodic solutions of the system of linear equations

$$x'U(\phi) = 0, \tag{8.8}$$

where x' is a row vector.

Indeed, according to (8.6), system (8.8) is equivalent to the system

$$x'O(\phi) = 0, \tag{8.9}$$

and the matrix $O_1(\phi)$ which extends $O(\phi)$ to an orthonormal basis in E^n, satisfies the identity

$$O_1^*(\phi)O(\phi) = 0, \tag{8.10}$$

where $*$ denotes the matrix transpose. Consequently, the columns of the matrix $O_1(\phi)$ define a system of linearly independent periodic solutions of equation (8.9).

Conversely, after making any $n-r$ linearly independent solutions of equation (8.9) belonging to the space $C^l(T_m)$ orthonormal, they form a matrix $O_1(\phi)$ satisfying identity (8.10). Consequently, they form a matrix which extends $O(\phi)$ to an orthonormal periodic basis in E^n.

Lemma 3. *Suppose that for any integer p satisfying inequality*

$$n - r + 1 \leq p \leq n \tag{8.11}$$

and any vector $V(\phi) = (v_1(\phi), \ldots, v_p(\phi))$, $\|V(\phi)\| = 1$, *belonging to* $C^l(\mathcal{T}_m)$, *the equation*

$$v_1(\phi)x_1 + \ldots + v_p(\phi)x_p = 0 \tag{8.12}$$

has $p - 1$ *linearly independent solutions belonging to* $C^l(\mathcal{T}_m)$. *Then any r-system of vectors* $U(\phi) = (u_1(\phi), \ldots, u_r(\phi))$ *in* $C^l(\mathcal{T}_m)$ *can be extended to a periodic basis in* E^n.

To prove Lemma 3 it is sufficient to find $n - r$ linearly independent periodic solutions of system (8.8) or the equivalent system (8.9). We denote the orthonormal columns of the matrix $O(\phi)$ by $e_j(\phi)$ $(j = 1, \ldots, r)$ and rewrite system (8.9) in the form

$$x'e_1(\phi) = 0, \ x'e_2(\phi) = 0, \ldots, \ x'e_r(\phi) = 0. \tag{8.13}$$

Consider the first equation of (8.13). It follows from our assumptions that this equation has $p - 1 = n - 1$ linearly independent solutions belonging to the space $C^l(\mathcal{T}_m)$. We denote them by $v_2(\phi), \ldots, v_n(\phi)$ and make the system of vectors $e_1(\phi), v_2(\phi), \ldots, v_n(\phi)$ orthonormal. This gives the orthonormal matrix

$$\Phi_1(\phi) = (e_1(\phi), V_1(\phi)),$$

which forms a periodic basis in E^n. Using the matrix $\Phi_1(\phi)$ we make a change of variables in (8.13) by introducing $y' = (y_1, z')$, $z' = (z_1, \ldots, z_{n-1})$ according to the formulae $x' = y'\Phi_1^*(\phi)$.

Since the vector $e_1(\phi)$ is orthogonal to the vectors $e_j(\phi)$ $(j = 2, \ldots, r)$ and to the vectors which form the matrix $V_1(\phi)$, the system of equations (8.13) in the new coordinates will take the form

$$y_1 = 0, \quad r = 1 \tag{8.14}$$

or, equivalently,

$$y_1 = 0, \quad z'U_1(\phi) = 0, \ r \geq 2, \tag{8.15}$$

where the matrix $U_1(\phi)$ defined by the expression

$$U_1(\phi) = (V_1^*(\phi)e_2(\phi), \ldots, V_1^*(\phi)e_r(\phi))$$

is an $(n-1) \times (r-1)$-matrix of rank $r-1$ belonging to the space $C^l(\mathcal{T}_m)$.

The unit basis vectors $(0,1,0,\ldots,0),\ldots,(0,\ldots,0,1)$ clearly form $n-r = n-1$ (for $r=1$) linearly independent solutions of system (8.14). This proves the lemma for $r=1$.

To prove Lemma 3 for $r \geq 2$ consider the system

$$z'U_1(\phi) = 0 \qquad (8.16)$$

which consists of $r-1$ equations with $n-1$ unknowns.

Making the matrix $U_1(\phi)$ orthonormal, we obtain a system of equations of the form (8.9), equivalent to (8.16):

$$z'O_1(\phi) = 0. \qquad (8.17)$$

Since the conditions of the lemma apply to the first equation of this system, it has $p-1 = n-2$ linearly independent solutions belonging to the space $C^l(\mathcal{T}_m)$. Using them, we transform system (8.17) to the form similar to (8.14):

$$\bar{y}_2 = 0, \quad r-1 = 1,$$

or to the form similar to (8.15):

$$\bar{y}_2 = 0, \quad \bar{z}'U_2(\phi) = 0, \quad r-1 \geq 2,$$

where $U_2(\phi)$ is an $(n-2) \times (r-2)$ matrix belonging to $C^l(\mathcal{T}_m)$ or rank $r-2$.

The composite of the above transformations reduces the initial system of equations (8.8) to the system

$$y_1 = 0, \quad \bar{y}_2 = 0, \quad r = 2 \qquad (8.18)$$

or to the system

$$y_1 = 0, \quad \bar{y}_2 = 0, \quad \bar{z}'U_2(\phi) = 0, \quad r \geq 3.$$

In the first case the unit basis vectors $(0,0,1,0,\ldots,0),\ldots,(0,0,\ldots,0,1)$ form $n-r = n-2$ linearly independent solutions of system (8.18), and this proves the lemma for $r=2$. In the second case, the transformations similar to the ones performed for system (8.16) lead to the system of equations

$$\bar{z}'U_2(\phi) = 0.$$

Continuing this transformation process applied to system (8.8) r times, we reduce it to the system

$$y_1 = 0, \ \bar{y}_2 = 0, \ldots, \ \bar{y}_r = 0. \tag{8.19}$$

The unit basis vectors

$$(0, \ldots, 0, 1, 0, \ldots, 0), \ (0, \ldots, 0, 0, 1, 0, \ldots, 0), \ldots, \ (0, \ldots, 1)$$

clearly form $n - r$ linearly independent solutions of system (8.19). This proves the lemma for an arbitrary r, $1 \le r \le n$.

Lemma 4. *Suppose that the set of values of the function*

$$u(\phi) = (u_1(\phi), \ldots, u_n(\phi)), \ \|u(\phi)\| = 1, \ u(\phi) \in C^l(T_m)$$

is not everywhere dense on the unit sphere $S_1(0)$:

$$\|u\|^2 = u_1^2 + u_2^2 + \ldots + u_n^2 = 1.$$

Then the equation

$$u_1(\phi)x_1 + \ldots + u_n(\phi)x_n = 0 \tag{8.20}$$

has $n - 1$ linearly independent solutions belonging to the space $C^l(T_m)$.

The proof of Lemma 4 follows from the statements of Lemma 1. Indeed, if the set of values of the function $u(\phi)$ is not everywhere dense on $S_1(0)$, then there is a δ-neighbourhood of a point $p \in S_1(0)$ containing no values of the function $u(\phi)$. Without loss of generality we can assume that the point p has the coordinates $(0, 0, \ldots, 0, 1)$. Under this assumption the coordinate $u_n(\phi)$ of the vector $u(\phi)$ satisfies the inequality

$$-1 \le u_n(\phi) \le \alpha, \quad \phi \in T_m,$$

where $\alpha = \text{const} < 1$. Thus the coordinates $u_{n-1}(\phi)$ and $u_n(\phi)$ of the vector $u(\phi)$ satisfy the condition

$$u_{n-1}^2(\phi) + (u_n(\phi - c))^2 \ge K^2 > 0 \tag{8.21}$$

for an arbitrary $c \in (\alpha, 1)$ and some $K = K(c) > 0$.

Let

$$(\psi_1(\phi), \psi_2(\phi), \ldots, \psi_{n-1}(\phi)) = \psi(\phi)$$

denote the spherical coordinates of the vector $u(\phi)$ defined on the sphere with centre $(0, 0, \ldots, c)$ and radius

$$R = R(\phi) = \sqrt{u_1^2(\phi) + u_2^2(\phi) + \ldots + u_{n-1}^2(\phi) + (u_n(\phi) - c)^2}.$$

In view of inequality (8.21) we can apply the theorem from §1.7 to $\psi(\phi)$ and thus we see that the coordinates of this function are appropriately smooth and periodic. The transition formulae relating Cartesian and spherical coordinates allow one to write the vector $u(\phi)$ in the form

$$u(\phi) = R(\phi) e_1(\psi(\phi)) + c e_0, \qquad (8.22)$$

where $e_1(\phi)$ is the vector (8.2) and $e_0 = (0, \ldots, 0, 1)$ is a unit basis vector. Using representation (8.22) we now give linearly independent solutions of equation (8.20) belonging to $C^l(\mathcal{T}_m)$. Let us start with the following system of vectors:

$$u(\phi), e_2(\psi(\phi)), \ldots, e_n(\psi(\phi)), \qquad (8.23)$$

where the $e_j(\phi)$ are the orthonormal vectors (8.3).

We show that the system (8.23) forms a periodic basis in E^n. Since the functions $\psi_j(\phi)$ satisfy conditions (7.3) (if in (7.3) we replace θ and r by ψ and l respectively), the functions (8.23) belong to the space $C^l(\mathcal{T}_m)$. It remains to show that vectors (8.23) are linearly independent. To see this, we evaluate the determinant of the matrix

$$\Phi = (u(\phi), e_2(\psi(\phi)), \ldots, e_n(\psi(\phi))).$$

Taking into account (8.22), we have

$$\det \Phi = R(\phi) + c \det(e_0, e_2(\psi(\phi)), \ldots, e_n(\psi(\phi))).$$

Taking into account the form of the vector $e_j(\phi)$ and continuing the calculation we find that

$$\det(e_0, e_2(\psi(\phi)), \ldots, e_n(\psi(\phi))) = \sin \psi_1(\phi) \sin \psi_2(\phi) \ldots \sin \psi_{n-1}(\phi).$$

Using formula (8.22) and the form of the vector $e_1(\phi)$ we now have

$$\det \Phi = R(\phi) + c \sin \psi_1(\phi) \sin \psi_2(\phi) \ldots \sin \psi_{n-1}(\phi) =$$

$$= R(\phi) + \frac{c}{R(\phi)}(u_n(\phi) - c) = \frac{R^2(\phi) + cu_n(\phi) - c^2}{R(\phi)} = \frac{1 - cu_n(\phi)}{R(\phi)}.$$

From the last relation it follows that

$$\det \Phi \geq (1 - c^2)/R(\phi) > 0, \quad \phi \in T_m.$$

Making the vectors (8.23) orthogonal we obtain $n - 1$ linearly independent solutions of equation (8.20) belonging to the space $C^l(T_m)$.

The following lemma will be used to verify whether the conditions of Lemma 4 hold.

Lemma 5. *If $u(\phi) = (u_1(\phi), \ldots, u_n(\phi)) \in C^l(T_m)$, $\|u(\phi)\| = 1$, $l \geq 1$ and the inequality*

$$n > m + 1 \tag{8.24}$$

holds, then the set of values of the function $u(\phi)$ is not everywhere dense on the unit sphere $S_1(0)$.

Indeed, the function $u(\phi)$ defines a smooth map U of the m-dimensional torus $T_m : 0 \leq \phi_j \leq 2\pi$ $(j = 1, \ldots, m)$ into the $n-1$-dimensional sphere $S_1(0)$. Because of (8.24), U is a map of a manifold of smaller dimension to a manifold of larger dimension. Under such a map, all the image points UT_m are "critical" values of the map U and their measure equals zero (see Sard's theorem [129]). Being closed, the set UT_m cannot "cover" the entire sphere $S_1(0)$.

The above statements enable one to prove the following result on the extendability of a periodic r-system of vectors to a periodic basis in E^n.

Theorem. *Let $n \times r$-matrix $U(\phi) = (u_1(\phi), \ldots, u_r(\phi))$ be a member of the space $C^l(T_m)$ and have rank r for all $\phi \in T_m$. If the inequality*

$$n > m + r \tag{8.25}$$

holds, then the matrix $U(\phi)$ can be extended to a periodic basis in E^n.

To prove the theorem, first suppose that $l \geq 1$. Since it follows from (8.25) that for any integer p satisfying the condition

$$n - r + 1 \leq p \leq n,$$

the inequality

$$p > m + 1,$$

holds, the equation

$$v_1(\phi)x_1 + \ldots + v_p(\phi)x_p = 0$$

(where, according to Lemmas 4 and 5, the coefficients $v(\phi) = (v_1(\phi), \ldots, v_p(\phi))$ belong to the space $C^l(T_m)$) has exactly $p - 1$ linearly independent solutions which belong to the space $C^l(T_m)$. Then Lemma 3 asserts that the r-system of vectors $U(\phi) = (u_1(\phi), \ldots, u_r(\phi))$ in the space $C^l(T_m)$ can be extended to a periodic basis of E^n.

Now let $l = 0$. We represent $U(\phi)$ in the form (8.6):

$$U(\phi) = O(\phi)T(\phi)$$

where $0(\phi)$ is an $n \times r$ orthonormal matrix belonging to the space $C(T_m)$.

Applying the approximation theorem to the matrix $O(\phi)$, we construct an orthogonal matrix $O_1(\phi)$ belonging to the space $C^1(T_m)$ and arbitrarily close to $O(\phi)$ with respect to the metric of $C(T_m)$. Let

$$\|O(\phi) - O_1(\phi)\| < \delta, \quad \phi \in T_m,$$

where δ is a positive number of suitable magnitude.

As was proved above, the matrix $O_1(\phi)$ can be extended to an orthonormal periodic basis in E^n. Consequently, there exists a matrix $V_1(\phi) \in C^1(T_m)$ such that

$$\det(O_1(\phi), V_1(\phi)) = 1.$$

The matrix $V_1(\phi)$ extends $O(\phi)$ to a periodic basis in E^n since

$$\det(O(\phi), V_1(\phi)) \geq 1 - \epsilon(\delta) \geq 1/2$$

for sufficiently small $\delta > 0$. This is sufficient to enable us to extend $U(\phi)$ to a periodic basis in E^n.

Note that the theorem given above cannot be improved, since if inequalities (8.25) fail to hold, then this can result in the inability to extend a periodic r-system of vectors from $C^l(T_m)$ to a periodic basis in E^n. The reasons for this and examples are given in [34], [35]. Note also that the problem on a periodic basis in E^n has a well known interpretation in topology, and the fact stated in the theorem has, in the latter context, a more elegant proof than the one given above.

1.9. Logarithm of a matrix in $C^l(T_m)$. Sibuja's theorem

Let $P(\phi)$ be an $n \times n$-matrix whose elements are functions in $C^l(T_m)$. A matrix $X = X(\phi)$ that satisfies the equation

$$e^X = P(\phi), \tag{9.1}$$

is called a *logarithm* of the matrix $P(\phi)$. Since the exponential of any matrix is non-singular, the condition

$$\det P(\phi) \neq 0 \tag{9.2}$$

is necessary for a logarithm of the matrix $P(\phi)$ to exist.

Suppose that condition (9.2) holds, let

$$\mathcal{J}(\phi) = \{\lambda_1(\phi)E_1 + Z_1, \ldots, \lambda_q(\phi)E_q + Z_q\}$$

be a Jordan form of the matrix $P(\phi)$, and let $T(\phi)$ be a matrix that reduces $P(\phi)$ to this form:

$$P(\phi) = T(\phi)\mathcal{J}(\phi)T^{-1}(\phi).$$

Let $C(\phi)$ be an arbitrary non-singular matrix commuting with the matrix $\mathcal{J}(\phi)$.

A general formula giving all solutions of equation (9.1), that is, giving all the logarithms of the matrix $P(\phi)$ has the form [39]:

$$X(\phi) = T(\phi)C(\phi)\ln \mathcal{J}(\phi)C^{-1}(\phi)T^{-1}(\phi), \tag{9.3}$$

where the matrix $\ln \mathcal{J}(\phi)$ defined by the expression:

$$\ln \mathcal{J}(\phi) = \{\ln(\lambda_1(\phi)E_1 + Z_1), \ldots, \ln(\lambda_q(\phi)E_q + Z_q)\}$$

is quasi-diagonal. The diagonal blocks of the matrix $\ln \mathcal{J}(\phi)$ can be found using the formula:

$$\ln(\lambda_j(\phi)E_j + Z_j) = \ln \lambda_j E_j + \frac{1}{\lambda_j}Z_j + \ldots =$$

$$= f(\lambda_j)E_j + \frac{f'(\lambda_j)}{1!}Z_j + \frac{f''(\lambda_j)}{2!}Z_j + \ldots + \frac{f^{(p_j-1)}(\lambda_j)}{(p_j - 1)!}Z_j^{p_j-1},$$

where

$$f(\lambda) = \ln \lambda, \quad Z_j = \begin{bmatrix} 0 & 0 & \cdots & 0 & 0 \\ 1 & 0 & \cdots & 0 & 0 \\ \cdot & \cdot & \cdot & \cdot & \cdot \\ 0 & 0 & \cdots & 1 & 0 \end{bmatrix}$$

and p_j is the dimension of the matrix Z_j.

In the above formulae q and p_j $(j = 1, \ldots, q)$ are functions of ϕ.

As one can see from (9.3), the logarithm of a non-singular matrix is multi-valued. We shall be interested in the real branch of the logarithm of a non-singular real matrix $P(\phi)$. As follows from [63], such a branch exists if and only if the matrix $P(\phi)$ either does not have elementary divisors which correspond to negative eigenvalues, or each such divisor is repeated an even number of times. The latter is true if, in particular,

$$P(\phi) = Q^2(\phi)$$

where $Q(\phi)$ is a real matrix.

Suppose that the matrix $P(\phi)$ has a real logarithm. We denote it by

$$X(\phi) = \ln P(\phi)$$

and find out when we can choose it so that it will belong to the space $C^l(T_m)$.

The example of the matrix

$$P_0(\phi) = \begin{bmatrix} \cos \phi & \sin \phi \\ -\sin \phi & \cos \phi \end{bmatrix} = \frac{1}{2} \begin{bmatrix} -i & i \\ 1 & 1 \end{bmatrix} \begin{bmatrix} e^{i\phi} & 0 \\ 0 & e^{-i\phi} \end{bmatrix} \begin{bmatrix} i & 1 \\ -i & 1 \end{bmatrix} \qquad (9.4)$$

shows that this cannot always be achieved. Indeed, formula (9.3) defines the real branch of the logarithm of matrix (9.4):

$$\ln P_0(\phi) = \begin{bmatrix} 0 & \phi - 2k\pi \\ -\phi + 2k\pi & 0 \end{bmatrix}, \qquad (9.5)$$

where k is an arbitrary number. For a fixed number k, matrix (9.5) is not periodic with respect to ϕ; for k equal to the integer part of $\phi/(2\pi)$, matrix (9.5) is periodic with respect to ϕ but it is not continuous. Thus for the matrix $P(\phi)$ given by (9.4), its logarithm does not belong to $C(T_1)$.

To find out the conditions for a matrix $\ln P(\phi)$ to belong to the space $C^l(T_m)$, we write down Cauchy's integral formula for $\ln P(\phi)$ [42]:

$$\ln P(\phi) = \frac{1}{2\pi i} \int_\Gamma (\lambda E - P(\phi))^{-1} \ln \lambda d\lambda, \qquad (9.6)$$

where Γ is a closed contour in the complex λ-plane bounding a simply connected region G containing the eigenvalues $\lambda_1(\phi), \ldots, \lambda_n(\phi)$ of $P(\phi)$ and not containing the point $\lambda = 0$, and $\ln \lambda$ is a branch of the logarithm that is regular in G. Since the boundary Γ of the region G in (9.6) can be chosen separately for each ϕ, we see that formula (9.6) is a local one.

The totality of values of the eigenvalues $\lambda_1(\phi), \ldots, \lambda_n(\phi)$ of the matrix $P(\phi)$ when ϕ covers $\mathcal{T}_m$ is called the *spectrum* of the matrix $P(\phi)$. We shall say that the spectrum of the matrix $P(\phi)$ *does not contain the origin* if it is contained in a simply connected region G of complex plane that does not enclose the origin. For the matrix $P(\phi)$ the contour Γ in (9.6) can be chosen to be the same for all $\phi \in \mathcal{T}_m$ if its spectrum does not contain the origin. In this case, formula (9.6) becomes global since it defines an integral over Γ which is the same for all ϕ; therefore the integral (9.6) preserves the properties of the matrix $P(\phi)$ considered as a function in the space $C^l(\mathcal{T}_m)$. The following result summarizes the above remarks.

Theorem 1. *Let $P(\phi)$ be a non-singular matrix in $C^l(\mathcal{T}_m)$ with spectrum not containing the origin. Then a branch of logarithm of the matrix $P(\phi)$ can be chosen in such a way that it belongs to the space $C^l(\mathcal{T}_m)$.*

Note that the above branch of the logarithm turns out to be real when $P(\phi)$ is real and its eigenvalues satisfy the conditions given above for the logarithm to be real.

Having finished our discussion of matrices in $C^l(\mathcal{T}_m)$ we now give a result on the block diagonalization of a matrix $P(\phi) \in C^l(\mathcal{T}_1)$ due to Y. Sibuja [36], [169].

Theorem 2. *Let $P(\phi)$ be an $n \times n$ square matrix of a scalar variable ϕ, periodic with respect to ϕ with the period 2π and l times continuously differentiable. Suppose that the characteristic polynomial $p(\phi, \lambda)$ of the matrix $P(\phi)$ can be represented in the form*

$$p(\phi, \lambda) = p_1(\phi, \lambda) \cdot p_2(\phi, \lambda),$$

where p_1 and p_2 are polynomials in λ of the form

$$p_1(\phi, \lambda) = (-\lambda)^{m_1} + q_1(\phi)\lambda^{m_1-1} + \ldots + q_{m_1}(\phi),$$
$$p_2(\phi, \lambda) = (-\lambda)^{m_2} + r_1(\phi)\lambda^{m_2-1} + \ldots + r_{m_2}(\phi)$$

with l times continuously differentiable periodic coefficients with period 2π. Suppose that the polynomials $p_1(\phi, \lambda)$ and $p_2(\phi, \lambda)$ have no common roots for any $\phi \in T_1$. Then there exists a non-singular matrix $T(\phi)$ that is periodic with respect to ϕ with period 2π, l times continuously differentiable and such that

$$P(\phi) = T(\phi)\text{diag}\{E(\phi), F(\phi)\}T^{-1}(\phi),$$

where $E(\phi)$ and $F(\phi)$ are the corresponding $(m_1 \times m_1)$ and $(m_2 \times m_2)$ square matrices with characteristic polynomials $p_1(\phi, \lambda)$ and $p_2(\phi, \lambda)$. If the matrix $P(\phi)$ and coefficients of the polynomials $p_1(\phi, \lambda)$ and $p_2(\phi, \lambda)$ are real then the matrix $T(\phi)$ can be chosen to be real and periodic with period 4π.

The example considered in [169] of the matrix

$$P(\phi) = P_0\left(-\frac{\phi}{2}\right)DP_0\left(\frac{\phi}{2}\right), \quad P(\phi) \in C^\infty(T_1),$$

where $P_0(\phi)$ is matrix (9.4), $D = \text{diag}\{d_1, d_2\}$, d_1, d_2 are different, shows that the doubling of the period of the matrix $T(\phi)$ in choosing it to be real is essential for Theorem 2 to hold. We also note that the analogue of Sibuja's theorem is not known for $m > 1$.

1.10. Gårding's inequality

Let

$$L_2 = -\sum_{\nu,\mu} A_{\nu\mu}(\phi)\frac{\partial^2}{\partial\phi_\nu\partial\phi_\mu} + \sum_\nu a_\nu(\phi)\frac{\partial}{\partial\phi_\nu} + b(\phi)$$

be a differential operator, the coefficients of which are square matrices $A_{\nu\mu}, a_\nu, b$ of order n belonging to $C^1(T_m)$, $C(T_m)$, $C(T_m)$ respectively. We require that the operator L_2 be *elliptic*, that is, the inequality

$$\sum_{\nu,\mu}\langle A_{\nu\mu}, \eta_\nu, \eta_\mu\rangle \geq \gamma_1\sum_{\nu=1}^m \|\eta_\nu\|^2 \tag{10.1}$$

must hold for any collection $\eta_1, \ldots, \eta_m$ of n-dimensional vectors $\eta_\nu = (\eta_\nu^1, \ldots, \eta_\nu^n)$ for some $\gamma_1 > 0$.

The ellipticity L_2 enables us to prove an important inequality for the inner product $(L_2u, u)_p$. In the literature it is known as Gårding's inequality [14].

Theorem. *If $A_{\nu\mu}, a_\nu, b \in C^p(T_m)$ then for any function $u \in H^{p+1}(T_m)$*

$$(L_2u, u)_p \geq \gamma\|u\|_{p+1}^2 - \delta\|u\|_0^2, \tag{10.2}$$

where γ and δ are positive constants which depend on the coefficients of L_2 but do not depend on u.

Proof. Let $u \in H^\infty(\mathcal{T}_m)$. Then $L_2 u \in C^p(\mathcal{T}_m) \subset H^p(\mathcal{T}_m)$ and the expression on the left hand side of (10.2) makes sense. We now derive inequality (10.2).

Let $p = 0$. Then

$$(L_2 u, u)_0 = -\sum_{\nu,\mu} \Big(\frac{\partial}{\partial \phi_\mu} \Big(A_{\nu\mu} \frac{\partial u}{\partial \phi_\nu} \Big), u \Big)_0 + \sum_{\nu,\mu} \Big(\frac{\partial A_{\nu\mu}}{\partial \phi_\mu} \frac{\partial u}{\partial \phi_\nu}, u \Big)_0 +$$

$$+ \sum_\nu \Big(a_\nu \frac{\partial u}{\partial \phi_\nu}, u \Big)_0 + (bu, u)_0 = \sum_{\nu,\mu} \Big(A_{\nu\mu} \frac{\partial u}{\partial \phi_\nu}, \frac{\partial u}{\partial \phi_\mu} \Big)_0 + \Phi_0.$$

We make an estimate for the functional ϕ_0. Using the estimation technique given in 1.3 we find that

$$|\Phi_0| \leq \sum_{\nu,\mu} \Big\| \frac{\partial A_{\nu\mu}}{\partial \phi_\mu} \frac{\partial u}{\partial \phi_\nu} \Big\|_0 \|u\|_0 + \sum_\nu \Big\| a_\nu \frac{\partial u}{\partial \phi_\nu} \Big\|_0 \|u\|_0 + \|bu\|_0 \|u\|_0 \leq$$

$$\leq 2c \Big(\sum_{\nu,\mu} \Big| \frac{\partial A_{\nu\mu}}{\partial \phi_\mu} \Big|_0 + \sum_\nu |a_\nu|_0 + |b|_0 \Big) \|u\|_1 \|u\|_0 \leq c' \Big(\delta_1^2 \|u\|_1^2 + \frac{1}{\delta_1^2} \|u\|_0^2 \Big) \quad (10.3)$$

where $c' = c \Big(\sum_{\nu,\mu} \big| \frac{\partial A_{\nu\mu}}{\partial \phi_\mu} \big|_0 + \sum_\nu |a_\nu|_0 + |b|_0 \Big)$, δ_1 is an arbitrarily small number, and c is a positive constant which does not depend either on u or on the coefficients of L_2.

We make an estimate of the product $\sum_{\nu,\mu} \Big(A_{\nu\mu} \frac{\partial u}{\partial \phi_\nu}, \frac{\partial u}{\partial \phi_\mu} \Big)_0$. Using inequality (10.1) we obtain the estimate:

$$\sum_{\nu,\mu} \Big(A_{\nu\mu} \frac{\partial u}{\partial \phi_\nu}, \frac{\partial u}{\partial \phi_\mu} \Big)_0 = \sum_{\nu,\mu} \frac{1}{(2\pi)^m} \int_0^{2\pi} \cdots \int_0^{2\pi} \Big\langle A_{\nu\mu} \frac{\partial u}{\partial \phi_\nu}, \frac{\partial u}{\partial \phi_\mu} \Big\rangle d\phi_1 \ldots d\phi_m \geq$$

$$\geq \gamma_1 \sum_{\nu=1}^m \frac{1}{(2\pi)^m} \int_0^{2\pi} \cdots \int_0^{2\pi} \Big\| \frac{\partial u}{\partial \phi_\nu} \Big\|^2 d\phi_1 \ldots d\phi_m =$$

$$= \gamma_1 \sum_{\nu=1}^m \Big\| \frac{\partial u}{\partial \phi_\nu} \Big\|_0^2 \geq \gamma_1 (c_1 \|u\|_1^2 - c_2 \|u\|_0^2), \quad (10.4)$$

where c_1 and c_2 are positive constants independent of u and $A_{\nu\mu}$.

Using both inequalities (10.3) and (10.4) we arrive at inequality (10.2) for $p = 0$:

$$(L_2 u, u)_0 \geq (\gamma_1 c_1 - c' \delta_1^2) \|u\|_1^2 - \Big(\gamma_1 c_2 - \frac{c'}{\delta_1^2} \Big) \|u\|_0^2 = \gamma \|u\|_1^2 - \delta \|u\|_0^2.$$

Let $p = 1$. Then

$$(L_2 u, u)_1 = (L_2 u, u)_0 + (L_2 u, -\Delta u)_0 \geq \sum_{i=1}^{m} \left(\frac{\partial}{\partial \phi_i} L_2 u, \frac{\partial u}{\partial \phi_i} \right)_0 + \gamma \|u\|_1^2 - \delta \|u\|_0^2 =$$

$$= \sum_{i=1}^{m} \left(L_2 \frac{\partial u}{\partial \phi_i}, \frac{\partial u}{\partial \phi_i} \right)_0 + \sum_{i=1}^{m} \left(\sum_{\nu,\mu} \frac{\partial A_{\nu\mu}}{\partial \phi_i} \frac{\partial^2 u}{\partial \phi_\nu \partial \phi_\mu} + \right.$$

$$\left. + \sum_{\nu} \frac{\partial a_\nu}{\partial \phi_i} \frac{\partial u}{\partial \phi_\nu} + \frac{\partial b}{\partial \phi_i} u, \frac{\partial u}{\partial \phi_i} \right)_0 + \gamma \|u\|_1^2 - \delta \|u\|_0^2 \geq$$

$$\geq \gamma \sum_{i=1}^{m} \left\| \frac{\partial u}{\partial \phi_i} \right\|_1^2 - \delta \sum_{i=1}^{m} \left\| \frac{\partial u}{\partial \phi_i} \right\|_0^2 + \sum_{i=1}^{m} \left(\Phi_i u, \frac{\partial u}{\partial \phi_i} \right)_0 + \gamma \|u\|_1^2 - \delta \|u\|_0^2 \geq$$

$$\geq \gamma c_1 \|u\|_2^2 - \gamma \bar{c}_2 \|u\|_0^2 - \delta \|u\|_1^2 + \gamma \|u\|_1^2 - \delta \|u\|_0^2 + \sum_{i=1}^{m} \left(\Phi_i u, \frac{\partial u}{\partial \phi_i} \right)_0 \geq$$

$$\geq \gamma c_1 \|u\|_2^2 - \delta c_2 \|u\|_1^2 + \gamma \|u\|_1^2 - \delta \|u\|_0^2 + \sum_{i=1}^{m} \left(\Phi_i u, \frac{\partial u}{\partial \phi_i} \right)_0,$$

where c_1, c_2 are positive constants independent of u and L_2.

We now estimate the product $(\Phi_i u, \partial u/\partial \phi_i)_0$. We have:

$$\left| \left(\Phi_i u, \frac{\partial u}{\partial \phi_i} \right)_0 \right| \leq \left[\sum_{\nu,\mu} \left\| \frac{\partial A_{\nu\mu}}{\partial \phi_i} \frac{\partial^2 u}{\partial \phi_\nu \partial \phi_\mu} \right\|_0 + \right.$$

$$\left. + \sum_{\nu} \left\| \frac{\partial a_\nu}{\partial \phi_i} \frac{\partial u}{\partial \phi_\nu} \right\|_0 + \left\| \frac{\partial b}{\partial \phi_i} u \right\|_0 \right] \cdot \left\| \frac{\partial u}{\partial \phi_i} \right\|_0 \leq$$

$$\leq c \left[\sum_{\nu,\mu} |A_{\nu\mu}|_1 + \sum_{\nu} |a_\nu|_1 + |b|_1 \right] \|u\|_2 \|u\|_1 = \frac{c'}{m} \|u\|_2 \|u\|_1, \qquad (10.5)$$

where c is a positive constant independent of u and L_2.

Taking into account estimate (10.5) we obtain the following inequality:

$$(L_2 u, u)_1 \geq \gamma c_1 \|u\|_2^2 - (\delta c_2 - \gamma + c') \|u\|_2 \|u\|_1 - \delta \|u\|_0^2,$$

which, in view of the inequalities

$$2\|u\|_2 \|u\|_1 \leq \delta_1^2 \|u\|_2^2 + \frac{1}{\delta_1^2} \|u\|_1^2 \leq \delta_1^2 \|u\|_2^2 + \frac{1}{\delta_1^2} \|u\|_2 \|u\|_0 \leq$$

$$\leq \delta_1^2 \|u\|_2^2 + \frac{1}{2\delta_1^2} \left(\delta_1^4 \|u\|_2^2 + \frac{1}{\delta_1^4} \|u\|_0^2 \right) = \frac{3}{2} \delta_1^2 \|u\|_2^2 + \frac{1}{2\delta_1^6} \|u\|_0^2,$$

takes the form of inequality (10.2):

$$(L_2 u, u)_1 \geq \left[\gamma c_1 - \delta_1^2 \frac{3}{2} (\delta c_2 - \gamma + c') \right] \|u\|_2^2 -$$

$$- \left[\delta + \frac{\delta c_2 - \gamma + c'}{2\delta_1^6} \right] \|u\|_0^2 = \gamma_{(1)} \|u\|_2^2 - \delta_{(1)} \|u\|_0^2,$$

where $\gamma_{(1)}$ and $\delta_{(1)}$ are positive constants which depend only on $\sum_{\nu,\mu} |A_{\nu\mu}|_1 + \sum_\nu |a_\nu|_1 + |b|_1$, and do not depend on u.

Now suppose that inequality (10.2) has been proved for $p \leq r$. Consider the product $(L_2 u, u)_{r+1}$. We have

$$(L_2 u, u)_{r+1} = (L_2 u, u)_r + (L_2 u, -\Delta u)_r \geq$$

$$\geq \sum_{i=1}^m \left(\frac{\partial}{\partial \phi_i} L_2 u, \frac{\partial u}{\partial \phi_i} \right)_r + \gamma \|u\|_{r+1}^2 - \delta \|u\|_0^2 =$$

$$= \sum_{i=1}^m \left(L_2 \frac{\partial u}{\partial \phi_i}, \frac{\partial u}{\partial \phi_i} \right)_r + \sum_{i=1}^m \left(\Phi_i u, \frac{\partial u}{\partial \phi_i} \right)_r + \gamma \|u\|_{r+1}^2 - \delta \|u\|_0^2 \geq$$

$$\geq \gamma c_1 \|u\|_{r+2}^2 - \delta c_2 \|u\|_1^2 + \gamma \|u\|_{r+1}^2 - \delta \|u\|_0^2 + \sum_{i=1}^m \left(\phi_i u, \frac{\partial u}{\partial \phi_i} \right)_r \geq$$

$$\geq \gamma c_1 \|u\|_{r+2}^2 - \left(\frac{\delta c_2}{2} \right) \cdot 2 \|u\|_{r+2} \|u\|_{r+1} - \delta \|u\|_0^2 + \sum_{i=1}^m \left(\Phi_i u, \frac{\partial u}{\partial \phi_i} \right)_r.$$

Making an estimate of $(\Phi_i u, \partial u/\partial \phi_i)_r$, we have:

$$\left| \left(\Phi_i u, \frac{\partial u}{\partial \phi_i} \right)_r \right| \leq \left[\sum_{\nu,\mu} \left\| \frac{\partial A_{\nu\mu}}{\partial \phi_i} \frac{\partial^2 u}{\partial \phi_\nu \partial \phi_\mu} \right\|_r + \sum_\nu \left\| \frac{\partial a_\nu}{\partial \phi_i} \frac{\partial u}{\partial \phi_\nu} \right\|_r + \right.$$

$$\left. + \left\| \frac{\partial b}{\partial \phi_i} u \right\|_r \right] \cdot \left\| \frac{\partial u}{\partial \phi_i} \right\|_r \leq 2c \left[\sum_{\nu,\mu} |A_{\nu\mu}|_{r+1} + \sum_\nu |a_\nu|_{r+1} + |b|_{r+1} \right] \times$$

$$\times \|u\|_{r+2} \|u\|_{r+1} = \frac{2c'}{m} \|u\|_{r+2} \|u\|_{r+1}. \qquad (10.6)$$

But since by (3.2)

$$2\|u\|_{r+2} \|u\|_{r+1} \leq \delta_1^2 \|u\|_{r+2}^2 + \frac{1}{\delta_1^2} \|u\|_{r+1}^2 \leq \delta_1^2 \|u\|_{r+2}^2 + \frac{1}{\delta_1^2} \|u\|_{r+2} \|u\|_{r+1} \leq$$

$$\leq \delta_1^2 \|u\|_{r+2}^2 + \frac{\|u\|_{r+2}}{\delta_1^2} \left(\delta_1^4 \|u\|_{r+2} + \frac{1}{\delta_1^{4(r+1)}} \|u\|_0 \right) =$$

$$= 2\delta_1^2\|u\|_{r+2}^2 + \frac{1}{\delta_1^{2+4(r+1)}}\|u\|_{r+2}\|u\|_0 \le$$

$$\le 2\delta_1^2\|u\|_{r+2}^2\frac{1}{2\delta_1^{2+4(r+1)}}\left(\delta_1^{4(r+2)}\|u\|_{r+2}^2 + \frac{1}{\delta_1^{4(r+2)}}\|u\|_0^2\right) =$$

$$= \frac{5}{2}\delta_1^2\|u\|_{r+2}^2 + \frac{1}{2\delta_1^{2[3+4(r+1)]}}\|u\|_0^2,$$

it follows from (10.6) that

$$\left|\left(\Phi_i u, \frac{\partial u}{\partial \phi_i}\right)_r\right| \le \frac{c'}{m}\left[\frac{5}{2}\delta_1^2\|u\|_{r+2}^2 + \frac{\|u\|_0^2}{2\delta_1^{2[3+4(r+1)]}}\right]. \tag{10.7}$$

Using inequality (10.7) we make an estimate of the product $(L_2 u, u)_{r+1}$ as follows:

$$(L_2 u, u)_{r+1} \ge \left[\gamma c_1 - \frac{5}{2}\left(\frac{\delta c_2}{2} + c'\right)\delta_1^2\right]\|u\|_{r+2}^2-$$

$$-\left[\left(\frac{\delta c_2}{2} + c'\right)\frac{1}{2\delta_1^{2[3+4(r+1)]}} + \delta\right]\|u\|_0^2 \ge \gamma_{(r+1)}\|u\|_{r+2}^2 - \delta_{(r+1)}\|u\|_0^2,$$

where $\gamma_{(r+1)}$ and $\delta_{(r+1)}$ are positive constants which are dependent only on $\sum_{\nu,\mu}|A_{\nu\mu}|_{r+1} + \sum_\nu |a_\nu|_{r+1} + |b|_{r+1}$ but do not depend on u.

By assuming that inequality (10.2) holds for $p \le r$ we have proved it for $p = r + 1$. Hence by induction, inequality (10.2) holds for any p and $u \in H^\infty(\mathcal{T}_m)$.

Now let $u \in H^{p+1}(\mathcal{T}_m)$. Then $L_2 u \in H^{p-1}(\mathcal{T}_m)$ and the product $(L_2 u, u)_p$ makes sense. For $u \in H^{p+1}(\mathcal{T}_m)$, inequality (10.2) follows from the following chain of inequalities

$$(L_2 u, u)_p = (L_2(u - u_k), u)_p + (L_2 u_k, u - u_k)_p + (L_2 u_k, u_k)_p \ge$$

$$\ge \gamma\|u_k\|_{p+1}^2 - \delta\|u_k\|^2 - \|L_2(u - u_k)\|_{p-1}\|u\|_{p+1}-$$

$$-\|L_2 u_k\|_{p-1}\|u - u_k\|_{p+1}, \tag{10.8}$$

where u_k is a sequence of functions in $H^\infty(\mathcal{T}_m)$ convergent to u in $H^{p+1}(\mathcal{T}_m)$. Indeed, passing to the limit as $k \to \infty$ in (10.8) we have

$$(L_2 u, u)_p \ge \gamma\|u\|_{p+1}^2 - \delta\|u\|_0^2 - \lim_{k\to\infty}\|L_2(u - u_k)\|_{p-1}\|u\|_{p+1}-$$

$$- \lim_{k\to\infty}\|L_2 u_k\|_{p-1} \times \lim_{k\to\infty}\|u - u_k\|_{p+1} = \gamma\|u\|_{p+1}^2 - \delta\|u\|_0^2,$$

where the constants γ and δ depend only on $\sum_{\nu,\mu}|A_{\nu\mu}|_p + \sum_\nu |a_\nu|_p + |b|_p$, but do not depend on u.

Chapter 2. Invariant sets and their stability

2.1. Preliminary notions and results

We shall be looking at the following system of differential equations:

$$\frac{dx}{dt} = X(x), \tag{1.1}$$

the right hand side of which is defined, continuous and satisfies a Lipschitz condition in a domain D of n-dimensional Euclidean space. Under these assumptions, for every $x_0 \in D$ there exists a unique solution $x(t, x_0)$ of system (1.1) such that $x(0, x_0) = x_0$. This solution is defined and belongs to the domain D for a maximal interval (T^-, T^+) in $\mathbf{R}$, where $-\infty \le T^- < 0 < T^+ \le +\infty$, $T^- = T^-(x_0)$, $T^+ = T^+(x_0)$. Since the system of equations (1.1) is autonomous, the function $x(t, x_0)$ satisfies the group condition with respect to t:

$$x(t + \tau, x_0) = x(t, x(\tau, x_0))$$

for any $x_0 \in D$ and any t, τ and $t + \tau$ in (T^-, T^+).

Following conventional terminology, the solution $x(t, x_0)$ for a fixed x_0 will be called the *motion*. The set $\{x(t, x_0), \; t \in (T^-, T^+)\}$ will be called the *trajectory* of the motion $x(t, x_0)$, and when $T^- = -\infty, T^+ = +\infty$ it will be denoted by $x(t, x_0; \mathbf{R})$. The set $\{x(t, x_0), t \in [0, T^+)\}$ for $T^+ = +\infty$ will be called the *positive semi-trajectory* and the set $\{x(t, x_0), \; t \in (T^-, 0]\}$ for $T^- = -\infty$ will be called the *negative semi-trajectory* of the motion $x(t, x_0)$ and will be denoted by $x(t, x_0; \mathbf{R}^+)$ and $x(t, x_0; \mathbf{R}^-)$ respectively. Finally, the set of points $\{x(t, x_0), \; T_1 \le t \le T_2\}$ for $T^- < T_1 < T_2 < T^+$ will be called a *finite arc* of the trajectory of the motion $x(t, x_0)$ with the time length $T_2 - T_1$ which we denote by $x(t, x_0; [T_1, T_2])$.

A set $M \subset D$ is called an *invariant set* of system (1.1) if it consists of points of the trajectories of this system.

A set $D_1 \subset D$ is said to be *compact* if any infinite sequence of points of it contains a convergent subsequence and the limit point of this subsequence belongs to D.

A set $\Omega \subset E^n$ is called *connected* if it cannot be represented as a union of two non-empty sets in such a way that neither set contains limit points of the other.

Let

$$0 \leq t_1 < t_2 < \ldots < t_n < \ldots, \qquad \lim_{n \to \infty} t_n = +\infty \qquad (1.2)$$

be a sequence of values of t such that

$$\lim_{n \to \infty} x(t_n, x_0) = x^0 \in D. \qquad (1.3)$$

Then x^0 is called an ω-limit point of the semi-trajectory $x(t, x_0; \mathbf{R}^+)$. Similarly, one defines an α-limit point of a semi-trajectory $x(t, x_0; \mathbf{R}^-)$.

The set of all ω-limit points of a semi-trajectory $x(t, x_0; \mathbf{R}^+)$ is denoted by Ω_{x_0}, the similar set of α-limit points of a semi-trajectory $x(t, x_0; \mathbf{R}^-)$ is denoted by A_{x_0}.

For subsets A and B of E^n we define the distance $\rho(A, B)$ between them by setting

$$\rho(A, B) = \inf_{\substack{x \in A \\ y \in B}} \rho(x, y) = \inf_{\substack{x \in A \\ y \in B}} \|x - y\|.$$

The common properties of the limit sets Ω_{x_0} and A_{x_0} of compact semi-trajectories are known from the qualitative theory of differential equations [17], [94] and are given for Ω_{x_0} in the following theorem.

Theorem. *If $x(t, x_0; \mathbf{R}^+)$ is a compact semi-trajectory of system* (1.1), *then Ω_{x_0} is a closed invariant compact connected set satisfying the condition*

$$\lim_{t \to +\infty} \rho(x(t, x_0), \Omega_{x_0}) = 0. \qquad (1.4)$$

To prove the theorem we shall follow the reasoning of [94]. Let $x^0 \in \Omega_{x_0}$. We choose a sequence (1.2) such that relation (1.3) holds and consider the sequence of functions $x(t + t_n, x_0)$ $(n = 1, 2, \ldots)$, each of which is defined for $t \in [-t_n, +\infty)$.

It follows from the continuity of the solution $x(t, x_0)$ with respect to x_0 that

$$\lim_{n \to \infty} x(t + t_n, x_0) = \lim_{n \to \infty} x(t, x(t_n, x_0)) = x(t, x^0) \qquad (1.5)$$

for each $t \in \mathbf{R}$. Because the semi-trajectories $x(t, x_0; \mathbf{R}^+)$ are compact, the values of the functions $x(t+t_n, x_0)$, $n = 1, 2, \ldots$, belong to some closed bounded subset D_1 of the domain D. But then $x(t, x^0) \in D_1$ for any $t \in \mathbf{R}$. Hence, in view of (1.5), we have $x(t, x^0) \subset \Omega_{x_0}$ for all $t \in \mathbf{R}$.

By passing to the limit in the equality

$$x(t + t_n, x_0) = x(t_n, x_0) + \int_0^t X(x(t + t_n, x_0))dt \qquad (1.6)$$

we prove that the function $x(t, x^0)$ defines a solution of system (1.1). This is sufficient for the set Ω_{x_0} to be invariant.

To show that the set Ω_{x_0} is closed we take a sequence of points x_n^0 ($n = 1, 2, \ldots$), $x_n^0 \in \Omega_{x_0}$, $\lim_{n \to \infty} x_n^0 = \bar{x}^0$, and show that $\bar{x}^0 \in \Omega_{x_0}$. Because $\bar{x}^0$ is a limit point of the sequence x_n^0, for any $\epsilon > 0$ there exists $n \geq 1$ such that $\rho(x_n^0, \bar{x}^0) < \epsilon/2$. Since x_n^0 is an ω-limit point of the semi-trajectory $x(t, x_0; \mathbf{R}^+)$, it follows that there exists $t_n > 0$ such that $\rho(x(t_n, x_0), x_n^0) < \epsilon/2$. But then $\rho(x(t_n, x_0), \bar{x}^0) < \epsilon$, and therefore $\bar{x}^0$ is an ω-limit point of the semi-trajectory $x(t, x_0; \mathbf{R}^+)$. This proves that the set Ω_{x_0} is closed. It is obvious that Ω_{x_0} is compact.

We show that the set Ω_{x_0} is connected. Assume the contrary. Then there exist two non-empty disjoint closed sets A and B such that $\Omega_{x_0} = A \cup B$ and $\rho(A, B) = d > 0$. Since $A \subset \Omega_{x_0}$ and $B \subset \Omega_{x_0}$, one can find t_n' and t_n'' arbitrarily large such that $x(t_n', x_0)$ lies in a $(d/3)$-neighbourhood of the set A and $x(t_n'', x_0)$ lies in a $(d/3)$-neighbourhood of the set B. Choosing subsequences $\{t_n'\}$ and $\{t_n''\}$ such that the inequalities

$$0 < t_1' < t_1'' < t_2' < t_2'' < \ldots < t_n' < t_n'' < t_{n+1}' < \cdots$$

hold, we find that

$$\rho(x(t_n', x_0), A) < d/3,$$
$$\rho(x(t_n'', x_0), A) \geq \rho(A, B) - \rho(x(t_n'', x_0), B) > 2d/3. \qquad (1.7)$$

Since $\rho(x(t, x_0), A)$ is a continuous function with respect to t, we can find $t_n' < \tau_n < t_n''$ such that

$$\rho(x(\tau_n, x_0), A) = d/2. \qquad (1.8)$$

Because the set $x(t, x_0; \mathbf{R}^+)$ is compact in D, we can choose a subsequence from the sequence $x(\tau_n, x_0)$ ($n = 1, 2, \ldots$) that converges in D. According to the definition, the limit point of this subsequence x^0 belongs to the set Ω_{x_0}. On the other hand, by passing to the limit in relation (1.8) we obtain

$$\rho(x^0, A) = d/2, \quad \rho(x^0, B) \geq \rho(A, B) - \rho(A, x^0) = d/2,$$

that is, $\Omega_{x_0} \neq A \cup B$. This contradiction shows that Ω_{x_0} is connected.

Finally, we prove limit equality (1.4). Suppose that it does not hold. Then one can find a sequence of positive numbers $\{t_n\}$, $\lim\limits_{n\to\infty} t_n = +\infty$ and a number $\alpha > 0$ such that

$$\rho(x(t_n, x_0), \Omega_{x_0}) \geq \alpha. \tag{1.9}$$

Because the semi-trajectory $x(t, x_0; \mathbf{R}^+)$ is compact in D, the subsequence $\{x(t_n, x_0)\}$ has a limit point x^0 in D. On passing to the limit in (1.9), we obtain for this point the inequality $\rho(x^0, \Omega_{x_0}) \geq \alpha$. The latter contradicts the definition of the set Ω_{x_0}, according to which the point x^0 must lie in Ω_{x_0}.

This contradiction proves relation (1.4) and this finishes the proof of the theorem.

2.2. One-sided invariant sets and their properties

A set $M \subset D$ consisting of positive semi-trajectories of system (1.1) is called a *positively invariant* set of this system. Similarly, a set consisting of negative semi-trajectories of system (1.1) is called a *negatively invariant* set of this system. Both of these sets are called *one-sided invariant* sets of system (1.1).

A closed compact *positively* invariant set M of system (1.1) is called *stable* if for any $\epsilon > 0$ there exists $\delta = \delta(\epsilon) > 0$ such that $x(t, U_\delta) \subset (U_\epsilon \cup M)$ for all $t \geq 0$.

If M is stable and satisfies the limit relation:

$$\lim_{t \to +\infty} \rho(x(t, x_0), M) = 0$$

for all $x_0 \in U_{\delta_0}$ and some $\delta_0 > 0$, then the set M is said to be *asymptotically stable* for system (1.1). If M is not stable, then it is called *unstable*. The term "unstable" will also sometimes be used for sets that are not positively invariant.

Here and in what follows, $U_\epsilon = U_\epsilon(M)$ will denote the ϵ-neighbourhood of the set M consisting of all the points $x \in E^n$ for which $0 < \rho(x, M) < \epsilon$.

Let D_1 be a bounded region that is contained in D together with a neighbourhood of it, let ∂D_1 be the boundary of D_1, and $\overline{D}_1 = D_1 \cup \partial D_1$ the closure of D_1. A function $V(x)$ that is defined and has continuous partial derivatives in some region containing $\overline{D}_1$ is said to be continuously differentiable in $\overline{D}_1$.

Let $V = V(x)$ be a function that is continuously differentiable in $\overline{D}_1$. We say that V is *of constant sign* on D_1 if for all $x \in \overline{D}_1$ the non-zero values of $V(x)$ have the same sign.

Consequently, if a function V is of constant sign on D_1 then either $V(x) \geq 0$ or $V(x) \leq 0$ for all $x \in \overline{D}_1$. A function V of constant sign on D_1 is said to be *sign-definite* on D_1 if the set of zeros of the function V is not empty and is compact in D_1.

Consequently, if a function V is sign-definite, then the set N_0 of all $x \in D_1$ such that $V(x) = 0$ satisfies the conditions: $N_0 \neq \varnothing$, $N_0 \cap \partial D_1 = \varnothing$, and the values of $V(x)$ have the same sign on the complement of N_0.

A sign-definite function on D_1 taking the value zero at a single point $x = 0$ is a sign-definite *Lyapunov function* [68].

We denote by V_μ the set of points $\overline{D}_1$ for which

$$0 < |V(x)| \leq \mu,$$

and by N_0 the set of zeros of the function V in $\overline{D}_1$.

Lemma. *Let $V = V(x)$ be a sign-definite function on D_1. Then for any $\epsilon > 0$ there exist $\mu = \mu(\epsilon) > 0$ and $\delta = \delta(\epsilon) > 0$ such that*

$$U_\epsilon(N_0) \supset V_\mu \supset U_\delta(N_0). \tag{2.1}$$

Indeed, for a sign-definite function V on D_1 the function $|V|$ is positive-definite on D_1 and the set V_μ is not empty for any $\mu > 0$.

Suppose that the left inclusion in relation (2.1) does not hold for some sufficiently small $\epsilon > 0$. We can suppose that ϵ is so small that $U_\epsilon(N_0)$ is contained in D_1. This can always be achieved because N_0 is compact in D_1. Under these assumptions, one can find a point x_n in $V_{1/n}$ that does not belong to $U_\epsilon(N_0) \cup N_0$ for any $n = 1, 2, \ldots$. For the sequence $\{x_n\}$ the following inequalities hold:

$$0 < V(x_n) < 1/n, \quad \rho(x_n, N_0) \geq \epsilon, \quad n = 1, 2, \ldots.$$

For a limit point x^0 of the sequence $\{x_n\}$ we have then the following relations

$$V(x^0) = 0, \quad \rho(x^0, N_0) \geq \epsilon, \quad x^0 \in \overline{D}_1.$$

But they contradict each other because of the definition of the set N_0. This contradiction shows that for any $\epsilon > 0$ there exists $\mu = \mu(\epsilon) > 0$ such that the left inclusion in (2.1) holds.

Suppose that for some $\mu = \mu(\epsilon) > 0$ there does not exist $\delta = \delta(\epsilon) > 0$ such that the right inclusion in (2.1) holds. Then there exists a point x_n in $U_{1/n}(N_0)$ that does not belong to $V_\mu \cup N_0$ for any $n = 1, 2, \ldots$. For the sequence $\{x_n\}$ the following relations hold:

$$\mu \leq |V(x_n)|, \quad \rho(x_n, N_0) < 1/n, \quad n = 1, 2, \ldots.$$

Then for a limit point x^0 of the sequence $\{x_n\}$ we have

$$\mu \leq |V(x^0)|, \quad \rho(x^0, N_0) = 0, \quad x^0 \in \overline{D}_1,$$

which, as follows from the definition of N_0, contradict each other. The contradiction shows that for each $\mu = \mu(\epsilon) > 0$ we can find a $\delta = \delta(\mu) = \delta(\epsilon) > 0$ such that the right inclusion in (2.1) holds. This finishes the proof of the lemma.

Remark. The value of $\mu = \mu(\epsilon)$ played no role in the proof of the right-hand inclusion in (2.1). Hence it follows from this proof that for any $\mu > 0$ there exists $\delta = \delta(\mu) > 0$ such that

$$V_\mu \subset U_\delta(N_0).$$

The function

$$\dot{V}(x) = \frac{\partial V(x)}{\partial x} X(x) = \sum_{\nu=1}^{n} \frac{\partial V}{\partial x_\nu} X_\nu(x)$$

is called the *total derivative of the function* $V(x)$ along the solutions of system (1.1).

Theorem 1. *If* $V = V(x)$ *is a sign-definite function on* D_1 *whose total derivative* $\dot{V} = \dot{V}(x)$ *along the solutions of system* (1.1) *is sign-definite, then the set*

$$V(x) = 0, \quad x \in D_1 \tag{2.2}$$

is a one-sided invariant set of system (1.1). *More precisely, this set is negatively invariant if the signs of* V *and* $\dot{V}$ *are the same and is positively invariant (and stable) if the signs of* V *and* $\dot{V}$ *are different).**

---

* We regard the case when $\dot{V}(x) = 0$ for all $x \in \overline{D}_1$ as the case of different signs of V and $\dot{V}$.

To prove Theorem 1, we assume that $V(x) \geq 0$ for $x \in \overline{D}_1$. Let $\dot{V}(x) \leq 0$ for $x \in \overline{D}_1$. We claim that the set N_0 consisting of points of D_1 such that $V(x) = 0$ is a positively invariant stable set of system (1.1). To see this, we consider a solution $x(t, x_0)$ of system (1.1) passing through $x_0 \in N_0$ when $t = 0$. Since N_0 is compact in D_1, it follows that $x(t, x_0) \in D_1$ for t in some interval (h_1, h_2), where $h_1 < 0$ and $h_2 > 0$. By hypothesis, $dV(x(t, x_0))/dt = \dot{V}(x(t, x_0)) \leq 0$ for $t \in (h_1, h_2)$. But then the function $V(x(t, x_0))$ is non-increasing for $t \in [0, h_2)$, and this leads to the inequality

$$V(x(t, x_0)) \leq V(x_0) = 0, \ t \in [0, h_2),$$

which proves that $x(t, x_0) \in N_0$ for all $t \in [0, h_2)$.

Since there is a neighbourhood of N_0 lying inside D_1, so that there is a "gap" between N_0 and the boundary ∂D_1 of the region D_1, it follows that the point $x(t, x_0)$ in changing its position continuously in the region D_1 cannot leave the set N_0 for $t > 0$. This means that N_0 consists of positive semi-trajectories of system (1.1) and therefore N_0 is a positively invariant set of system (1.1).

We now prove that the set N_0 is stable. Let ϵ be a sufficiently small positive number such that $U_\epsilon = U_\epsilon(N_0) \subset \overline{D}_1$. By the lemma we can choose $\mu = \mu(\epsilon) > 0$ and $\delta = \delta(\epsilon) > 0$ such that the inclusions

$$U_\epsilon \supset V_\mu \supset U_\delta \qquad (2.3)$$

hold, where $U_\delta = U_\delta(N_0)$.

Taking into account (2.3) we see that the solution $x(t, x_0)$ of system (1.1), which starts at a point $x_0 \in U_\delta$ for $t = 0$, belongs to the set $V_\mu \cup N_0$ for all t in some interval (h_1, h_2), where $h_1 < 0$, $h_2 > 0$. Because the function $\dot{V}(x)$ is of constant sign in D_1, we have the inequality:

$$V(x(t, x_0)) \leq V(x_0) \leq \mu \qquad (2.4)$$

for all $t \in [0, h_2)$. Hence $x(t, x_0) \in V_\mu \cup N_0$ for all $t \in [0, h_2]$.

Let h_2 be the largest value in $\mathbf{R}^+ \cup \{+\infty\}$ for which inequality (2.4) holds. If h_2 is finite, then we immediately reach a contradiction. Thus $x(t, x_0) \in V_\mu$ for all $t \geq 0$. Since $U_\epsilon \supset V_\mu$ (see (2.3)), we have $x(t, x_0) \in U_\epsilon$ for all $t \geq 0$. This proves that $x(t, U_\delta) \subset U_\epsilon$ for $t \in \mathbf{R}^+$. And this means that the set N_0 is stable.

Now let $\dot{V} = \dot{V}(x) \geq 0$ for $x \in \overline{D}_1$. The function $-\dot{V} = -\dot{V}(x)$ is the total derivative of the function $V(x)$ along solutions of the system

$$\frac{dx}{dt} = -X(x) \tag{2.5}$$

and satisfies the inequality $-\dot{V}(x) \leq 0$ on $\overline{D}_1$.

Applying the above result to the system (2.5), we see that N_0 is a positively invariant stable set of system (2.5). Because system (2.5) was obtained from (1.1) by changing t to $-t$, N_0 is a negatively invariant set of system (1.1). This completes the proof.

It should be noted that it is not sufficient that the signs of V and $\dot{V}$ in the conditions of Theorem 1 be the same in order that the set (2.2) be unstable. However, if the signs of V and $\dot{V}$ are the same then set (2.2) can be stable only if it is invariant and contained in a "larger" invariant set of system (1.1); the closure of the set $x(t, U_{\delta_0} \cup N_0)$ for some $\delta_0 > 0$ and all $t \in \mathbf{R}$ is an example.

Theorem 2. *If the function $V = V(x)$ and its total derivative $\dot{V} = \dot{V}(x)$ along solutions of system* (1.1) *are sign-definite on D_1 and their zero sets in D_1 are the same, then the one-sided invariant set* (2.2) *is asymptotically stable if the signs of V and $\dot{V}$ are different and is unstable otherwise.*

To prove the theorem we suppose that $V(x) \geq 0$ for $x \in D_1$. Let $\dot{V} = \dot{V}(x) \leq 0$ for $x \in D_1$. It follows from Theorem 1 that the zero set N_0 in D_1 of the function $V(x)$ is a positively invariant stable set of system (1.1). Consequently, for a sufficiently small $\epsilon > 0$ there exists $\delta = \delta(\epsilon) > 0$ such that $x(t, U_\delta) \subset (U_\epsilon \cup N_0) \subset D_1$ for $t \geq 0$, where $U_\delta = U_\delta(N_0)$, $U_\epsilon = U_\epsilon(N_0)$. We now show that

$$\lim_{t \to +\infty} \rho(x(t, x_0), N_0) = 0 \tag{2.6}$$

for any point $x_0 \in U_{\delta_0} = U_{\delta_0}(N_0)$, where δ_0 is a sufficiently small number.

Indeed, let $\delta = \delta(\epsilon)$ be chosen such that $x(t, U_\delta) \subset (U_\epsilon \cup N_0) \subset D_1$ for $t \geq 0$. For a point $x_0 \in U_\delta$ we consider the semi-trajectory $x(t, x_0; \mathbf{R}^+)$. If $x(\tau, x_0) \in N_0$ for some $\tau \in \mathbf{R}^+$, then the semi-trajectory $x(t, x(\tau, x_0); \mathbf{R}^+)$ lies inside N_0 for $t \geq 0$ because the set N_0 is positively invariant, and so relation (2.6) holds for $x(t, x_0)$.

Let $x(t, x_0) \notin N_0$ for all $t \geq 0$. Then $dV(x(t, x_0))/dt = \dot{V}(x(t, x_0)) < 0$ for $t \geq 0$ and, consequently, the function $V(x(t, x_0))$, which is monotonically

decreasing, tends to a limit $\mu_0 \geq 0$ as $t \to +\infty$. We claim that $\mu_0 = 0$. For
suppose that $\mu_0 > 0$. Then $V(x(t, x_0)) \geq \mu_0$ for $t \geq 0$. Using the remark made
in the proof of the lemma, we take $\delta_0 = \delta_0(\mu_0) > 0$ such that the inclusion
$V_{\mu_0} \supset U_{\delta_0}$ holds. From the definition of the set V_{μ_0} and the above inclusion it
follows that $x(t, x_0) \in U_\epsilon$ and $x(t, x_0) \notin U_{\delta_0}$ for $t \geq 0$. But then $\delta_0 < \epsilon$ and
$x(t, x_0) \in U_\epsilon \backslash U_{\delta_0}$ for $t \geq 0$. We set $\sup\limits_{x \in U_\epsilon \backslash U_{\delta_0}} \dot{V}(x) = -\alpha$. Since $\dot{V}(x)$ takes only
negative values on the closure of the set $U_\epsilon \backslash U_{\delta_0}$, it follows that $\alpha > 0$. But
then $dV(x(t, x_0))/dt \leq -\alpha$ for $t \geq 0$. This leads to the inequality

$$V(x(t, x_0)) \leq V(x_0) - \alpha t, \quad t \in \mathbf{R}^+. \tag{2.7}$$

Inequality (2.7) contradicts the fact that the function $V(x)$ is positive definite
on D_1. This contradiction shows that $\mu_0 = 0$. Thus, if $x(t, x_0) \notin N_0$ for $t \geq 0$,
then

$$\lim_{t \to +\infty} V(x(t, x_0)) = 0. \tag{2.8}$$

Consider the limit set Ω_{x_0} of the semi-trajectory $x(t, x_0; \mathbf{R}^+)$. Now since
$\lim\limits_{t \to +\infty} \rho(x(t, x_0), \Omega_{x_0}) = 0$, to obtain (2.6) it is sufficient to prove that $\Omega_{x_0} \subset N_0$.
Suppose that $\Omega_{x_0} \not\subset N_0$. Let $y \in \Omega_{x_0} \backslash N_0$. Since y is an ω-limit point of the
semi-trajectory $x(t, x_0; \mathbf{R}^+)$, we can find a sequence $0 < t_1 < t_2 < \ldots < t_n <$
$\ldots$, $\lim\limits_{n \to \infty} t_n = +\infty$ such that $\lim\limits_{n \to \infty} x(t_n, x_0) = y$. But then $\lim\limits_{n \to \infty} V(x(t_n, x_0)) =$
$V(y) \neq 0$, which contradicts (2.8). This contradiction shows that $\Omega_{x_0} \subset N_0$ for
any $x_0 \in U_\delta$. This leads to relation (2.6) for all $x_0 \in U_\delta$ and therefore proves
that the set N_0 is asymptotically stable.

Now let $\dot{V}(x) \geq 0$ for $x \in \overline{D}_1$. Then $-\dot{V}(x) \leq 0$ for $x \in \overline{D}_1$ and the system
of equations (2.5) satisfies the conditions of Theorem 2 which ensure that the
set N_0 is asymptotically stable as an invariant set of system (2.5). This leads
to the relation

$$\lim_{t \to -\infty} \rho(x(t, x_0), N_0) = 0$$

for all $x_0 \in U_{\delta_0}$, where δ_0 is a positive number. But then the limit set A_{x_0}
of any semi-trajectory $x(t, x_0; \mathbf{R}^-)$, $x_0 \in U_{\delta_0}$, belongs to the set N_0. Let
$\mu_1 > 0$ be so small that $V_{\mu_1} \subset U_{\delta_0}$. Consider the semi-trajectory $x(t, x_0; \mathbf{R}^-)$
for $x_0 \in U_{\delta_0} \backslash V_{\mu_1}$. Since $A_{x_0} \subset N_0$, we can find a sequence of numbers

$$0 < t_1 < t_2 < \ldots < t_n < \ldots, \quad \lim_{n \to \infty} t_n = +\infty$$

such that

$$\lim_{n \to \infty} \rho(x(-t_n, x_0), N_0) = 0.$$

This means that there is at least one point of the sequence

$$x(-t_1, x_0), \ x(-t_2, x_0), \ldots, \ x(-t_n, x_0), \ldots \qquad (2.9)$$

in any neighbourhood of N_0. Each solution $x(t, x(-t_n, x_0))$ $(n = 1, 2, \ldots)$ takes the value $x(t_n, x(-t_n, x_0)) = x_0 \notin V_{\mu_1}$ for $t = t_n > 0$. If now we fix $\epsilon_0 > 0$ such that the inclusion $U_{\epsilon_0} \subset V_{\mu_1}$ holds, then in any neighbourhood of N_0 one can find a point of (2.9) that leaves the ϵ_0-neighbourhood of N_0 when $t > 0$. Thus the set N_0 cannot be a stable set of system (1.1) and this completes the proof of Theorem 2.

It should be noted that the parts of the given theorem that relate to the stability of the set N_0 are similar to the results on stability theory of invariant sets of system (1.1) [46], [47] and, in particular, are generalizations of the well-known Lyapunov theorems on stability of equilibrium points of system (1.1) when N_0 consists of a single point [68], [72].

We end our discussion of one-sided invariant sets and their stability by giving one further result [47] which is a quantitative characterization of trajectories of system (1.1). It has an obvious extension to positively invariant sets of system (1.1).

Theorem 3. *In order that a closed invariant set of system (1.1) having a a sufficiently small compact neighbourhood not containing entire trajectories be stable, it is necessary and sufficient that there does not exist a motion* $x(t, x_0)$, $x_0 \notin M$, *having α-limit points in M.*

The proof of the theorem is based on the fact that if the conditions of the theorem are satisfied, then the quantity

$$\delta(\epsilon) = \inf_{\substack{t \leq 0 \\ \rho(x_0, M) = \epsilon}} \rho(x(t, x_0), M)$$

is positive for sufficiently small $\epsilon > 0$. And by virtue of the definition of stability, this is sufficient for the set M to be stable.

2.3. Locally invariant sets. Reduction principle

A set $M \subset D$ is called a *locally invariant* set of system (1.1) if it consists of arcs of trajectories of this system. In other words, M is a locally invariant set of system (1.1) if for any point $x_0 \in M$ there exists $T_1 = T_1(x_0) < T_2 = T_2(x_0)$, $T_1 T_2 \leq 0$ such that $x(t, x_0; [T_1, T_2]) \subset M$.

Any arc of a trajectory and any region D_1 contained in D are obvious examples of invariant sets of system (1.1).

General criteria for finding locally invariant sets of system (1.1) are given in the following theorem.

Theorem 1. *If a function $V = V(x)$ and its total derivative $\dot{V} = \dot{V}(x)$ along solutions of system* (1.1) *are of constant sign on D_1 then the set*

$$V(x) = 0, \quad x \in D_1 \tag{3.1}$$

is a locally invariant set of system (1.1) *if it is non-empty.*

To prove the theorem, suppose, for example, that $V(x) \geq 0$ and $\dot{V}(x) \leq 0$ for $x \in D_1$. We denote set (3.1) by N.

Let $x_0 \in N$. Then $x_0 \in D_1$ and consequently, $x(t, x_0) \in D_1$ for $t \in (h_1, h_2)$, where $h_1 < 0 < h_2$. Then the function $V(x(t, x_0))$ satisfies the inequality $V(x(t, x_0)) \leq V(x_0) = 0$ for all $t \in [0, h_2)$. This shows that the arc $x(t, x_0; [0, T_2])$ belongs to the set N for $T_2 < h_2$. This is sufficient for N to be a locally invariant set of system (1.1).

Necessary conditions for the set (3.1) to be a locally invariant set of system (1.1) are given in the following theorem.

Theorem 2. *If a function $V = V(x)$ is continuously differentiable and its zero set* (3.1) *in D_1 is not empty and is locally invariant with respect to system* (1.1), *then its total derivative $\dot{V} = \dot{V}(x)$ along solutions of system* (1.1) *takes zero values on the set* (3.1).

Corollary. *If the functions $V = V(x)$ and $\dot{V} = \dot{V}(x) = (\partial V(x)/\partial x) X(x)$ are of constant sign on D_1 then in D_1 the zero set of the function V is contained in the zero set of the function $\dot{V}$.*

The statement of the Corollary follows from Theorem 1 and Theorem 2.

Under certain additional conditions, a locally invariant set of system (1.1) contains an invariant set, or is itself an invariant set of this system. An example of these conditions is given in the theorem given below.

Theorem 3. *Let $V = V(x)$ be a sign-definite function on D_1 the total derivative $\dot{V} = \dot{V}(x)$ of which along solutions of system* (1.1) *is of constant sign. If $V(x)$ and $X(x)$ are analytic on D_1, then the set* (3.1) *is an invariant set of system* (1.1).

Thus an additional condition that the functions $V(x)$ and $X(x)$ be analytic is one that guarantees that the locally invariant set (3.1) is in fact an invariant set of system (1.1).

Turning to the proof of the theorem, we denote set (3.1) by N and consider a solution $x(t, x_0)$ of system (1.1) for $x_0 \in N$. According to Theorem 1 of §2.2, this solution is defined on the semi-axis and takes values in N for $t \in \mathbf{R}^+$. We suppose for definiteness that $x(t, x_0; \mathbf{R}^+) \subset N$. Then

$$V(x(t, x_0)) = 0 \qquad (3.2)$$

for all $t \geq 0$. Because the functions $V(x)$ and $X(x)$ are analytic on D_1, we have the decomposition

$$V(x(t, x_0)) = \sum_{k=0}^{\infty} \frac{V_k t^k}{k!}, \quad V_k = \left. \frac{d^k V(x(t, x_0))}{dt^k} \right|_{t=0}, \qquad (3.3)$$

which converges for t in some interval $|t| < \epsilon$.

It follows from equalities (3.2) and (3.3) that

$$V_k = 0, \quad k = 0, 1, 2, \ldots.$$

But then

$$V(x(t, x_0)) \equiv 0 \text{ for all } \quad |t| < \epsilon.$$

This leads to the inclusion $x(t, x_0; [-\epsilon, \epsilon]) \subset N$. Since $x(-\epsilon, x_0)$ belongs to N, some arc that starts at this point, namely, the arc $x(t, x(-\epsilon, x_0); [-\epsilon_1, 2\epsilon])$ for some ϵ also belongs to N. Continuing in this manner we see that the entire semi-trajectory $x(t, x_0; \mathbf{R}^-)$ belongs to the set N. Since x_0 is an arbitrary point of N, this means that $x(t, N) = N$ for all $t \in (-\infty, \infty)$. The theorem is proved.

Let N be a locally invariant set of system (1.1), M a positively invariant set of this system such that $N \supset M$. We shall say that *M is stable in N* if for any neighbourhood U_ϵ of M there exists a neighbourhood U_δ of it such that $x(t, U_\delta \cap N) \subset (U_\epsilon \cup M)$ for all $t \geq 0$. We say that M *is asymptotically stable in N* if M is stable in N and there is a $\delta_0 > 0$ such that

$$\lim_{t \to +\infty} \rho(x(t, x_0), M) = 0 \tag{3.4}$$

for all $x_0 \in U_{\delta_0} \cap N$.

It is clear that stability of a set M in N does not always imply that this set is stable; however, if M is unstable in N it follows that M is unstable. Thus there arises the problem of giving conditions under which stability of M will follow from its stability in a locally invariant set N. This problem has been solved by V.A. Pliss [98] when M is a point. His result is known in the literature on stability theory as the *reduction principle*. The result for the general case given in Theorem 4 is similar to this principle and is due to the author [108].

Theorem 4. *Suppose that a locally invariant set N of system (1.1) contains a closed positively invariant set M that is asymptotically stable in N. If N is a set of type (3.1) for some function $V = V(x)$ of constant sign on D_1, the total derivative of which $\dot{V} = \dot{V}(x)$ along solutions of system (1.1) is of constant sign and has the opposite sign of V, then M is a stable set of system (1.1). If, in addition, the zero sets in D_1 of V and $\dot{V}$ are the same, then M is an asymptotically stable set of system (1.1).*

Before going to the proof of the theorem we establish some auxiliary results.

Lemma 1. *Under the conditions of Theorem 4, the limit relation (3.4) is uniform with respect to $x_0 \in U_{\delta_1} \cap N$, where δ_1 is a positive number.*

Suppose that the lemma does not hold. Then for any $\delta_1 > 0$ one can find a number $\mu > 0$ such that there is a sequence of numbers t_n and a convergent sequence of points x_n in $U_{\delta_1} \cap N$ satisfying

$$\rho(x_n, M) < \delta_1, \quad x(t, x_n; [0, t_n]) \subset D_1,$$
$$\rho(x(t_n, x_n), M) \geq \mu, \quad t_n \to +\infty. \tag{3.5}$$

Since M is closed and is contained in the bounded region D_1, some neighbourhood of M is contained in D_1. The number δ_1 can be chosen so small that the limit of the sequence x_n belongs to the δ_0-neighbourhood of the set M, where δ_0 is chosen in such a way that both the points $x(t, U_{\delta_0} \cap N)$ belong to N for $t \geq 0$ and limit relation (3.4) holds for points of $U_{\delta_0} \cap N$. Let

$$\lim_{n \to \infty} x_n = x^0.$$

Since $x^0 \in D_1$ and $V(x_n) = 0$, it follows that $x^0 \in N$ and consequently, $x^0 \in (U_{\delta_0} \cup M) \cap N$. But then

$$\lim_{t \to +\infty} \rho(x(t, x^0), M) = 0,$$

and we can find $T > 0$ such that

$$\rho(x(t, x^0), M) < \delta(\mu)/2 \tag{3.6}$$

for all $t \geq T$, where $\delta(\mu)$ is a positive constant such that $x(t, U_{\delta(\mu)} \cap N) \subset (U_{\mu/2} \cup M)$ for all $t \geq 0$.

Since the solutions of system (1.1) depend continuously on the initial data, we can find $N > 0$ such that the inequality

$$\rho(x(t, x_n), x(t, x^0)) < \delta(\mu)/2 \tag{3.7}$$

holds for all $t \in [0, T]$ and $n > N$. We can suppose that N is so large that

$$t_n > T \text{ when } n > N. \tag{3.8}$$

It then follows from inequalities (3.6) and (3.7) that

$$\rho(x(T, x_n), M) \leq \rho(x(T, x^0), M) +$$
$$+ \rho(x(T, x_n), x(T, x^0)) < \delta(\mu)/2 + \delta(\mu)/2 = \delta(\mu).$$

The last inequality means that $x(T, x_n) \in (U_{\delta(\mu)} \cup M)$ for $n > N$. Since the function $\dot{V}$ has the opposite sign of V for all $x \in D_1$ and, according to (3.5) and (3.8), $x(t, x_n, [0, T)) \subset D_1$, it follows that $V(x(T, x_n)) = 0$ and $x(T, x_n) \in (U_{\delta(\mu)} \cup M) \cap N$ for $n > N$. The last inclusion shows that the value of the function $x(t, x(T, x_n))$ belongs to the set $U_{\mu/2} \cup M$ for all $t \geq 0$ and $n > N$, which yields the inequality

$$\rho(x(t, x(T, x_n)), M) < \mu/2 \tag{3.9}$$

for $t \geq 0$ and $n > N$. We set $t = t_n - T$ in (3.9). By virtue of inequality (3.8) this is always possible. As a result we obtain

$$\rho(x(t_n, x_n), M) < \mu/2 \text{ when } n > N. \tag{3.10}$$

Inequality (3.10) contradicts one of the inequalities (3.5). The contradiction proves the lemma.

Remark. By setting $V \equiv 0$ in the conditions of Theorem 4, it will follow from Lemma 1 that the limit (3.4) is uniform for $x_0 \in U_{\delta_1}(M)$ and for any closed compact asymptotically stable positively invariant set M of system (1.1). This property for an invariant set is well known [47]. It is the basis of the definition of *uniform asymptotical stability* of an invariant set.

We denote by $V_{\delta,\mu}$ the set of points $x \in D_1$, that belong to the δ-neighbourhood of the set M and satisfying the inequality

$$|V| \leq \mu.$$

Lemma 2. *If the conditions of Theorem 4 are satisfied then $V_{\delta,\mu}$ contains some $\delta_1 = \delta_1(\mu)$-neighbourhood $U_{\delta_1(\mu)}$ of M which contracts to the set $V_{\delta,0} = U_\delta \cap N$ as $\mu \to 0$, so that*

$$\lim_{\mu \to 0} \sup_{y \in U_{\delta_1(\mu)}} \rho(y, V_{\delta,0}) = 0. \tag{3.11}$$

First we suppose that $V_{\delta,\mu}$ does not contain a δ_1-neighbourhood of M for any $\delta_1 > 0$. Then one can find a convergent sequence of points x_n such that

$$\lim_{n \to \infty} \rho(x_n, M) = 0, \quad |V(x_n)| \geq \mu.$$

Since the functional ρ is continuous and the set M is closed, the limit of the sequence x_n,

$$x^0 = \lim_{n \to \infty} x_n$$

belongs to M. But then

$$V(x^0) = 0,$$

which contradicts the relation

$$|V(x^0)| = \lim_{n \to \infty} |V(x_n)| \geq \mu.$$

Now suppose that the limit relation (3.11) does not hold. Then one can find convergent sequences μ_n and y_n such that

$$y_n \in V_{\delta,\mu_n}, \quad \rho(y_n, N) \geq \epsilon_0 > 0, \quad \mu_n \to 0.$$

The limit of the sequence y_n,

$$y_0 = \lim_{n \to \infty} y_n$$

satisfies the inequalities

$$|V(y_0)| = \lim_{n \to \infty} |V(y_n)| \leq \lim_{n \to \infty} \mu_n = 0, \quad \rho(y_0, N) \geq \epsilon_0,$$

which are contradictory since the first one means that $y_0 \in N$ and the second that $y_0 \notin N$.

We now turn to the proof of Theorem 4. Using the statements of Lemma 1 and asymptotical stability of M in N we choose positive numbers ϵ and $\delta = \delta(\epsilon)$ such that the relations

$$\rho(x(t, x_0), M) < \epsilon/2, \quad (U_\epsilon \cup M) \subset D_1$$

hold for all $t \geq 0$ and all $x_0 \in V_{\delta,0}$ and such that the limit

$$\lim_{t \to +\infty} \rho(x(t, x_0), M) = 0 \tag{3.12}$$

is uniform with respect to $x_0 \in V_{\delta,0}$.

Let T be a positive number which is determined so that the condition

$$\rho(x(T, x_0), M) < \delta/2$$

holds for all $x_0 \in V_{\delta,0}$. We can choose such T because the limit relation (3.12) is uniform. Now choose $\mu_0 = \mu_0(T, \epsilon) > 0$ in such a way that the trajectory arc $x(t, x_0; [0,T])$ belongs to the set $U_\epsilon \cup M$ for any $x_0 \in V_{\delta,\mu_0}$. Such a choice of μ_0 is possible because of the following considerations.

Suppose that μ_0 cannot be chosen in the indicated manner. Then one can find convergent sequences $\{t_n\}$, $\{y_n\}$, $\{\mu_n\}$ such that $0 \leq t_n \leq T$, $\lim_{n \to \infty} t_n = \tau$, $\lim_{n \to \infty} y_n = y_0$, $\lim_{n \to \infty} \mu_n = 0$ and

$$y_n \in V_{\delta,\mu_n}, \quad , x(t, y_0; [0, t_n)) \subset (U_\epsilon \cup M), \quad x(t_n, y_n) \in \partial U_\epsilon.$$

The last relation is equivalent to the equality

$$\rho(x(t_n, y_n), M) = \epsilon,$$

from which, on passing to the limit, we obtain

$$\rho(x(\tau, y_0), M) = \epsilon. \qquad (3.13)$$

According to Lemma 2, the set V_{δ,μ_n} contracts to $V_{\delta,0}$ as $\mu_n \to 0$. Therefore $y_0 \in V_{\delta,0}$ and, as follows from the choice of δ,

$$\rho(x(\tau, y_0), M) < \epsilon/2.$$

We have arrived at a contradiction with equality (3.13). Since solutions of system (1.1) depend continuously on initial conditions, we have the inequality

$$\rho(x(t, x_0), x(t, x_1)) < \delta/2, \quad t \in [0, T] \qquad (3.14)$$

for any two points $x_0, x_1 \in (V_{\delta,\mu_0} \cup M)$ such that

$$\rho(x_0, x_1) \leq d, \qquad (3.15)$$

where d is sufficiently small.

According to Lemma 2, the set $V_{\delta,\mu}$ contracts to $V_{\delta,0}$ when $\mu \to 0$. So, one can always choose $\mu = \mu(\delta)$, $0 < \mu(\delta) < \mu_0$, such that for any point $x_1 \in V_{\delta,\mu}$ there will be a point x_0 in $V_{\delta,0}$ satisfying inequality (3.15). Suppose that μ has been chosen in this manner. Let y be an arbitrary point of the set $V_{\delta,\mu}$ and x^0 a point of $V_{\delta,0}$ for which

$$\rho(y, x^0) < d.$$

Inequality (3.14) considered for $x_0 = x^0 \in V_{\delta,0}$ and $x_1 = y$ leads to the relation

$$\rho(x(t, x^0), x(t, y)) < \delta/2 \leq \epsilon/2, \quad t \in [0, T], \qquad (3.16)$$

which proves that

$$\rho(x(t, y), M) \leq \rho(x(t, y), x(t, x^0)) + \rho(x(t, x^0), M) < \epsilon/2 + \epsilon/2 = \epsilon, \quad t \in [0, T].$$

Moreover, taking into consideration the choice of T, relation (3.16) leads to the inequality

$$\rho(x(T, y), M) \leq \rho(x(T, y), x(T, x^0)) + \rho(x(T, x^0), M) < \delta,$$

proving that

$$x(T, y) \in (U_\delta \cup M). \tag{3.17}$$

Since the functions V and $\dot{V}$ have different signs in D_1 and $x(t, y) \in (U_\delta \cup M) \subset D_1$ for $t \in [0, T]$, it follows that

$$|V(x(t, y))| \leq |V(y)| \leq \mu, \quad t \in [0, T].$$

The last inequality leads to the inclusion

$$x(t, y) \in (V_{\epsilon,\mu} \cup M), \quad t \in [0, T]$$

which, together with relation (3.17), proves that the trajectory which starts at the point $y \in V_{\delta,\mu}$ at the moment $t = T$ passes through a point $x(T, y)$ belonging to the set $(U_\delta \cup M) \cap (V_{\epsilon,\mu} \cup M) = V_{\delta,\mu} \cup M$ without leaving $V_{\epsilon,\mu} \cup M$. If $x(T, y) \epsilon M$, then $x(t, y) \in M$ for all $t \geq T$ because the set M is positively invariant. If $x(T, y) \notin M$, then $x(T, y) \in V_{\delta,\mu}$, so that $x(t, x(T, y)) = x(t + T, y) \in (V_{\epsilon,\mu} \cup M)$ for all $t \in [0, T]$. This leads to the inclusion

$$x(t, y) \in (V_{\epsilon,\mu} \cup M)$$

for $t \geq 0$. This proves that

$$x(t, V_{\delta,\mu}) \subset (V_{\epsilon,\mu} \cup M) \text{ for } t \geq 0. \tag{3.18}$$

According to Lemma 2, the set $V_{\delta,\mu}$ contains some $\delta_1 = \delta_1(\mu)$-neighbourhood of the set M. But then for any $\epsilon > 0$ one can find $\delta_1 = \delta_1(\mu) = \delta_1(\mu(\delta)) = \delta_1(\mu(\delta(\epsilon))) = \delta_2(\epsilon) > 0$ such that

$$x(t, U_{\delta_2(\epsilon)}) \subset (V_{\epsilon,\mu(\delta(\epsilon))} \cup M) \subset (U_\epsilon \cup M) \tag{3.19}$$

for all $t > 0$. Relation (3.19) proves that the set M is stable.

Assume now that the zero sets of the functions V and $\dot{V}$ coincide in D_1. Choose δ and $\mu = \mu(\delta)$ such that relation (3.18) holds. Then for a point $x_0 \in V_{\delta,\mu}$ either $x(t, x_0) \in M$ for $t = \tau$ or $x(t, x_0) \in V_{\epsilon,\mu}$ for all $t > 0$. In the second case $|V(x(t, x_0))|$ is monotonically decreasing and as $t \to +\infty$ tends to a limit which, according to considerations similar to those given in the proof of Theorem 2 of §2.2, is equal to zero: $\lim_{t \to +\infty} V(x(t, x_0)) = 0$. But then the limit set Ω_{x_0} of the semi-trajectory $x(t, x_0; \mathbf{R}^+)$ belongs to the set $\overline{V}_{\epsilon,0}$. We claim that it is possible to choose $\delta_0 > 0$ such that the limit set Ω_{x_0} lies inside M for any point $x_0 \in \overline{V}_{\delta_0, \mu(\delta_0)}$. This will prove the limit relation

$$\lim_{t \to +\infty} \rho(x(t, x_0), M) = 0$$

for points $x_0 \in \overline{V}_{\delta_0, \mu(\delta_0)}$ and hence the asymptotic stability of the set M will be established.

In fact, suppose that it is impossible to choose such a $\delta_0 > 0$. Then for any $\epsilon_0 > 0$ no matter how small, there will be a point y, $y \in \overline{V}_{\delta_0, \mu(\delta_0)}$, such that $\Omega_y \subset \overline{V}_{\epsilon_0, \mu(\delta_0)}$ and M does not contain Ω_y, where $\delta_0 = \delta_0(\epsilon) \le \epsilon_0$ is determined in such a way that $x(t, \overline{V}_{\delta_0, \mu(\delta_0)}) \subset (V_{\epsilon_0, \mu(\delta_0)} \cup M)$ for all $t \ge 0$. Let $z \in \Omega_y$, $z \notin M$. Then because the set Ω_y is invariant it follows that $x(t, z) \in \Omega_y$ for all $t \in (-\infty, +\infty)$. Thus the semi-trajectory $x(t, z; \mathbf{R}^-)$ lies in $\overline{V}_{\epsilon_0, 0}$. But this contradicts the fact that M is stable in N.

Indeed, because $x(t, z; \mathbf{R}^-) \subset \overline{V}_{\epsilon_0, 0}$ and $\epsilon_0 > 0$ is arbitrarily small, the α-limit points of the semi-trajectory $x(t, z; \mathbf{R}^-)$ lie either in M or in $(N \cap \overline{U}_{\epsilon_0}) \backslash M$. In either case, we can take a point on the semi-trajectory $x(t, z; \mathbf{R}^-)$ so close to the set A_z that the time of passage from this point to the point z for $t > 0$ is greater than a preassigned fixed number T. Since this point belongs to $N \cap \overline{U}_{\epsilon_0}$ and the limit relation (3.4) is uniform for points in $N \cap \overline{U}_{\epsilon_0}$, it follows, in view of the arbitrary smallness of ϵ_0, that the time of passage from a given point to the point z for $t > 0$ is determined by the distance from z to M and cannot be greater than the number T. This contradiction proves relation (3.18). Thus the asymptotic stability of the set M has been proved.

We make two remarks concerning Theorem 4.

Remark 1. If the condition on asymptotic stability of M on N is dropped this may lead to instability of the set M. The following system is an example of this:

$$\frac{dx_1}{dt} = 0, \quad \frac{dx_2}{dt} = x_1 x_2. \tag{3.20}$$

Here the equilibrium in $M = \{(0,0)\}$ is unstable. By taking $V = x_1^2$ we see that the set M is stable in N and that all the conditions of Theorem 4 are satisfied except for the condition of asymptotic stability of M in N.

Remark 2. If we drop the condition that the zero sets of V and $\dot{V}$ in D_1 coincide then it may happen that the set M will only be stable but not asymptotically stable.

The system

$$\frac{dx_1}{dt} = 0, \quad \frac{dx_2}{dt} = x_1 - x_2,$$

gives an example of the situation where the equilibrium position $M = \{(0,0)\}$ is stable but not asymptotically stable. For the same function V as in the previous example all the conditions of Theorem 4 are fulfilled.

2.4. Behaviour of an invariant set under small perturbations of the system

Let M_0 be a closed invariant set of system (1.1). We shall find out the behaviour of M_0 under small perturbations of system (1.1). The magnitude of a small perturbation will be defined by a positive parameter μ. Using it, we write the perturbed system of equations in the form

$$\frac{dx}{dt} = X(x) + \mu Y(x). \tag{4.1}$$

We shall assume that the functions X and Y are defined, continuous and satisfy Lipschitz condition for all x in D.

Denote the *region of attraction* of M_0 by $\Pi(M_0)$. It consists of all the points $x_0 \in D$ such that $\Omega_{x_0} \subset M$.

Theorem 1. *If M_0 is a closed compact asymptotically stable invariant set of system* (1.1) *then one can find $\delta > 0$ and $\mu_0 = \mu_0(\delta) > 0$ such that for all $\mu < \mu_0$ the system of equations* (4.1) *has a closed asymptotically invariant set $M = M(\mu)$ for which $\lim_{\mu \to 0} \rho(M_0, M) = 0$ and $U_\delta(M_0) \subset \Pi(M)$.*

Theorem 1 determines the behaviour of an asymptotically stable set M_0 under small perturbations. It establishes that small perturbations merely "reconstruct" the trajectories of the perturbed system in a neighbourhood of M_0 by forming from them an asymptotically stable set $M(\mu)$ which is drawn to M_0 for $\mu \to 0$ and that the region of attraction of any of these sets contains some neighbourhood of M_0 common to all the sets $M(\mu)$ for $\mu < \mu_0$.

Turning to the proof of the theorem, we denote by $x(t, x_0, \mu)$ the solution of the system of equations (4.1) for which $x(0, x_0, \mu) = x_0$. We set $x(t, x_0, 0) = x(t, x_0)$. Since $x(t, x_0, \mu)$ is continuous with respect to all the arguments t, x_0, μ for $t \in [0, T]$, $x_0 \in \overline{U}_\delta(M_0)$ and $\mu \in [0, \mu_0]$ for any $T > 0$ and small δ and μ_0, it follows that the relation

$$\lim_{\mu \to 0} \rho(x(t, x_0, \mu), \ x(t, x_0)) = 0 \tag{4.2}$$

holds uniformly with respect to $(t, x_0) \in [0, T] \times \overline{U}_\delta(M_0)$.

Since the set M_0 is asymptotically stable, the limit

$$\lim_{t \to +\infty} \rho(x(t, x_0), M_0) = 0 \tag{4.3}$$

is uniform for all $x_0 \in \overline{U}_\delta(M_0)$ for sufficiently small $\delta > 0$.

Equalities (4.2) and (4.3) enable us, using a sufficiently small $\delta_1 > 0$, to choose $\delta = \delta(\delta_1) > 0$, $T = T(\delta_1) > 0$, $\mu(\delta_1) > 0$, $\lim_{\delta_1 \to 0} \mu(\delta_1) = 0$ monotonically, such that the relations

$$x(t, x_0, \mu) \in \overline{U}_{\delta_1}(M_0) \subset D, \quad x(T, x_0, \mu) \in \overline{U}_\delta(M_0) \tag{4.4}$$

hold for all $x_0 \in \overline{U}_\delta(M_0)$, $\mu < \mu(\delta_1)$, $t \in [0, +\infty)$.

Indeed, fixing $\delta_1 > 0$ and using (4.3) we choose the constants $\delta = \delta(\delta_1) > 0$ and $T = T(\delta) > 0$ such that

$$\begin{aligned} \rho(x(t, x_0), M_0) &< \frac{\delta_1}{2}, \quad t \geq 0, \\ \rho(x(t, x_0), M_0) &< \frac{\delta}{2}, \quad t > T \end{aligned} \tag{4.5}$$

for all $x_0 \in \overline{U}_\delta(M_0)$. Taking into consideration (4.2) and using the above δ_1, δ and T we find $\mu(\delta_1) = \mu(\delta_1, \delta, T) > 0$ such that

$$\rho(x(t, x_0, \mu), \ x(t, x_0)) < \frac{\delta}{2}, \quad t \in [0, T] \tag{4.6}$$

for all $x_0 \in \overline{U}_\delta(M_0)$ and all $\mu < \mu(\delta_1)$.

Inequalities (4.5) and (4.6) lead to the estimates

$$\rho(x(t, x_0, \mu), M_0) \leq \rho(x(t, x_0, \mu), \ x(t, x_0)) + \rho(x(t, x_0), M_0) < \delta_1,$$

$$\rho(x(T, x_0, \mu), M_0) < \delta$$

for all $x_0 \in \overline{U}_\delta(M_0)$, $\mu < \mu(\delta_1)$, $t \in [0, T]$. The last inequality means that the motion $x(t, x_0, \mu)$ starting at the point $x_0 \in \overline{U}_\delta(M_0)$ remains in $\overline{U}_{\delta_1}(M_0)$ all the time for $t \in [0, T]$ and at $t = T$ comes back to a point in $\overline{U}_\delta(M_0)$. This is sufficient for a positive semi-trajectory of the motion $x(t, x_0, \mu)$ to belong to the set $\overline{U}_{\delta_1}(M_0)$ and for the point $x(T, x_0, \mu)$ to belong to the set $\overline{U}_\delta(M_0)$ for all $x_0 \in U_\delta(M_0)$, $\mu < \mu(\delta_1)$.

Let M_1 be union of all positive semi-trajectories of the motion $x(t, x_0, \mu)$ for which $x_0 \in \overline{U}_\delta(M_0)$, and let $\overline{M}_1$ be its closure. Both M_1 and $\overline{M}_1$ are positively invariant sets of system (4.1) and the following inclusion relations hold for them:

$$\overline{U}_{\delta_1}(M_0) \supset \overline{M}_1 \supset M_1 \supset \overline{U}_\delta(M_0). \tag{4.7}$$

We define the set M to be the union of all trajectories of system (4.1) lying in $\overline{M}_1$. It is clear that this set is a closed compact invariant set of system (4.1). Applying (4.7) to M we find that $\overline{U}_{\delta_1}(M_0) \supset M$. In view of this, the inequality $\rho(M_0, M) \leq \delta_1$ holds, which proves the limit relation $\lim_{\mu \to 0} \rho(M_0, M) = 0$ since δ_1 was chosen arbitrarily. We now prove the inclusion $\overline{U}_\delta(M_0) \subset \Pi(M)$. To do this, we choose $\epsilon = \epsilon_1(\delta_1) > 0$, $T_1 = T_1(\delta_1) > 0$ such that the relations

$$x(t, x_0, \mu) \in \overline{U}_{\epsilon_1}(M_0) \subset D, \quad x(T_1, x_0, \mu) \in \overline{U}_\delta(M_0) \tag{4.8}$$

analogous to (4.4) hold for sufficiently small $\delta_1 > 0$ and all $x_0 \in \overline{U}_{2\delta_1}(M_0)$, $\mu < \mu(\delta_1)$, $t \in [0, +\infty)$.

The set M_2 formed by the positive semi-trajectories of the motion $x(t, x_0, \mu)$, where $x_0 \in \overline{U}_{2\delta_1}(M_0)$, satisfy the inclusions

$$\overline{U}_{\epsilon_1}(M_0) \supset \overline{M}_2 \supset M_2 \supset \overline{U}_{2\delta_1}(M_0)$$

analogous to (4.7).

Moreover, since $x(T_1, x_0, \mu) \in \overline{U}_\delta(M_0)$ for all $x_0 \in \overline{U}_{2\delta_1}(M_0)$, it follows that $x(t, x_0, \mu) \in \overline{U}_{\delta_1}(M_0)$ for all $t \geq T_1$. Therefore $\overline{U}_{\delta_1}(M_0)$ contains an ω-limit set of the motion $x(t, x_0, \mu)$ for any $x_0 \in \overline{U}_{2\delta_1}(M_0)$, $\mu < \mu(\delta_1)$. Because of the inclusion $x(T_1, x_0, \mu) \in \overline{U}_\delta(M_0)$, this set is an ω-limit set of the motion $x(t, x(T_1, x_0, \mu), \mu)$ which starts at a point of the set $\overline{U}_\delta(M_0)$. Therefore it belongs to the set M which by definition contains all the trajectories of motions which start at a point in $\overline{U}_\delta(M_0)$. This proves that

$$\overline{U}_{2\delta_1}(M_0) \subset \Pi(M)$$

for all $\mu < \mu(\delta_1)$, so that $\overline{U}_{\delta_1}(M_0) \subset \Pi(M)$ for $\mu < \mu(\delta_1)$. The required inclusion is proved.

We now prove that the set M is asymptotically stable. Since $\rho(M_0, M) \to 0$ as $\mu \to 0$, we can find $\delta_0 > 0$ and $\bar{\mu}_0 > 0$ such that the δ_0-neighbourhood

$U_{\delta_0}(M)$ of the set M is contained in the set $\overline{U}_\delta(M_0)$ for any $\mu < \bar{\mu}_0$. But then for $\mu < \min(\bar{\mu}_0, \mu(\delta_1))$ we have the inclusion $U_{\delta_0}(M) \subset \Pi(M)$, which proves that

$$\lim_{t \to +\infty} \rho(x(t, x_0, \mu), M) = 0$$

for all $x_0 \in U_{\delta_0}(M)$, $\mu < \mu_0 < \min(\bar{\mu}_0, \mu(\delta_1))$.

Now to show that the set M is asymptotically stable for $\mu < \mu_0$ it is sufficient to prove that this set is stable.

Since $U_{\delta_0}(M) \subset \Pi(M)$, there are no whole trajectories of the motions $x(t, x_0, \mu)$ of system (4.1) that start in $U_{\delta_0}(M)$ and remain entirely in $U_{\delta_0}(M)$. Otherwise, these trajectories would belong to M. But this is impossible since $U_{\delta_0}(M)$ does not contain points of the set M. So, by applying Theorem 3 of §2.2 to show that the set M is stable it is sufficient to prove that there are no α-limit points in M of motions $x(t, x_0, \mu)$ that start at points $x_0 \in U_{\delta_0}(M)$ for $\mu < \mu_0$.

We now prove that M does not contain any α-limit points of motions $x(t, x_0, \mu)$ starting in $\overline{U}_{2\delta_1}(M_0)$ outside the set M. Thus let $x_0 \in \overline{U}_{2\delta_1}(M_0)\backslash M$. Then either $x_0 \in \overline{U}_{2\delta_1}(M_0)\backslash M_1$ or $x \in M_1\backslash M$. Consider the motion $x(t, x_0, \mu)$ for $t \leq 0$.

First suppose that $x_0 \in \overline{U}_{2\delta_1}(M_0)\backslash M_1$. We show that this motion leaves $\overline{U}_{2\delta_1}(M_0)$ for some $t = -h$, where $0 \leq h \leq T_1$. Indeed, let $x(t, x_0, \mu) \in \overline{U}_{2\delta_1}(M_0)$ for all $t \leq 0$. Then $x(t, x_0, \mu) \in \overline{U}_{2\delta_1}(M_0)\backslash M_1$ for $t \leq 0$, since otherwise $x(-t_1, x_0, \mu) \in M_1$ for $t_1 > 0$ and $x(t, x(-t_1, x_0, \mu), \mu) \in M_1$ for all $t \geq 0$, which is impossible in view of the equality $x(t_1, x(-t_1, x_0, \mu), \mu) = x_0$ and the relation $x_0 \notin M_1$. We take a point $x(-T, x_0, \mu)$ of the motion $x(t, x_0, \mu)$. Because it belongs to the set $\overline{U}_{2\delta}(M_0)$, according to (4.8) it must at time T_1 belong to the set $\overline{U}_\delta(M_0)$. But this is impossible since $x(T_1, x(-T_1, x_0, \mu), \mu) = x_0$ and $x_0 \notin M_1 \supset \overline{U}_\delta(M_0)$. This contradiction shows that $x(t, x_0, \mu) \notin \overline{U}_{2\delta_1}(M_0)$ for $t = -h$, where $h > 0$. But then the motion $x(t, x_0, \mu)$ does not have α-limit points in the set $\overline{U}_{2\delta_1}(M_0)$.

Now let $x_0 \in M_1\backslash M$. If the motion $x(t, x_0, \mu)$ leaves M_1 for $t < 0$, then it enters $\overline{U}_{2\delta}(M_0)$ and leaves it at some time $t = -h < 0$. Consequently, the motion $x(t, x_0, \mu)$ does not have α-limit points in $\overline{U}_{2\delta_1}(M_0)$.

If the motion $x(t, x_0, \mu) \in M_1$ for all $t \leq 0$, then the entire trajectory of this motion lies in M_1, which contradicts the assumption that $x_0 \notin M_1$. This contradiction shows that the inclusion $x(t, x_0, \mu) \in M_1$ for all $t \leq 0$ is impossible.

It follows from the above remarks that the set M does not contain α-limit points of the motions $x(t, x_0, \mu)$ starting in $\overline{U}_{2\delta_1}(M_0)$ for sufficiently small μ. Consequently, the set M is asymptotically stable for sufficiently small μ. Theorem 1 is proved.

Theorem 1 enables us to investigate the behaviour of the stable set M_0 of system (1.1) at the moment when it loses stability as a result of small perturbations of the system. Namely, we have the following theorem.

Theorem 2. *Suppose that the system of equations* (4.1) *has the same closed compact invariant set M_0 for all sufficiently small $\mu \in [0, \mu_0]$. If M_0 is asymptotically stable for $\mu = 0$ and unstable for $\mu \in (0, \mu_0]$, then for sufficiently small μ_0 the system of equations* (4.1) *has an invariant set M_μ for all $\mu \in (0, \mu_0]$ such that $M_0 \cap M_\mu = \varnothing$, $\lim_{\mu \to 0} \rho(M_0, M_\mu) = 0$.*

In other words, Theorem 2 states that when an asymptotically stable set M_0 becomes unstable a new invariant set M_μ, $\mu \in (0, \mu_0]$, of system (4.1) is "born".

To prove Theorem 2 it is sufficient to set M_μ equal to

$$M_\mu = M \backslash M_0, \tag{4.9}$$

where $M = M(\mu)$ is the asymptotically stable set of system (4.1) of Theorem 1. Since M contains all the trajectories that start in $\overline{U}_{\delta_1}(M_0)$, it follows that $M_0 \subset M$. Because M is an asymptotically stable set and M_0 is an unstable set of system (4.1) when $\mu \in (0, \mu_0]$, we have that $M_0 \neq M$. But then set (4.9) is non-empty, invariant, and together with M satisfies the limit relation $\lim_{\mu \to 0} \rho(M_0, M_\mu) = 0$. Theorem 2 is proved.

It should be noted that small perturbations of system (1.1) may have a severe "destructive" effect on the invariant set M_0 of Theorem 1. For example, for equation (4.1) with X and Y defined by the expressions

$$X = \begin{cases} -(x+1), & x \leq -1, \\ 0, & -1 \leq x \leq +1, \ Y = 1, \\ -(x-1), & x \geq +1, \end{cases}$$

the set M_0 consists of the interval $-1 \leq x \leq +1$, while $M = M(\mu)$ consists merely of the point $x_\mu = 1 + \mu$. The set M_μ born from M_0 may also differ significantly from M_0. For example, in the system

$$\frac{dx_1}{dt} = x_2 - \tfrac{1}{2}x_1(\rho^2 - \mu), \quad \frac{dx_2}{dt} = -x_1 - \tfrac{1}{2}x_2(\rho^2 - \mu),$$

where $\rho^2 = x_1^2 + x_2^2$, the set M_0 consists of the single point $x_1 = 0, x_2 = 0$, while M_μ consists of all the points of the disc $\rho^2 \le \mu$ without the centre.

Note that a change from stability to instability of the set M_0 does not necessarily lead to the "birth" from M_0 of a new invariant set of system (4.1) for $\mu \in (0, \mu_0]$. This is the case, for example, for the point $M_0 = \{0\}$ in the equation

$$\frac{dx}{dt} = x^3 \sin^2 \frac{1}{x} + \mu x.$$

2.5. Quasi-periodic motions and their closure

Suppose that the system of equations (1.1) has a quasi-periodic motion

$$x = x(t, x_0) = f(\omega t), \quad f(\phi) \in C(\mathcal{T}_m), \tag{5.1}$$

the trajectory $x(t, x_0; \mathbf{R})$ of which is compact in D. The closure of the trajectory $x(t, x_0; \mathbf{R})$ belongs to the region D and consists of the points defined by the equation

$$x = f(\phi), \quad \phi \in \mathcal{T}_m. \tag{5.2}$$

According to equation of motion (1.1) we have

$$f(\omega t) = f(0) + \int_0^t X(f(\omega s))ds, \quad t \in \mathbf{R},$$

so that, passing to the limit in the sequence of functions $f(\omega t + \omega t_n)$, $\lim_{n\to\infty} \omega t_n = \phi \bmod 2\pi$, we obtain the identity

$$f(\omega t + \phi) = f(\phi) + \int_0^t X(f(\omega s + \phi))ds, \tag{5.3}$$

which proves that

$$x(t, f(\phi)) = f(\omega t + \phi), \quad t \in \mathbf{R}, \ \phi \in \mathcal{T}_m.$$

Consequently, the set (5.2) is an invariant set of system (1.1), the trajectories of which are quasi-periodic with the same frequency basis.

We denote set (5.2) by M and determine its properties as a subset of E^n.

Lemma 1. *If $x(t, x_0) \in C^s(\omega)$ and $x(t, x_0) \in C(\Omega)$, then $x(t, x_0) \in C^s(\Omega)$.*

According to the lemma, the number s, describing smoothness of M, is an invariant of the motion $x(t, x_0)$.

Turning to the proof of the lemma, we write down the identities

$$
\begin{aligned}
x(t, x_0) &= f(\omega t), \quad f \in C^s(T_m), \\
x(t, x_0) &= f_1(\Omega t), \quad f_1 \in C(T_m),
\end{aligned}
\tag{5.4}
$$

where m and m_1 are the dimensions of the bases ω and Ω.

According to the definition of a basis, we can choose a basis Ω, Ω', using the numbers (Ω, ω) by extending Ω to a basis with numbers from ω. With such a choice

$$
\omega = \frac{1}{d}(P_1\Omega + P_2\Omega'), \quad \text{rank}[P_1, P_2] = m,
\tag{5.5}
$$

where d is an integer, P_1 and P_2 are integer matrices. From identities (5.4) it follows that

$$
\sup_{t \in \mathbf{R}} \left\| f\left(\frac{1}{d}P_1\Omega t + \frac{1}{d}P_2\Omega' t\right) - f_1(\Omega t) \right\| =
$$

$$
= \max_{(\psi, v) \in T_p} \| f(P_1\psi + P_2 v) - f_1(d\psi) \| = 0,
$$

that is, we have the identity

$$
f_1(d\psi) = f(P_1\psi + P_2 v), \quad (\psi, v) \in T_p,
\tag{5.6}
$$

where p is the dimension of the basis Ω, Ω'. But then

$$
f_1(\psi) = f\left(\frac{1}{d}P_1\psi\right), \quad \psi \in T_{m_1}.
\tag{5.7}
$$

Relation (5.7) proves that $f_1 \in C^s(T_{m_1})$, hence $x(t, x_0) \in C^s(\Omega)$.

The number m such that $x(t, x_0) \in C(\omega)$ for some basis $\omega = (\omega_1, \ldots, \omega_m)$ and $x(t, x_0) \notin C(\Omega)$ for any basis $\Omega = (\Omega_1, \ldots, \Omega_{m_1})$ with $m_1 < m$ will be called the *real dimension* of the frequency basis of the quasi-periodic function $x(t, x_0)$. The real dimension of the frequency basis describes the dimension of the set M at each point of it.

This fact can easily be expressed in mathematical terms when $x(t, x_0) \in C^s(\omega)$ for $s \geq 1$. In fact, we have the following lemma.

Lemma 2. *Let $x(t, x_0) = f(\omega t) \in C^s(\omega)$, where $s \geq 1$, $\omega = (\omega_1, \ldots, \omega_p)$. Then*

$$\text{rank } \frac{\partial f(\phi)}{\partial \phi} = m, \quad \phi \in T_p, \tag{5.8}$$

where m is the real dimension of the frequency basis of the $x(t, x_0)$.

To prove the lemma, first suppose that $p = m$, that is, the dimension of the frequency basis ω is real. Suppose that relation (5.8) does not hold for some $\phi = \phi_0 \in T_m$, so that

$$\text{rank } \frac{\partial f(\phi_0)}{\partial \phi} = m_1 < m.$$

Differentiating (5.3) we see that

$$\frac{\partial f(\omega t + \phi_0)}{\partial \phi_i} = \frac{\partial f(\phi_0)}{\partial \phi_i} + \int_0^t \frac{\partial X(f(\omega s + \phi_0))}{\partial x} \frac{\partial f(\omega s + \phi_0)}{\partial \phi_i} ds$$

for all $t \in \mathbf{R}$ $(i = 1, \ldots, m)$. This proves that the functions

$$\frac{\partial f(\omega t + \phi_0)}{\partial \phi_1}, \ldots, \frac{\partial f(\omega t + \phi_0)}{\partial \phi_m} \tag{5.9}$$

are solutions of the same system of linear differential equations

$$\frac{dy}{dt} = \frac{\partial X(f(\omega t + \phi_0))}{\partial x} y, \tag{5.10}$$

which consists of n equations.

When $t = 0$ the functions (5.9) are linearly dependent, therefore they are linearly dependent for all $t \in \mathbf{R}$. Consequently, we can find constants $c_1, \ldots, c_m$, $\sum_{i=1}^m |c_i| \neq 0$ such that

$$\sum_{i=1}^m \frac{\partial f(\omega t + \phi_0)}{\partial \phi_i} c_i = 0, \quad t \in \mathbf{R}. \tag{5.11}$$

Without loss of generality we can assume that $c_m \neq 0$. Because (5.11) is homogeneous we can make c_m equal to ω_m. Let $c_m = \omega_m$ in (5.11). Differentiating (5.3) and subtracting identity (5.11) we find that

$$\sum_{i=1}^{m-1} \frac{\partial f(\omega t + \phi_0)}{\partial \phi_i} \Omega_i = X(f(\omega t + \phi_0)), \quad t \in \mathbf{R}, \tag{5.12}$$

where $\Omega_i = \omega_i - c_i$ $(i = 1, \ldots, m-1)$. On passing to the limit we obtain from (5.12) the new identity:

$$\sum_{i=1}^{m-1} \frac{\partial f(\phi)}{\partial \phi_i} \Omega_i = X(f(\phi)), \ \phi \in T_m,$$

from which it follows that $X = f(\Omega_1 t, \ldots, \Omega_{m-1} t, \phi_m)$ is a solution of the system of equations (1.1). By virtue of the uniqueness of the solution of the Cauchy problem for system (1.1), we now have

$$f(\omega t) = f(\Omega t, 0), \ t \in \mathbf{R}$$

where we use the notation $\Omega = (\Omega_1, \ldots, \Omega_{m-1})$. This contradicts the assumption that m is the real dimension of the frequency basis of the $x(t, x_0) = f(\omega t)$. This contradiction proves relation (5.8) for $p = m$.

Now let $p > m$. Then $x(t, x_0) \in C(\Omega)$, where $\Omega = (\Omega_1, \ldots, \Omega_m)$. By Lemma 1, $x(t, x_0) \in C^s(\Omega)$ and the $f_1 \in C^s(T_m)$ satisfying the equality $x(t, x_0) = f_1(\Omega t)$ is related to $f \in C^s(T_p)$ by formula (5.7). Upon differentiating this relation we find that

$$m = \text{rank} \frac{\partial f_1(\psi)}{\partial \psi} \le \min\left(\text{rank} \frac{\partial f(\phi)}{\partial \phi}, \ \text{rank} P_1\right).$$

Thus, rank $\frac{\partial f(\phi)}{\partial \phi} \ge m$.

On the other hand, we have the relation

$$f(\phi) = f_1\left(\frac{1}{d_1}\overline{P}_1 \phi\right), \ \phi \in T_p$$

analogous to (5.7). But then

$$\text{rank} \frac{\partial f(\phi)}{\partial \phi} \le \min\left(\text{rank} \frac{\partial f_1(\psi)}{\partial \psi}, \ \text{rank} \overline{P}_1\right) = \min(m, \text{rank} \overline{P}_1).$$

From this and the previous inequalities it follows that

$$\text{rank} \frac{\partial f(\phi)}{\partial \phi} = m, \ \phi \in T_p.$$

We denote the image of the set $U_\delta(\phi_0) = \{\phi | \|\phi - \phi_0\| < \delta\}$ under the map $f : \phi \to f(\phi)$ by $M_\delta(f(\phi_0))$. The following statement is the analogue of Lemma 2 in the general case.

Lemma 3. *Let $x(t, x_0) = f(\omega t) \subset C(\omega)$, $\omega = (\omega_1, \ldots, \omega_m)$ and let m be the real dimension of the frequency basis of the $x(t, x_0)$. Then there exists $\delta > 0$ such that the map*

$$f : \overline{U}_\delta(\phi_0) \to \overline{M}_\delta(f(\phi_0)) \tag{5.13}$$

is a homeomorphism for any $\phi_0 \in T_m$.

By [1], the map (5.13) is a homeomorphism if it is one-to-one. Suppose that the map (5.13) is not one-to-one. Then there exist sequences of points $\phi_\nu'' \in T_m$ and $\phi_\nu' \in T_m$ $(\nu = 1, 2, \ldots)$ such that $\phi_\nu'' \neq \phi_\nu'$ and

$$f(\phi_\nu'') = f(\phi_\nu'), \quad \|\phi_\nu'' - \phi_\nu'\| \to 0, \text{ as } \nu \to \infty.$$

Then $x(t, f(\phi_\nu'')) = x(t, f(\phi_\nu'))$ for all $t \in \mathbf{R}$ $(\nu = 1, 2, \ldots)$, which leads to the system of identities

$$f(\omega t + \phi_\nu'') = f(\omega t + \phi_\nu'), \quad \nu = 1, 2, \ldots, t \in \mathbf{R}. \tag{5.14}$$

Passing to the limit in (5.14) we obtain the identities

$$f(\phi + \phi_\nu'') = f(\phi + \phi_\nu'), \quad \nu = 1, 2, \ldots,$$

which lead to a similar system of identities for the Fourier coefficients of the function $f(\phi)$:

$$f_k(e^{i(k, \phi_\nu'' - \phi_\nu')} - 1) = 0, \quad \nu = 1, 2, \ldots. \tag{5.15}$$

Considering (5.15) on the spectrum of the $f(\phi)$, that is, for those k such that $f_k \neq 0$, we arrive at the equalities:

$$(k, \phi_\nu'' - \phi_\nu') = 0 \bmod 2\pi, \quad \nu = 1, 2, \ldots. \tag{5.16}$$

Since $\|\phi_\nu'' - \phi_\nu'\| \to 0$ as $\nu \to \infty$, there exists $\nu_0 = \nu_0(k) > 0$ such that

$$(k, \phi_\nu'' - \phi_\nu') = 0, \quad \nu \geq \nu_0. \tag{5.17}$$

Dividing each equality in (5.17) by $\|\phi_\nu'' - \phi_\nu'\| \neq 0$ and passing to the limit as $\nu \to \infty$ we obtain the equality

$$(k, \xi) = 0, \tag{5.18}$$

where ξ is the limit of the sequence $(\phi_\nu'' - \phi_\nu')/\|\phi_\nu'' - \phi_\nu'\|$ $(\nu = 1, 2, \ldots)$. Because $\|\xi\| = 1$ we can assume that the mth coordinate ξ_m of the vector $\xi = (\xi_1, \ldots, \xi_m)$ is non-zero. Solving relation (5.18), we find that

$$k_m = -\frac{k_1\xi_1 + \ldots + k_{m-1}\xi_{m-1}}{\xi_m}. \tag{5.19}$$

Using equality (5.19) we prove that

$$f(\phi_1, \ldots, \phi_m) = f\left(\phi_1 - \frac{\xi_1}{\xi_m}\phi_m, \ldots, \phi_{m-1} - \frac{\xi_{m-1}}{\xi_m}\phi_m, 0\right) \tag{5.20}$$

for all $\phi \in \mathcal{T}_m$. To do this we set

$$P_N(\phi) = \sum_{\|k\| \le N} f_k e^{i(k,\phi)},$$

where $f(\phi) \simeq \sum_k f_k e^{i(k,\phi)}$ is the Fourier series of the function $f(\phi)$. Because relation (5.19) holds for the values of k in the spectrum of $f(\phi)$, it also holds on the spectrum of $P_N(\phi)$. This leads to the identity

$$P_N(\phi) = P_N\left(\phi' - \frac{\xi'}{\xi_m}\phi_m, 0\right), \quad \phi \in \mathcal{T}_m,$$

where $\phi' = (\phi_1, \ldots, \phi_{m-1})$, $\xi' = (\xi_1, \ldots, \xi_{m-1})$.

But then

$$J = \frac{1}{(2\pi)^m}\int_0^{2\pi}\cdots\int_0^{2\pi}\left\|f(\phi) - f\left(\phi' - \frac{\xi'}{\xi_m}\phi_m, 0\right)\right\|^2 d\phi \le$$

$$\le \|f(\phi) - P_N(\phi)\|_0^2 + \frac{1}{(2\pi)^m}\int_0^{2\pi}\int_0^{2\pi}\left\|P_N\left(\phi' - \frac{\xi'}{\xi_m}\phi_m, 0\right) -\right.$$

$$\left. - f\left(\phi' - \frac{\xi'}{\xi_m}\phi_m, 0\right)\right\|^2 d\phi' d\phi_m \le \|f(\phi) - P_N(\phi)\|_0^2 +$$

$$+ \frac{1}{(2\pi)^m}\int_0^{2\pi}\int_{(\xi'/\xi_m)\phi_m}^{2\pi+(\xi'/\xi_m)\phi_m}\|P_N(\phi', 0) - f(\phi', 0)\|^2 d\phi' d\phi_m \le$$

$$\le \|f(\phi) - P_N(\phi)\|_0^2 + \|f(\phi', 0) - P_N(\phi', 0)\|_0^2,$$

and passing to the limit as $N \to \infty$ we see that $J = 0$. Consequently, $\|f(\phi) - f(\phi' - (\xi'/\xi_m)\phi_m, 0)\| = 0$ almost everywhere on $\mathcal{T}_m$. And because $f(\phi)$

is continuous for $\phi \in \mathcal{T}_m$, this equality holds for all $\phi \in \mathcal{T}_m$. Relation (5.20) is proved.

It follows from (5.20) that

$$f(\omega t) = f(\Omega t, 0), \quad t \in \mathbf{R},$$

where $\Omega = (\Omega_1, \ldots, \Omega_{m-1})$, $\Omega_j = \omega_j - (\xi_j/\xi_m)\omega_m$. But then the real dimension of the frequency basis of the function $x(t, x_0) = f(\omega t)$ does not exceed $m - 1$ which contradicts the assumption.

The map $f : \phi \to f(\phi)$ does not always define a homeomorphism between the torus $\mathcal{T}_m = S_1(\phi_1) \times \ldots \times S_1(\phi_m)$ and the manifold M, where $S_1(\phi_j)$ is a circle with the angular coordinate system. This becomes evident if we take a function $f \in C(\mathcal{T}_m)$ that is periodic with respect to one of the variables ϕ_ν with period $2\pi/d$, where d is an integer, $d \geq 2$. Then every point of M is the image of d different points of $\mathcal{T}_m$, namely, the points $\phi + l\theta$ ($l = 1, \ldots, d$), where $\theta_1 = \ldots = \theta_{\nu-1} = \theta_{\nu+1} = \ldots = \theta_m = 0$, $\theta_\nu = 2\pi/d$. The non-uniqueness of the inverse image of a point $f(\phi_0) \in M$ under the map $f : \mathcal{T}_m \to M$ is related to the fact that it is possible to choose a frequency basis of a quasi-periodic $x(t, x_0)$ in different ways.

For $x(t, x_0) \in C(\omega)$ we introduce the notion of *maximal frequency basis* by defining a frequency basis $\omega = (\omega_1, \ldots, \omega_m)$ to be maximal if any other frequency basis $\Omega = (\Omega_1, \ldots, \Omega_{m_1})$ of this is related to ω via the relation

$$\omega = P\Omega, \tag{5.21}$$

where P is an integer matrix of rank m.

From this definition it becomes clear that the number of frequencies in a maximal basis is equal to the dimension of the real frequency basis of the function $x(t, x_0)$, and two maximal bases ω and Ω satisfy (5.21), where both P and P^{-1} are integer-valued. A maximal basis of a periodic function of one variable is equal to the largest frequency of the function $x(t, x_0)$ calculated from the smallest of its periods.

Theorem 1. *Let* $x(t, x_0) = f(\omega t) \in C(\omega)$. *The map* $f : \mathcal{T}_m \to M$ *is a homeomorphism if and only if the frequency basis* $\omega = (\omega_1, \ldots, \omega_m)$ *is maximal.*

To prove the theorem first assume that the mapping $f : \mathcal{T}_m \to M$ is a homeomorphism. Then $f(\phi) \neq f(0)$ if $\phi \neq 0 \bmod 2\pi$. Let $\Omega = (\Omega_1, \ldots, \Omega_{m_1})$

be a basis of the function $x(t, x_0)$. Then

$$x(t, x_0) = f_1(\Omega t), \quad t \in \mathbf{R},$$

where $f_1(\psi) = f_1(\psi_1, \ldots, \psi_{m_1}) \in C(T_{m_1})$. It follows from the arguments of the proof of Lemma 1 that the bases ω and Ω satisfy equality (5.5) and the functions f and f_1 satisfy identities (5.6) and (5.7). Because the mapping $f : T_m \to M$ is one-to-one, the matrix $P_2 = 0$ in equality (5.5).

Indeed, if $P_2 \neq 0$ then it follows from identity (5.6) that $f_1(0) = f(P_2 v) = f(0)$ for all $v \in T_{p-m_1}$, where $p > m_1$. For small v this contradicts the inequality $f(0) \neq f(\phi)$, where $\phi \neq 0 \bmod 2\pi$. Thus the bases ω and Ω satisfy the equality:

$$\omega = \frac{1}{d} P_1 \Omega, \tag{5.22}$$

where d is an integer, P_1 is an integer matrix of rank m and the functions f and f_1 satisfy identity (5.7). Since f_1 is periodic, it follows from (5.7) that

$$f(0) = f\left(\frac{2\pi}{d} p_j\right), \quad j = 1, \ldots, m_1, \tag{5.23}$$

where p_j is the jth column of the matrix P_1. Because $f(0) \neq f(\phi)$ for $\phi \neq 0 \bmod 2\pi$, equalities (5.23) can hold only if

$$p_j = 0 \bmod d, \quad j = 1, \ldots, m_1.$$

Then the matrix $(1/d)P_1 = P$ is an integer matrix of rank m and equality (5.22) is equivalent to equality (5.21). Thus, any basis of the function $x(t, x_0)$ different from ω is related to ω by relation (5.21) and consequently, the basis ω is maximal.

Suppose now that the frequency basis $\omega = (\omega_1, \ldots, \omega_m)$ of $x(t, x_0) = f(\omega t) \in C(\omega)$ is maximal. Suppose that the map $f : T_m \to M$ is not a homeomorphism. Since $f(\phi)$ is continuous and periodic with respect to ϕ_ν ($\nu = 1, \ldots, m$) with period 2π, the assumption made is equivalent to the property that the inverse image under the map f of a point of the set M is not unique. Consequently, there exist two points $\phi_1 \in T_m$ and $\phi_2 \in T_m$, $\phi_1 \neq \phi_2$ such that $f(\phi_1) = f(\phi_2)$. It follows from the arguments given in the proof of Lemma 3 that

$$f(\phi + \phi_1) = f(\phi + \phi_2), \quad \phi \in T_m. \tag{5.24}$$

Given a vector ϕ and a number α, we shall denote by $\phi \bmod \alpha$ the vector ϕ_0 with non-negative components of the smallest length such that $\phi - \phi_0 = \alpha p$, where p is an integer vector. We set $\delta = (\phi_2 - \phi_1)/(2\pi) \bmod 1$. Then it follows from identity (5.24) that

$$f(\phi) = f(\phi + 2\pi\delta), \quad \phi \in T_m. \tag{5.25}$$

Equality (5.25) shows that the function $F(t) = f(\delta t)$ is periodic with respect to t with period 2π. From this it necessarily follows that

$$\delta = p/d, \tag{5.26}$$

where d is an integer, $p = (p_1, \ldots, p_m)$ is an integer vector, $0 \le p_j < d$, $j = 1, \ldots, m$, $p \ne 0$.

Indeed, by choosing the basis $(1, \Omega')$ from the numbers $(1, \delta)$ we obtain the following representation for δ:

$$\delta = \frac{P_1}{d} + \frac{P_2}{d}\Omega', \tag{5.27}$$

where d is an integer, P_1 is an integer vector, P_2 is a matrix of integers, $\mathrm{rank}[P_1, P_2] \ge m_1 - 1$, m_1 being the number of frequencies of the basis $(1, \Omega')$.

From (5.27) and the identity $F(dt) = f(P_1 t + P_2\Omega' t)$, $t \in \mathbf{R}$, it follows that

$$F(d\psi) = f(P_1\psi + P_2 v), \quad (\psi, v) \in T_{m_1}. \tag{5.28}$$

For $m_1 > 1$ the vector $d\delta$ cannot be integer-valued for any integer d and so it follows from (5.27) that $P_2 \ne 0$. But if $P_2 \ne 0$, the identity (5.28) leads to the equality

$$f(0) = f(P_2 v), \quad v \in T_{m_1 - 1}. \tag{5.29}$$

Since $f(\omega t)$ has a frequency basis $\omega = (\omega_1, \ldots, \omega_m)$ of real dimension, the map $f : \overline{U}_{\delta_0}(0) \to \overline{M}_{\delta_0}(f(0))$ is a homeomorphism for small $\delta_0 > 0$. This contradicts equality (5.29) considered for those v satisfying $P_2 v \in \overline{U}_{\delta_0}(0)$. The contradiction shows that $m_1 = 1$. Consequently the basis of numbers $(1, \delta)$ consists of the number 1, and representation (5.27) has the form (5.26), where $P_1 = p$. The inequalities for the coordinates of the vector p follow from similar inequalities for the coordinates of the vector δ. This finishes the proof of (5.26).

Suppose that the jth coordinate p_j of the vector p is non-zero. We define the numbers $\Omega = (\Omega_1, \ldots, \Omega_m)$ by the relations:

$$\omega_1 = \frac{p_1}{d}\Omega_j + (-1)^{\nu_1}\Omega_1, \ldots, \quad \omega_j = \frac{p_j}{d}\Omega_j, \ldots, \omega_m = \frac{p_m}{d}\Omega_j + (-1)^{\nu_m}\Omega_m,$$

where the integers ν_i, $i \neq j$, are chosen so that the numbers Ω_i $(i = 1, \ldots, m)$ are positive. Since the numbers ω are incommensurable, the numbers Ω are also incommensurable.

Consider the function

$$f_1(\psi) = f\left(\frac{p_1}{d}\psi_j + (-1)^{\nu_1}\psi_1, \ldots, \frac{p_j}{d}\psi_j, \ldots, \frac{p_m}{d}\psi_j + (-1)^{\nu_m}\psi_m\right).$$

This is periodic with respect to the variables ψ_ν, $\nu \neq j$, with period 2π since f is periodic, and it is periodic with respect to ψ_j with period 2π since it follows from (5.24) that

$$f_1(\psi_1, \ldots, \psi_j + 2k_j\pi, \ldots, \psi_m) = f\left(\frac{p_1}{d}\psi_j + (-1)^{\nu_1}\psi_1 +\right.$$

$$+ 2\pi\frac{p_1}{d}k_j, \ldots, \frac{p_j}{d}\psi_j + 2\pi\frac{p_j}{d}k_j, \ldots, \frac{p_m}{d}\psi_j + (-1)^{\nu_m}\psi_m + 2\pi\frac{p_m}{d}k_j\right) =$$

$$= f_1(\psi_1, \ldots, \psi_j, \ldots, \psi_m)$$

for any integer k_j. Since $f_1(\psi)$ is periodic, the function $f_1(\Omega t)$ is quasi-periodic, and from the formulae relating ω and Ω, f and f_1 it follows that

$$f_1(\Omega t) = f(\omega t) = x(t, x_0), \quad t \in \mathbf{R}.$$

This shows that $x(t, x_0) \in C(\Omega)$.

We set $(1/d)P = ((-1)^{\nu_1}e_1, \ldots, p/d, \ldots, (-1)^{\nu_m}e_m)$, where

$$e_\nu = (0, \ldots, 1, \ldots, 0)$$

is the νth unit vector, $\nu \neq j$. The relations between ω and Ω can be expressed in terms of the matrix $(1/d)P$ in the form

$$\omega = (1/d)P\Omega. \tag{5.30}$$

Since the matrix $(1/d)P$ in (5.30) is not integer-valued, the basis ω of the function $x(t, x_0)$ is not maximal.

Thus we have arrived at a contradiction, which proves that the map $f : \mathcal{T}_m \to M$ is a homeomorphism. Theorem 1 is proved.

Next we consider the problem of the existence of a maximal frequency basis of the function $x(t, x_0) \in C(\omega)$. To do this we first study the properties of the map $f : \mathcal{T}_m \to M$ when $x(t, x_0) = f(\omega t) = C(\omega)$ and the basis ω is not a maximal basis of the function $x(t, x_0)$.

Let Γ_d be a subgroup of the group $\mathbf{Z}^m$ of all integer vectors $k = (k_1, \ldots, k_m)$ with the property that for a vector p the vector $kd + p$ also belongs to it for $\forall k \in \mathbf{Z}^m$, and let d be an integer ≥ 2. We define T'_d to be the set of all vectors $p \mod d$, where p runs over the subgroup Γ_d. We denote the set of all non-zero vectors of T'_d by T_d. If $T_d \neq \varnothing$, then we define the set T_d by prescribing all its elements $p_1, \ldots, p_N$, $p_\nu \neq 0$ for $\nu = 1, \ldots, N$, $N \geq 1$ and write it as $T_d = \{p_1, \ldots, p_N\}$.

The set of vectors $T_d = \{p_1, \ldots, p_N\}$ has the following property: if we arrange the positive numbers in the set of jth coordinates $p_{j\nu}$ of all vectors $p_\nu \in T_d$ in increasing order to form the sequence $p_j^{(1)}, p_j^{(2)}, \ldots, p_j^{(l_j)}$ ($p_j^{(i)} < p_j^{(i+1)}$, $i = 1, \ldots, l_j - 1$), then $p_j^{(1)}$ is a divisor of the number d and $p_j^{(i)} = i p_j^{(1)}$ for $i = 1, \ldots, l_j$ and $l_j + 1 = d/p_j^{(1)}$.

This property can be easily derived from the definition of the set T_d.

Lemma 4. *If the frequency basis* $\omega = (\omega_1, \ldots, \omega_m)$ *of the* $x(t, x_0) = f(\omega t) \in C(\omega)$ *is not maximal but* m *is the real dimension of the frequency basis, then for the function* $f(\phi)$ *there is a set* $T_d = \{p_1, \ldots, p_N\}$ *such that*

$$f\left(\phi + 2\pi \frac{p_\nu}{d}\right) = f(\phi), \quad \phi \in \mathcal{T}_m \tag{5.31}$$

for any $p_\nu \in T_d$ *and*

$$f\left(\phi + 2\pi \frac{p}{d}\right) \neq f(\phi), \quad \phi \in \mathcal{T}_m \tag{5.32}$$

for any p *such that* $p \mod d \notin T_d$.

To prove the lemma, we consider the set $\{\phi_1, \ldots, \phi_N\}$ of all the roots of the equation

$$f(\phi) = f(0), \quad \phi \in \mathcal{T}_m, \tag{5.33}$$

where $\phi_j = \phi_\nu$ if $j \neq \nu$, $\phi_j \neq 0$ for $j = 1, \ldots, N$. From the arguments in the proof of Lemma 3 it follows that this set is finite and from the arguments given in the proof of Theorem 1 one can see that $\phi_\nu = 2\pi\delta_\nu$ for $\nu = 1,\ldots,N$, where δ_ν has the form of (5.26) with $p = p'_\nu$ and $d = d_\nu$. By taking the l.c.m. of all denominators in the expression for δ_ν in the form (5.26) we find a representation $\delta_\nu = p_\nu/d$ ($\nu = 1,\ldots,N$) for δ_ν, where d is the l.c.m. Then, the vector $p_\nu = (p_{1\nu},\ldots,p_{m\nu})$ satisfies the inequality $0 \leq p_{j\nu} < d$ for $j = 1,\ldots,m$, $\nu = 1,\ldots,N$.

We set $T_d = \{p_1,\ldots,p_N\}$. It follows from identity (5.24) considered for $\phi_1 = 0$ and $\phi_2 = 2\pi p_\nu/d$ that identity (5.31) holds for $\nu = 1,\ldots,N$. The properties of the set T_d clearly follow from identities (5.31) because the function $f(\phi)$ is periodic with respect to ϕ_ν, $\nu = 1,\ldots,m$, with period 2π. Inequality (5.32) also follows because the set $\{\phi_1,\ldots,\phi_N\}$ contains all the roots of equation (5.33) and consequently T'_d contains all the vectors $p = p \bmod d$ for which equality (5.31) holds. Lemma 4 is proved.

Lemma 4 indicates a simple procedure for transforming an arbitrary basis $\omega = (\omega_1,\ldots,\omega_m)$ of the function $x(t,x_0) = f(\omega t) \in C(\omega)$ to a maximal one. We describe one step of this procedure. Suppose that m is the real dimension of the frequency basis of the function $x(t,x_0)$. Consider the set $T_d = \{p_1,\ldots,p_N\}$. Suppose that the first coordinate of at least one of the vectors from T_d is not zero. Then the first coordinates of the vectors p_ν ($\nu = 1,\ldots,N$) are multiples of the least positive number $p_1^{(1)}$. Without loss of generality we can assume that this is the vector $p_1 = (p_{11},\ldots,p_{m1})$ which has $p_1^{(1)}$ as its first coordinate. We define the numbers $(\Omega_1,\ldots,\Omega_m) = \Omega$ by setting

$$\omega_1 = \frac{p_1^{(1)}}{d}\Omega_1, \quad \omega_{j+1} = \frac{p_{j+1,1}}{d}\Omega_1 + (-1)^{\nu_{j+1}}\Omega_{j+1}, \quad j = 1,\ldots,m-1, \quad (5.34)$$

and choosing integers ν_{j+1} so that the numbers $\Omega_1,\ldots,\Omega_m$ are positive. Because ω is a basis, the numbers $\Omega_1,\ldots,\Omega_m$ are incommensurable.

Consider the function

$$f_1(\psi) = f\left(\frac{p_1^{(1)}}{d}\psi_1, \frac{p_{21}}{d}\psi_1 + (-1)^{\nu_2}\psi_2, \ldots, \frac{p_{m1}}{d}\psi_1 + (-1)^{\nu_m}\psi_m\right). \quad (5.35)$$

By identity (5.31) the function $f_1(\psi)$ turns out to be periodic with respect to ψ_ν ($\nu = 1,\ldots,m$) with period 2π. Thus $f_1(\Omega t) \in C(\Omega)$. It follows from formulae (5.34) and (5.35) that

$$f_1(\Omega t) = f(\omega t) = x(t,x_0), \quad t \in \mathbf{R},$$

therefore $x(t, x_0) \in C(\Omega)$.

We define the matrix P by setting

$$P/d = (p_1/d, (-1)^{\nu_2} e_2, \ldots, (-1)^{\nu_m} e_m), \tag{5.36}$$

where $e_\nu = (0, \ldots, 1, \ldots, 0)$ is the ν-unit vector taken as a column vector. According to formulae (5.34)–(5.36) we have the equalities

$$\omega = \frac{P}{d}\Omega, \quad f_1(\psi) = f\left(\frac{P}{d}\psi\right), \quad \psi \in T_m. \tag{5.37}$$

Consider the roots of the equation

$$f_1(\psi) = f_1(0), \quad \psi \in T_m. \tag{5.38}$$

According to formulae (5.37) these roots are defined in terms of the values $\phi_k = 2\pi k$ and $\phi_{\nu,k} = 2\pi(p_\nu/d + k)$ (where $k = (k_1, \ldots, k_m) \in \mathbf{Z}^m$, $p_\nu \in T_d$) from the formulae

$$2\pi k = \frac{P}{d}\psi_k, \quad 2\pi\left(\frac{p_\nu}{d} + k\right) = \frac{P}{d}\psi_{\nu,k} = \frac{P}{d}\psi_\nu \tag{5.39}$$

as $\psi_k \bmod 2\pi$ and $\psi_\nu \bmod 2\pi$.

From (5.39) we select the equalities for the first coordinates of the vectors $\psi_k = (\psi_{1k}, \ldots, \psi_{mk})$ and $\psi_\nu = (\psi_{1\nu}, \ldots, \psi_{m\nu})$. We have

$$2\pi k_1 = \frac{p_1^{(1)}}{d}\psi_{1k}, \quad 2\pi\left(\frac{p_{1\nu}}{d} + k_1\right) = \frac{p_1^{(1)}}{d}\psi_{1\nu} \tag{5.40}$$

Since $p_1^{(1)}$ is a divisor of the numbers d and $p_{1\nu}$ it follows from (5.40) that

$$\psi_{1k} = 0 \bmod 2\pi, \quad \psi_{1\nu} = 0 \bmod 2\pi.$$

Hence $\psi_k \bmod 2\pi$ and $\psi_\nu \bmod 2\pi$ are vectors that have their first coordinates equal to zero for arbitrary $k \in \mathbf{Z}^m$ and any $\nu = 1, \ldots, N$. But then the set $T_{d'} = \{p_1', \ldots, p_{N'}'\}$ constructed using the function $f_1(\phi)$ consists of the vectors p_ν' $(\nu = 1, \ldots, N')$ with first coordinate equal to zero. This finishes the first step of the procedure of transforming a basis ω into a maximal basis Ω.

Suppose that the first coordinates of all the vectors in $T_d = \{p_1, \ldots, p_N\}$ are equal to zero and the second coordinate of at least one of the vectors from T_d is not equal to zero. In this case we have to carry out the same transformations with the function $f(\phi)$ and basis ω as in the case considered above except that the transformations must be applied to the variable $\phi' = (\phi_2, \ldots, \phi_m)$ and the basis $\omega' = (\omega_2, \ldots, \omega_m)$, leaving the coordinate ϕ_1 and the frequency ω_1 unchanged. In this way we find a new $f_1(\psi)$ and a new basis Ω such that $f_1(\Omega t) = f(\omega t) = x(t, x_0) \in C(\Omega)$, and the set $T_{d'} = \{p'_1, \ldots, p'_{N'}\}$ constructed using the function $f_1(\phi)$ consists of the vectors p'_ν $(\nu = 1, \ldots, N')$ with the first and second coordinates equal to zero.

Continuing this transformation process with the function $f(\phi)$ and basis ω we can construct in a finite number of steps a function f' and a basis Ω' such that the map $f' : T_m \to M$ is a homeomorphism and the basis Ω' is a maximal basis of the function $x(t, x_0) = f'(\Omega' t) = f(\omega t)$.

As a result we have the following statement.

Theorem 2. *If the motion $x(t, x_0) = f(\omega t)$ is quasi-periodic, then the function $x(t, x_0)$ has a maximal frequency basis.*

Theorems 1 and 2 together with Lemma 1 yield the following result.

Theorem 3. *If the motion $x(t, x_0)$ is quasi-periodic and $x(t, x_0) \in C^s(\omega)$ with $s \geq 0$ then the set $M = \overline{x(t, x_0; \mathbf{R})}$ is C^s-homeomorphic to an m-torus, where m is the real dimension of the frequency basis of the function $x(t, x_0)$.*

A set that is C^s-homeomorphic to an m-dimensional torus is usually called an m-dimensional *toroidal manifold* of smoothness s or simply an m-dimensional torus of smoothness s.

Theorem 3 therefore states that the closure of a quasi-periodic orbit of system (1.1) is an m-dimensional s-smooth torus.

Because M is homeomorphic to an m-dimensional torus and belongs to the space E^n it necessarily follows that $m < n$ [1]. At the same time the value $m = n - 1$ is possible; it is easy to construct an analytic submanifold of the space E^{m+1} that is analytically diffeomorphic to an m-dimensional torus.

2.6. Invariance equations of a smooth manifold and the trajectory flow on it

Let M be an m-dimensional smooth compact manifold, a neighbourhood $U(M)$ of which lies in a domain $D \subset E^n$. We form a finite covering of the set M:

$$M = \bigcup_{i=1}^{N} M_\delta(x_i)$$

by choosing a sufficiently dense net of points $x_i \in M$ $(i = 1, \ldots, N)$ and defining

$$M_\delta(x_i) = M \bigcap \amalg_\delta^n(x_i), \quad i = 1, \ldots, N,$$

where $\amalg_\delta^n(x_i)$ is an n-dimensional ball in E^n with centre at the point x_i and redius δ.

For a sufficiently small $\delta > 0$, the set $M_\delta(x_i)$ is diffeomorphic to an m-dimensional ball $\amalg_\delta^m(0)$ of the space E^m with a certain coordinate system $\phi = (\phi_1, \ldots, \phi_m)$ and consequently, it consists of the points $x \in E^n$ given by the equation

$$x = F_i(\phi), \quad \phi \in \amalg_\delta^m(0),$$

where the function $F_i(\phi)$ is continuously differentiable for $\phi \in \amalg_\delta^m(0)$ and

$$\text{rank}\, \frac{\partial F_i(\phi)}{\partial \phi} = m, \quad \phi \in \amalg_\delta^m(0).$$

We define a multivalued function $F(\phi)$ by setting

$$F(\phi) = \bigcup_{i=1}^{N} F_i(\phi), \quad \phi = \amalg_\delta^m(0),$$

and use it to write the equation of the manifold M in the form

$$x = F(\phi), \quad \phi \in \amalg_\delta^m(0). \tag{6.1}$$

We write down the invariance conditions for M with respect to the system of equations (6.1) by using the analytic expression for the equation of the manifold M.

Let $x_0 \in M$, so that $x_0 = F_i(\phi_0)$ for some i, $1 \le i \le N$, and $\phi_0 \in \amalg_\delta^m(0)$. The set M is an invariant set of system (1.1) if and only if the vector $X(x_0)$ defining the right side of this system at the point $x = x_0$ belongs to the tangent hyperplane to M at the point x_0. This leads to the equality

$$\frac{\partial F_i(\phi_0)}{\partial \phi} r_i = X(F_i(\phi_0)) \tag{6.2}$$

for some $r_i = r_i(\phi_0)$.

Upon multiplying (6.2) by the transpose $(\partial F_i(\phi_0)/\partial \phi)^*$ of the matrix $\partial F_i(\phi_0)/\partial \phi$, we obtain the equality

$$\Gamma_i(\phi_0)r_i = \left(\frac{\partial F_i(\phi_0)}{\partial \phi}\right)^* X(F_i(\phi_0)), \tag{6.3}$$

where we have set

$$\Gamma_i(\phi) = \left(\frac{\partial F_i(\phi)}{\partial \phi}\right)^* \frac{\partial F_i(\phi)}{\partial \phi}, \quad \phi \in \amalg_\delta^m(0). \tag{6.4}$$

Matrix (6.4) is the Gramm matrix of linearly independent vectors, therefore

$$\det \Gamma_i(\phi) \neq 0, \quad \phi \in \amalg_\delta^m(0).$$

The vector r_i is uniquely defined from equality (6.3):

$$r_i = \Gamma_i^{-1}(\phi_0)\left(\frac{\partial F_i(\phi_0)}{\partial \phi}\right)^* X(F_i(\phi_0)).$$

Substituting this value of r_i into equality (6.2) we find that the function $F_i(\phi)$ satisfies the relation

$$\frac{\partial F_i(\phi)}{\partial \phi}\Gamma_i^{-1}(\phi)\left(\frac{\partial F_i(\phi)}{\partial \phi}\right)^* X(F_i(\phi)) = X(F_i(\phi))$$

for every $\phi \in \amalg_\delta^m(0)$.

Consequently, the manifold M is an invariant set of system (1.1) if and only if the function $F(\phi)$ satisfies the equation

$$(F_\phi \Gamma^{-1} F_\phi^* - E)X(F) = 0 \tag{6.5}$$

for all $\phi \in \amalg_\delta^m(0)$. In (6.5) we have introduced the notation

$$F_\phi = \partial F/\partial \phi, \quad \Gamma = F_\phi^* F_\phi, \quad F_\phi^* = (\partial F/\partial \phi)^*,$$

and the expression "$F(\phi)$ satisfies" means that each function $F_i(\phi)$ for $i = 1, \dots, N$ satisfies equation (6.5).

Equation (6.5) will be called the *invariance equation* of the manifold M with respect to system of equations (1.1).

Suppose that the function $F(\phi)$ satisfies the invariance equation (6.5). We consider the differential equation

$$\frac{d\phi}{dt} = \Gamma_i^{-1}(\phi)F_{i\phi}^*(\phi)X(F_i(\phi)), \quad \phi \in \amalg_\delta^m(0). \tag{6.6}$$

For a sufficiently dense net of points x_i $(i = 1, \ldots, N)$ on M, every point $x_0 \in M$ is an interior point of the ball $\amalg_\delta^n(x_i)$, so that x_0 is not a boundary point of $M_\delta(x_i)$.

Because boundary points are mapped into boundary points under a diffeomorphism, $\phi_0 = F_i^{-1}(x_0)$ is an interior point of the ball $\amalg_\delta^m(0)$. Hence the solution $\phi = \phi_i(t, \phi_0)$ of equation (6.6) is defined for t in some interval $I_0 = I_0(\phi_0)$ containing the point $t = 0$. Differentiating the function $F_i(\phi_i(t, \phi_0))$ and using relations (6.5) and (6.6) we find that

$$\frac{dF_i(\phi_i(t, \phi_0))}{dt} = X(F_i(\phi_i(t, \phi_0))), \quad t \in I_0.$$

But then

$$x(t, x_0) = F_i(\phi_i(t, \phi_0)), \quad t \in I_0, \tag{6.7}$$

where $x(t, x_0)$ is the solution of system (1.1), $x(0, x_0) = x_0$. Using the standard notation, relation (6.7) means that

$$x(t, F_i(\phi_0)) = F_i(\phi_i(t, \phi_0)), \quad t \in I_0. \tag{6.8}$$

Consequently, any trajectory of system (1.1) that starts at a point of the set M, can be obtained by using equality (6.8) from equation (6.1) of the manifold and the solutions of the equation

$$\frac{d\phi}{dt} = \Gamma^{-1}(\phi)F_\phi^*(\phi)X(F(\phi)), \quad \phi \in \amalg_\delta^m(0) \tag{6.9}$$

understood as the set of equations (6.6) for $i = 1, \ldots, N$. Having in mind this possibility we shall say that equation (6.9) defines a *trajectory flow* for system (1.1) on the invariant manifold M, or that it is the equation of this flow.

We make some remarks concerning equations (6.5) and (6.9).

Let $B(\phi) = \bigcup_{i=1}^{N} B_i(\phi)$, $\phi \in \amalg_\delta^m(0)$, be an $n \times (n - m)$ matrix that extends the matrix $F_\phi(\phi)$ to a non-singular matrix $[F_\phi(\phi), B(\phi)] : \det[F_\phi(\phi), B(\phi)] \neq 0$, $\phi \in \amalg_\delta^m(0)$. Multiplying equation (6.5) by the matrix $[F_\phi(\phi), B(\phi)]^*$ we obtain the equation

$$B^*(\phi)(F_\phi \Gamma^{-1} F_\phi^* - E)X(F) = 0$$

equivalent to (6.5) and containing $n - m$ equations instead of the initial n equations.

Let $P(\phi) = \bigcup_{i=1}^{N} P_i(\phi)$, $\phi \in \text{III}_\delta^m(0)$, be an $m \times n$ matrix such that its product with the matrix $F_\phi(\phi)$ is non-singular:

$$\det P(\phi)F_\phi(\phi) \neq 0, \quad \phi \in \text{III}_\delta^m(0).$$

The following identity follows from invariance equation (6.5):

$$P(\phi)F_\phi(\phi)\Gamma^{-1}(\phi)F_\phi^*(\phi)X(F(\phi)) = P(\phi)X(F(\phi)), \quad \phi \in \text{III}_\delta^m(0),$$

and we can use it to write equation (6.9) in the following form:

$$\frac{d\phi}{dt} = (P(\phi)F_\phi(\phi))^{-1}P(\phi)X(F(\phi)), \quad \phi \in \text{III}_\delta^m(0),$$

where there is "certain degree of freedom" in choosing the right hand side.

Finally we note that the non-uniqueness involved in the description of the equation of M in the form of (6.1) disappears whenever the manifold equation is given by a single-valued function $F(\phi)$ for all ϕ in a compact set K. In this case each of the equations (6.5) and (6.9) is to be understood as a separate equation for all $\phi \in K$.

In particular, the invariance equation (6.5) for the m-dimensional torus M defined by equation (5.2) with the function $f(\phi)$ satisfying the conditions

$$f(\phi) \in C^{s+1}(T_m), \quad \text{rank}\, \frac{\partial f(\phi)}{\partial \phi} = m \quad \forall \phi \in T_m \tag{6.10}$$

has the form

$$\left(\frac{\partial f}{\partial \phi} \Gamma^{-1} \left(\frac{\partial f}{\partial \phi} \right)^* - E \right) X(f) = 0, \quad \phi \in T_m, \tag{6.11}$$

while the trajectory flow equation (6.9) for system (1.1) on M takes the form

$$\frac{d\phi}{dt} = \Gamma^{-1}(\phi) \left(\frac{\partial f(\phi)}{\partial \phi} \right)^* X(f(\phi)), \quad \phi \in T_m. \tag{6.12}$$

For example, for a torus defined by the equations

$$x_{2\nu-1} = \rho_\nu^0 \sin \phi_\nu, \quad x_{2\nu} = \rho_\nu^0 \cos \phi_\nu, \quad \nu = 1, \ldots, m, \tag{6.13}$$

where $\rho_\nu^0 = \text{const} > 0$ for $\nu = 1, \ldots, m$, invariance equation (6.11) has the form of the system

$$\sin \phi_\nu (\sin \phi_\nu X_{2\nu-1} + \cos \phi_\nu X_{2\nu}) = 0, \quad \phi \in T_m,$$

$$\cos \phi_\nu (\sin \phi_\nu X_{2\nu-1} + \cos \phi_\nu X_{2\nu}) = 0, \quad \nu = 1, \ldots, m$$

equivalent to the system

$$\sin \phi_\nu X_{2\nu-1} + \cos \phi_\nu X_{2\nu} = 0, \quad \nu = 1,\dots,m, \ \phi \in T_m, \qquad (6.14)$$

where X_j $(j = 1,\dots,2m)$ denotes the expression

$$X_j = X_j(\rho_1^0 \sin \phi_1, \rho_1^0 \cos \phi_1, \dots, \rho_m^0 \sin \phi_m, \ \rho_m^0 \cos \phi_m).$$

If system of equalities (6.14) is satisfied for all $\phi \in T_m$, then the torus
(6.13) is an invariant set of system (1.1) with $n = 2m$. In the latter case the
trajectory flow of system (1.1) on the torus (6.13) is defined by equation (6.12)
which takes the form of the system

$$\frac{d\phi_\nu}{dt} = \frac{1}{\rho_\nu^0}(\cos \phi_\nu X_{2\nu-1} - \sin \phi_\nu X_{2\nu}), \quad \nu = 1,\dots,m, \ \phi \in T_m.$$

2.7. Local coordinates in a neighbourhood of a toroidal manifold.
Stability of an invariant torus

Let M be a toroidal manifold defined by equation (6.1) with the function
$F(\phi) = f(\phi)$ satisfying conditions (6.10). Suppose that the matrix $\partial f(\phi)/\partial \phi$
can be extended to a periodic basis in E^n and that $B(\phi) \in C^{s+1}(T_m)$ is the
extending matrix so that

$$\det[\partial f(\phi)/\partial \phi, \ B(\phi)] \neq 0, \quad \phi \in T_m. \qquad (7.1)$$

In studying the trajectories of system (1.1) which start in a neighbourhood of the manifold M it is convenient to go from the Euclidean coordinates
$x = (x_1,\dots,x_n)$ to local coordinates $\phi = (\phi_1,\dots,\phi_m), \quad h = (h_1,\dots,h_{n-m})$
introduced in a neighbourhood of M in such a way that the manifold equation
(6.1) takes the form

$$h = 0, \ \phi \in T_m. \qquad (7.2)$$

Under the above assumptions concerning the manifold M, the local coordinates
ϕ, h can be introduced by giving equations relating the variables x with the
variables ϕ, h by means of equalities

$$x = f(\phi) + B(\phi)h. \qquad (7.3)$$

Lemma 1. *For each sufficiently small $\delta > 0$ one can find $\delta_1 = \delta_1(\delta) > 0$, $\delta_1(\delta) \to 0$ as $\delta \to 0$ such that every point x satisfying the condition*

$$\rho(x, M) < \delta, \tag{7.4}$$

has local coordinates ϕ, h such that

$$\|h\| < \delta_1, \quad \phi \in \mathcal{T}_m. \tag{7.5}$$

Indeed, since

$$\rho(x, M) = \inf_{y \in M} \rho(x, y) = \inf_{\phi \in \mathcal{T}_m} \|x - f(\phi)\| = \|x - f(\psi)\|$$

for some $\psi = \psi(x) \in \mathcal{T}_m$, inequality (7.4) means that

$$\|x - f(\psi)\| < \delta.$$

We set

$$x = f(\psi) + z$$

and show that for a sufficiently small $\delta > 0$ the equation

$$f(\psi) + z = f(\phi) + B(\phi)h \tag{7.6}$$

can be uniquely solved with respect to ϕ, h in the domain

$$\|\phi - \psi\| < \delta_1, \ \|h\| < \delta_1, \ \|z\| < \delta, \tag{7.7}$$

where $\delta_1 = \delta_1(\delta)$ is a sufficiently small number, $\delta_1(\delta) \to 0$ as $\delta \to 0$.

To do this we write equation (7.6) in the form

$$\frac{\partial f(\psi)}{\partial \phi}(\phi - \psi) + B(\psi)h = z - \left[f(\phi) - f(\psi) - \frac{\partial f(\psi)}{\partial \phi}(\phi - \psi) \right] - [B(\phi) - B(\psi)]h \tag{7.8}$$

and, taking into account inequality (7.1), solve (7.8) with respect to $\phi - \psi, h$. As a result we obtain the system of equations

$$\phi - \psi = L_1 \left\{ z - \left[f(\psi + (\phi - \psi)) - f(\psi) - \frac{\partial f(\psi)}{\partial \phi}(\phi - \psi) \right] - [B(\psi + (\phi - \psi)) - B(\psi)]h \right\},$$

$$h = L_2 \left\{ z - \left[f(\psi + (\phi - \psi)) - f(\psi) - \frac{\partial f(\psi)}{\partial \phi}(\phi - \psi) \right] - \right.$$

$$\left. - [B(\psi + (\phi - \psi)) - B(\psi)]h \right\}, \tag{7.9}$$

where $L_1 = L_1(\psi)$ and $L_2 - L_2(\psi)$ are matrices of appropriate diemensions.

Using this system of equations one can see how to fix $\delta_1 = \delta_1(\delta) > 0$ for a sufficiently small $\delta > 0$ in such a way that in the domain (7.7) the operator defined by the right hand side of this system will be contracting in the space of the vectors $\phi - \psi, h$. This is sufficient to solve uniquely equations (7.9) with respect to $\phi - \psi, h$ in the domain (7.7) for the corresponding choice of $\delta_1 = \delta_1(\delta) > 0$ and sufficiently small $\delta > 0$. But then any point x satisfying condition (7.4) has local coordinates ϕ, h satisfying inequality (7.5).

Consider the matrix

$$\Gamma_0(\phi) = B^*(\phi)B(\phi), \quad \phi \in \mathcal{T}_m.$$

It is the Gramm matrix of linearly independent vectors and so the eigenvalues of this matrix are positive for all $\phi \in \mathcal{T}_m$. Since the matrix $\Gamma_0(\phi)$ is periodic with respect to ϕ one can find upper and lower estimates γ_0, γ^0 for these numbers that are independent of ϕ. The quadratic form $\langle \Gamma_0(\phi)h, h \rangle$ can be estimated by the inequality

$$\gamma_0\|h\|^2 \leq \langle \Gamma_0(\phi)h, h \rangle \leq \gamma^0\|h\|^2 \tag{7.10}$$

for all $\phi \in \mathcal{T}_m$, $h \in E^{n-m}$. Inequality (7.10) leads to the estimate

$$\gamma_0\|h\|^2 \leq \|x - f(\phi)\|^2 \leq \gamma^0\|h\|^2 \tag{7.11}$$

which relates the Cartesian coordinates $x = (x_1, \ldots, x_n)$ and the local coordinates $\phi = (\phi_1, \ldots, \phi_m)$, $h = (h_1, \ldots, h_{n-m})$ of any point of a neighbourhood of the manifold M.

It is clear from Lemma 1 and inequality (7.11) that a δ-neighbourhood of M considered as a subset of points in E^n is a certain domain contained in a δ_1-neighbourhood of manifold (7.2) considered as a subset of points in $\mathcal{T}_m \times E^{n-m}$ and vice versa.

We use this fact to write in local coordinates the equations of the motion of system (1.1) which starts in a neighbourhood of M. To do this we differentiate relation (7.3) as a formula of change of variables in system (1.1) and we obtain instead of (1.1) the system

$$\left[\frac{\partial f(\phi)}{\partial \phi} + \frac{\partial B(\phi)h}{\partial \phi}\right]\frac{d\phi}{dt} + B(\phi)\frac{dh}{dt} = X(f(\phi) + B(\phi)h). \tag{7.12}$$

Condition (7.1) enables us to solve system (7.12) with respect to $d\phi/dt$, dh/dt for all h in the domain (7.5) for sufficiently small $\delta_1 > 0$. This leads to the system of equations:

$$\frac{d\phi}{dt} = L_1(\phi, h)X(f(\phi) + B(\phi)h)$$
$$\frac{dh}{dt} = L_2(\phi, h)X(f(\phi) + B(\phi)h)$$

(7.13)

where $L_1(\phi, h)$, $L_2(\phi, h)$ denote the blocks of the matrix

$$L(\phi, h) = \text{column}(L_1(\phi, h), L_2(\phi, h))$$

which is the inverse of the matrix $(\partial f(\phi)/\partial\phi + \partial B(\phi)h/\partial\phi, B(\phi))$.

The right hand side of the system of equations (7.13) is defined on the domain (7.5), is periodic with respect to ϕ_ν ($\nu = 1, \ldots, m$) with period 2π and has degree of smoothness with respect to ϕ, h one less than the degree of smoothness of the manifold M and the smoothness of the right hand side of system (1.1) in a neighbourhood of M. The system of equations (7.13) is equivalent to system of equations (1.1) considered on the inverse image of the domain (7.5) under the mapping $x \to (\phi, h)$ given by the change of variables formula (7.3).

Note that the expressions for the matrices $L_1(\phi, h)$ and $L_2(\phi, h)$ in terms of the matrices $f_\phi = \partial f(\phi)/\partial\phi$ and $B = B(\phi)$ can be found by using the Frobenius formula [39] for inverting a four-block matrix. This leads, for example, to the following values for $L_1(\phi, h)$ and $L_2(\phi, h)$:

$$L_1(\phi, h) = \left[\left(f_\phi + \frac{\partial Bh}{\partial\phi}\right)^*(E - B\Gamma_0^{-1}B^*)\left(f_\phi + \frac{\partial Bh}{\partial\phi}\right)\right]^{-1} \times$$

$$\times \left(f_\phi + \frac{\partial Bh}{\partial\phi}\right)^*(E - B\Gamma_0^{-1}B^*),$$

$$L_2(\phi, h) = \left\{B^*\left[E - \left(f_\phi + \frac{\partial Bh}{\partial\phi}\right)\Gamma_1^{-1}\left(f_\phi + \frac{\partial Bh}{\partial\phi}\right)^*\right]B\right\}^{-1} \times$$

$$\times B^*\left\{E - \left(f_\phi + \frac{\partial Bh}{\partial\phi}\right)\Gamma_1^{-1}\left(f_\phi + \frac{\partial Bh}{\partial\phi}\right)^*\right\},$$

where

$$\Gamma_1 = \Gamma_1(\phi, h) = \left(f_\phi + \frac{\partial Bh}{\partial\phi}\right)^*\left(f_\phi + \frac{\partial Bh}{\partial\phi}\right).$$

Suppose that the manifold M is an invariant set for system (1.1). Then set (7.2) which is an m-dimensional torus in $T_m \times E^{n-m}$ is an invariant set of system (7.13). Estimate (7.11) shows that the nature of the stability of the manifold M is uniquely determined by that of the torus (7.2): if the manifold M is asymptotically stable, stable or unstable then torus (7.2) is asymptotically stable, stable or unstable, and conversely.

Consider in equations (7.13) the terms linear with respect to h. The invariance condition for the torus (7.2) is expressed by the identity

$$L_2(\phi,0)X(f(\phi)) = 0, \quad \phi \in T_m, \tag{7.14}$$

therefore, to within terms of order $\|h\|^2$, equations (7.13) have the form

$$\frac{d\phi}{dt} = a(\phi) + \Phi(\phi)h, \quad \frac{dh}{dt} = P(\phi)h. \tag{7.15}$$

To find the values of the function $a(\phi)$ and matrices $\Phi(\phi)$, $P(\phi)$ it is enough to substitute the expressions (7.15) into equation (7.12) and equate the terms of order $\|h\|^0$ and $\|h\|$ in the equalities so obtained. This leads us to equations relating $a = a(\phi)$, $\Phi = \Phi(\phi)$ and $P = P(\phi)$:

$$\frac{\partial f(\phi)}{\partial \phi}a = X(f(\phi)),$$

$$\frac{\partial f(\phi)}{\partial \phi}\Phi + B(\phi)P = \frac{\partial X(f(\phi))}{\partial x}B(\phi) - \frac{\partial B(\phi)}{\partial \phi}a(\phi). \tag{7.16}$$

The first equation of (7.16) defines $a(\phi)$ by the expression

$$a(\phi) = \Gamma^{-1}(\phi)\left(\frac{\partial f(\phi)}{\partial \phi}\right)^* X(f(\phi)), \quad \phi \in T_m. \tag{7.17}$$

The second defines $\Phi(\phi)$ and $P(\phi)$ as

$$\Phi(\phi) = L_1^0(\phi)\left(\frac{\partial X(f(\phi))}{\partial x}B(\phi) - \frac{\partial B(\phi)}{\partial \phi}a(\phi)\right),$$

$$P(\phi) = L_2^0(\phi)\left(\frac{\partial X(f(\phi))}{\partial x}B(\phi) - \frac{\partial B(\phi)}{\partial \phi}a(\phi)\right), \tag{7.18}$$

where $\Gamma(\phi) = f_\phi^* f_\phi$ and $L_1^0(\phi)$, $L_2^0(\phi)$ are the blocks of the inverse matrix $L^0(\phi)$ of the matrix $[f_\phi, B]$. In particular,

$$L_1^0(\phi) = [f_\phi^*(E - B\Gamma_0^{-1}B^*)f_\phi]^{-1}f_\phi^*(E - B\Gamma_0^{-1}B^*),$$

$$L_2^0(\phi) = [B^*(E - f_\phi\Gamma^{-1}f_\phi^*)B]^{-1}B^*(E - f_\phi\Gamma^{-1}f_\phi^*).$$

The system of equations

$$\frac{d\phi}{dt} = a(\phi), \quad \frac{dh}{dt} = P(\phi)h, \tag{7.19}$$

where $a(\phi)$ is the function (7.17), $P(\phi)$ is matrix (7.18), will be called the *system of variation equations* of the invariant torus (7.2). The first of equations (7.19) describes the "tangent" and the second the "normal" components of the trajectory flow of system (7.13) in a neighbourhood of torus (7.2). It is clear that the variation equations are defined subject to the condition that the function $X(x)$ has continuous derivatives with respect to x in a neighbourhood of the manifold M.

Theorem 1. *Suppose that system of equations* (1.1) *and the toroidal manifold M satisfy the above-mentioned smoothness and invariance conditions, and that local coordinates can be introduced. If there is a positive-definite quadratic form*

$$V = \langle S(\phi)h, h \rangle \geq \gamma \|h\|^2, \quad \gamma = \text{const} > 0, \tag{7.20}$$

defined by the symmetric matrix $S(\phi) \in C^1(\mathcal{T}_m)$ such that its total derivative given by virtue of variation equations (7.19) *as*

$$\dot{V} = \langle \hat{S}(\phi)h, h \rangle, \hat{S}(\phi) = S(\phi)P(\phi) + P^*(\phi)S(\phi) + \frac{\partial S(\phi)}{\partial \phi}a(\phi) \tag{7.21}$$

is negative definite:

$$\dot{V} \leq -\beta \|h\|^2, \quad \beta = \text{const} > 0, \tag{7.22}$$

then M is an asymptotically stable invariant set of system (1.1).

To prove the theorem it is sufficient to prove that the invariant torus (7.2) of system (7.13) is asymptotically stable. Let ϵ be a fixed number in the interval $(0, \delta_1)$. From identity (7.14) and the smoothness conditions for the function $X(x)$ it follows that

$$\|L_1(\phi, h)X(f(\phi) + B(\phi)h) - a(\phi)\| \leq M(\epsilon),$$
$$\|L_2(\phi, h)X(f(\phi) + B(\phi)h) - P(\phi)h\| \leq M(\epsilon)\|h\| \tag{7.23}$$

for all ϕ, h in the domain

$$\|h\| < \epsilon, \quad \phi \in \mathcal{T}_m, \tag{7.24}$$

where $M(\epsilon)$ is a positive constant, $M(\epsilon) \to 0$ as $\epsilon \to 0$.

Consider the derivative of the function V taken along the solutions of system (7.13).

We have

$$\frac{dV}{dt} = \dot{V} + \left\langle \frac{\partial S(\phi)}{\partial \phi}(L_1(\phi, h)X(f(\phi) + B(\phi)h) - a(\phi))h, h \right\rangle +$$

$$+ 2\langle S(\phi)(L_2(\phi, h)X(f(\phi) + B(\phi)h) - P(\phi)h), h \rangle,$$

where $\dot{V}$ is the function (7.21). From inequalities (7.22) and (7.23) we have the estimate for dV/dt:

$$\frac{d\dot{V}}{dt} \leq -(\beta - KM(\epsilon))\|h\|^2, \tag{7.25}$$

where K is a positive constant chosen to satisfy the condition

$$\max_{\phi \in T_m} \left\| \frac{\partial S(\phi)}{\partial \phi} \right\| + 2 \max_{\phi \in T_m} \|S(\phi)\| \leq K.$$

This estimate holds for all ϕ, h in the domain (7.24). For sufficiently small $\epsilon > 0$ it follows from estimate (7.25) that

$$\frac{dV}{dt} \leq -\frac{\beta}{2}\|h\|^2 \leq -\frac{\beta}{2\gamma_1}V, \tag{7.26}$$

where γ_1 is a constant determined from the inequality

$$\max_{\phi \in T_m, \|\xi\|=1} \langle S(\phi)\xi, \xi \rangle \leq \gamma_1.$$

Since the function V is positive definite we can write inequality (7.26) in the form

$$\gamma\|h_t\|^2 \leq V(t) \leq e^{-(\beta/(2\gamma_1))t}V(0) \leq \gamma_1 e^{-(\beta/(2\gamma_1))t}\|h_0\|^2, \quad t \in [0, T), \tag{7.27}$$

where (ϕ_t, h_t) is the solution of system (7.13) that takes the value (ϕ_0, h_0) when $t = 0$, $V(t)$ is the value of V when $\phi = \phi_t$, $h = h_t$, and T is a non-negative number defined to be the maximal value for which (ϕ_t, h_t) belongs to domain (7.24) for any $t \in [0, T)$. We choose ϕ_0, h_0 in the domain

$$\|h_0\| < (\gamma/\gamma_1)^{1/2}\epsilon, \quad \phi_0 \in T_m. \tag{7.28}$$

Then T will be positive and inequality (7.27) takes the form

$$\|h_t\| < \epsilon e^{-(\beta/(4\gamma_1))t}, \quad t \in [0, T). \tag{7.29}$$

Since T is positive, it follows from inequality (7.29) that the curve (ϕ_t, h_t) does not hit the boundary of the domain (7.24) at any finite $t \geq 0$. Hence it follows that $T = +\infty$ and

$$\|h_t\| < \epsilon e^{-(\beta/(4\gamma_1))t}, \quad t \in [0, +\infty). \tag{7.30}$$

The arbitrariness of the choice of ϵ and inequality (7.30) prove that any solution (ϕ_t, h_t) of system (7.13) with initial conditions in the domain (7.28) satisfies the conditions needed for the torus (7.2) to be asymptotically stable:

$$\|h_t\| < \epsilon, \quad t \in [0, +\infty), \quad \lim_{t \to +\infty} \|h_t\| = 0.$$

This proves the theorem.

We conclude this section by noting that it is always possible to introduce local coordinates ϕ, h in a neighbourhood of our manifold M. This follows from the theorem on periodic basis in E^n for $n = m + 1$ or $n > 2m$. Otherwise it is necessary to embed the system of equations (1.1) in an "extended" system with the required properties in order to introduce local coordinates. For example, system (1.1) and the equation

$$\frac{dy}{dt} = \lambda y, \quad y = (y_1, \dots, y_p), \tag{7.31}$$

where $p > 2m - n$, $\lambda = \mathrm{diag}\{\lambda_1, \dots, \lambda_p\}$, $\lambda_j = \mathrm{const} < 0 \ (j = 1, \dots, p)$, can be taken as the extended system of equations.

However it is possible to introduce local coordinates ϕ, h and write motion equations (1.1), (7.31) in the form of system (7.13) in a neighbourhood of the set formed by the points $x \in M$, $y = 0$. Here the stability characteristics of M considered as an invariant set of system (1.1) can be transferred to the manifold (7.2) considered as an invariant set of system (7.13). Consequently, to study it we can use the various statements proved for the manifold (7.2) and, in particular, the theorem on stability with respect to variation equations (7.19) given before.

2.8. Recurrent motions and multi-frequency oscillations

A motion $x(t, x_0)$ of system (1.1) is called *recurrent* if for a given $\epsilon > 0$ and any $t_0, \tau \in \mathbf{R}$, there exist $T = T(\epsilon) > 0$ and θ such that $t_0 < \theta < t_0 + T$ and

$$\|x(\tau, x_0) - x(\theta, x_0)\| < \epsilon.$$

According to the definition, recurrent motion has the property that any arc of time length $T = T(\epsilon)$ approximates the whole arc to within ϵ:

$$x(t, x_0; \mathbf{R}) \subset U_\epsilon(x(t, x_0; [t_0, t_0 + T])), \tag{8.1}$$

where $U_\epsilon(x(t, x_0; [t_0, t_0 + T]))$ is an ϵ-neighbourhood of the arc $x(t, x_0; [t_0, t_0 + T])$ of the recurrent trajectory $x(t, x_0; \mathbf{R})$.

Intuitively, the oscillations described by system (1.1) must be related to (8.1) by requiring that it should hold for the trajectories of the motion. This is the case, in particular, for periodic, quasi-periodic and almost periodic oscillations of system (1.1), the trajectories of which are recurrent.

We now explain the relation between recurrent motions and invariant sets of system (1.1).

A set $M \subset D$ is called a *minimal set* of system (1.1) if it is not empty, closed, invariant and does not have a proper subset with these properties.

Using Baire's theorem on the power of a totally ordered sequence of distinct closed sets ordered by inclusion [94], one can prove by contradiction the following theorem on a minimal set [94].

Theorem 1. *Every invariant closed compact set contains a minimal set.*

The following theorems due to Birkhoff [17] hold for trajectories of a minimal set.

Theorem 2. *Every motion of system (1.1) which starts in a compact minimal set is recurrent.*

Theorem 3. *If the recurrent motion of system (1.1) is compact, then the closure of its trajectory is a compact minimal set.*

Birkhoff's theorems can be proved by reductio ad absurdum: the assumption that the theorem does not hold necessarily contradicts the characteristic property of a minimal set by virtue of which any trajectory of a minimal set is everywhere dense in it and vice versa [94].

Suppose that the manifold M given by the equation

$$x = f(\phi), \quad \phi \in \mathcal{T}_m, \tag{8.2}$$

with the function $f(\phi) \in C^s(\mathcal{T}_m)$ defining a local homeomorphism from the torus $\mathcal{T}_m$ onto its image $f(\mathcal{T}_m) = M$, is an invariant set of system (1.1). Here, the property that $f : \mathcal{T}_m \rightarrow M$ be a local homeomorphism is ensured by the relation

$$\text{rank} \frac{\partial f(\phi)}{\partial \phi} = m, \quad \phi \in \mathcal{T}_m \tag{8.3}$$

for $s \geq 1$ and by the assumption that f satisfies conditions of Lemma 5.3 for $s = 0$.

As follows from the results of §2.5, the existence of such a manifold for system (1.1) is a necessary condition for its quasi-periodic oscillations with a frequency basis containing m frequencies to exist. We associate the multi-frequency oscillations of system (1.1) with the manifold M by defining each oscillation to be a recurrent motion the trajectory of which belongs to the minimal set M. The existence of minimal sets on M follows because the manifold M is compact and closed, and Birkhoff's theorems imply that the motions that start at the points of minimal sets are recurrent and they belong to a "certain family" of similar motions each of which "winds round" the minimal set.

With such a definition of a multi-frequency oscillation the number m charactrizes how many frequencies there are by giving an upper bound m for the dimension of a frequency basis for any quasi-periodic motion of system (1) contained in multi-frequency oscillations. Periodic oscillations whose trajectories form M as well as stationary states of system (1.1), the points of which lie in M and form there a closed set with connected components joined by separatrices of system (1.1) in M are considered as a "single frequency" oscillation.

The behaviour of "double frequency" oscillations corresponding to the value $m = 2$ is determined for $s \geq 2$ by the Poincaré-Denjoy theory [100] which gives a qualitative analysis of trajectories of a dynamical system on a two-dimensional torus $\mathcal{T}_2$. For $m = 2$ and $s \geq 2$ the trajectory flows of system (1.1) on M defined by the system of equations (6.12) is a smooth dynamical system on $\mathcal{T}_2$.

It follows from the results of Poincaré-Denjoy that the minimal sets on M are formed either by the trajectories of the quasi-periodic motions with a double frequency basis or by the trajectories of periodic motions or stationary states. In this case every trajectory of a quasi-periodic motion "sweeps" the manifold M, and the trajectories of periodic motions are either limit cycles or cover all of M.

The properties of multi-frequency oscillations of system (1.1) for $m \geq 3$ are not well studied. It is known from Pugh's closure lemma [102] that a recurrent motion of smooth system (1.1) can be turned into a periodic one by a small change of the right hand side of the system, and from the results in [5], [89], [103] it follows that if the right hand side of system (6.12) is sufficiently smooth and close to a constant, then quasi-periodic oscillations are "typical" of the multi-frequency oscillations of system (1.1).

Chapter 3. Some problems of the linear theory

3.1. Introductory remarks and definitions

We shall be considering a system of equations of the form

$$\frac{d\phi}{dt} = a(\phi), \quad \frac{dx}{dt} = P(\phi)x + f(\phi), \tag{1.1}$$

where $a, P, f \in C^r(T_m)$, $\phi = (\phi_1, \ldots, \phi_m)$, $x = (x_1, \ldots, x_n)$. Underlining its linearity with respect to the variable x we shall call it a *linear non-homogeneous system of equations* defined on the direct product of the m-dimensional torus T_m and the Euclidean space E^n. An invariant manifold of system of equations (1.1) of the form

$$x = u(\phi), \quad \phi \in T_m, \tag{1.2}$$

where $u \in C^s(T_m)$, will be called an m-dimensional s times continuously differentiable invariant torus of system (1.1). For $s = 0$ an invariant manifold of system (1.1) of the form (1.2) will be called an invariant torus of this system.

The invariant torus

$$x = 0, \quad \phi \in T_m \tag{1.3}$$

of the homogeneous system of equations

$$\frac{d\phi}{dt} = a(\phi), \quad \frac{dx}{dt} = P(\phi)x, \tag{1.4}$$

which is sometimes called a *linear extension* of a dynamical system on a torus [27] will be called *exponentially stable* if for any solution $\phi = \phi_t(\phi_0)$, $x = x_t(\phi_0, x_0)$ of system (1.4) satisfying the condition

$$\phi_0(\phi_0) = \phi_0, \quad x_0(\phi_0, x_0) = x_0,$$

the inequality

$$\|x_t(\phi_0, x_0)\| \le K e^{-\gamma(t-\tau)} \|x_\tau(\phi_0, x_0)\|$$

holds for all $t \geq \tau$, arbitrary $\phi_0 \in T_m$, $x_0 \in E^n$, $\tau \in \mathbf{R}$ and some positive K, γ that are independent of ϕ_0, x_0, τ.

A system of equations (1.4) for which the invariant torus (1.3) is exponentially stable will be called exponentially stable.

An invariant torus (1.3) of the system of equations (1.4) will be called *exponentially dichotomous* and the system of equations itself exponentially dichotomous whenever for any $\phi_0 \in T_m$ the Euclidean space E^n can be represented as a direct sum of two complementary subspaces E^+ and E^- of the dimensions r and $n - r$ in such a way that any solution $\phi_t(\phi_0)$, $x_t(\phi_0, x_0)$ of system (1.4) with $x_0 \in E^+$ satisfies the inequality

$$\|x_t(\phi_0, x_0)\| \leq K e^{-\gamma(t-\tau)} \|x_\tau(\phi_0, x_0)\|, \quad t \geq \tau,$$

and any solution $\phi_t(\phi_0)$, $x_t(\phi_0, x_0)$ of (1.4) with $x_0 \in E^-$ satisfies the inequality

$$\|x_t(\phi_0, x_0)\| \leq K_1 e^{\gamma_1(t-\tau)} \|x_\tau(\phi_0, x_0)\|, \quad t \leq \tau,$$

for arbitrary $\tau \in \mathbf{R}$, $\phi_0 \in T_m$ and some positive K, γ, K_1, γ_1 independent of ϕ_0, x_0, τ.

An invariant torus (1.2) of the non-homogeneous system of equations (1.1) will be called exponentially stable (or dichotomous) whenever the invariant torus (1.3) of the homogeneous system (1.4) corresponding to system (1.1) is exponentially stable (or dichotomous).

We shall say that the system of equations (1.4) is *block $C^S(T_m)$-decomposable* if there exists a non-singular matrix $\Phi \in C^s(T_m)$ such that the change of variables

$$x = \Phi(\phi)y$$

transforms this system into the system

$$\frac{d\phi}{dt} = a(\phi), \quad \frac{dy}{dt} = Q(\phi)y, \tag{1.5}$$

where the matrix Q is block-diagonal:

$$Q = \mathrm{diag}\{Q_1, Q_2\}.$$

Here the property of being block $C^0(T_m)$-decomposable means that systems (1.4) and (1.5) are topologically equivalent in the sense that

$$x_t(\phi_0, x_0) = \Phi(\phi_t(\phi_0))y_t(\phi_0, y_0)$$

for all $t \in \mathbf{R}$, $\phi_0 \in T_m$, $x_0 \in E^n$, $y_0 = \Phi^{-1}(\phi_0)x_0$.

The system of equations (1.1) is the simplest among the systems obtained by taking a linearization of the equation of motion of a dynamical system in the Euclidean space, considered in a neighbourhood of a smooth torus in a local coordinate system. This explains the deep relation between these systems with the theory of multi-frequency oscillations, the elements of which are considered here.

Some problems that we shall be dealing with in this chapter are finding existence conditions and methods of construction for the invariant tori of system (1.1), studying their exponential stability and dichotomous properties, finding the degree of smoothness, and finding conditions for the system (1.4) to be block $C^s(T_m)$-decomposable.

3.2. Adjoint system of equations.
Necessary conditions for the existence of an invariant torus

We shall be considering system of equations (1.1) under the assumption that $a \in C^{r+1}(T_m)$, $P \in C^r(T_m)$, $r \geq 0$. We write down the homogeneous system of equations

$$\frac{d\phi}{dt} = a(\phi), \quad \frac{dx}{dt} = P(\phi)x \tag{2.1}$$

corresponding to (1.1) and use it to construct the operator $L : H^1(T_m) \to H^0(T_m)$, by setting

$$Lu = \sum_{\nu=1}^{m} a_\nu(\phi)\frac{\partial u}{\partial \phi_\nu} + b(\phi)u, \tag{2.2}$$

where $(a_1(\phi), \ldots, a_m(\phi)) = a(\phi)$, $b(\phi) = -P(\phi)$.

The operator L is related to the *formal adjoint* operator $L^* : H^1(T_m) \to H^0(T_m)$ defined by the expression

$$L^*u = -\sum_{\nu-1}^{m} a_\nu(\phi)\frac{\partial u}{\partial \phi_\nu} + (b^*(\phi) - \mu(\phi))u, \tag{2.3}$$

where $b^*(\phi)$ is the matrix adjoint to $b(\phi)$ and

$$\mu(\phi) = \sum_{\nu=1}^{m} \frac{\partial a_\nu(\phi)}{\partial \phi_\nu} = \operatorname{tr}\frac{\partial a(\phi)}{\partial \phi}.$$

Lemma 1. *For any* $u, v \in H^1(T_m)$ *the following equality holds:*

$$(Lu, v)_0 = (u, L^*v)_0. \tag{2.4}$$

Indeed, according to the definition of the operator L,

$$(Lu, v)_0 = \frac{1}{(2\pi)^m} \int_0^{2\pi} \cdots \int_0^{2\pi} \left[\sum_{\nu=1}^m \langle a_\nu(\phi) \frac{\partial u(\phi)}{\partial \phi_\nu}, v(\phi) \rangle + \langle b(\phi)u(\phi), \ v(\phi) \rangle \right] d\phi,$$

where $\langle u, v \rangle = \sum_{i=1}^n u_i v_i$ is the usual inner product of vectors u, v considered as elements of n-dimensional Euclidean space. Since

$$\sum_{\nu=1}^m \langle a_\nu \frac{\partial u}{\partial \phi_\nu}, v \rangle + \langle bu, v \rangle = \sum_{\nu=1}^m \frac{\partial}{\partial \phi_\nu} \langle a_\nu u, v \rangle - \mu \langle u, v \rangle -$$

$$- \langle u, \sum_{\nu=1}^m a_\nu \frac{\partial v}{\partial \phi_\nu} \rangle + \langle u, b^*v \rangle = \sum_{\nu=1}^m \frac{\partial}{\partial \phi_\nu} \langle a_\nu u, v \rangle + \langle u, L^*v \rangle, \tag{2.5}$$

it follows from (2.5) that

$$(Lu, v)_0 = (u, L^*v)_0.$$

We write the system of ordinary differential equations

$$\frac{d\phi}{dt} = -a(\phi), \quad \frac{dx}{dt} = [P^*(\phi) + \mu(\phi)]x \tag{2.6}$$

in terms of L^* in such a way that the operator L^* can be written using system (2.6) in the same way as the operator L is written using system (2.1). We call system (2.6) the *adjoint* of system (2.1).

We denote by $C'(T_m)$ the subspace of $C(T_m)$ consisting of all those functions u for which $u(\phi_t(\phi_0))$ has continuous derivatives with respect to t such that
$$\frac{du(\phi_t(\phi_0))}{dt} = \dot{u}(\phi_t(\phi_0))$$
for all $t \in \mathbf{R}$, $\phi_0 \in T_m$ and some function $u(\phi) \in C(T_m)$.

Lemma 2. *Let* $w \in C'(T_m)$. *Then*

$$\int_0^{2\pi} \cdots \int_0^{2\pi} \dot{w}(\phi)d\phi_1 \ldots d\phi_m = - \int_0^{2\pi} \cdots \int_0^{2\pi} \mu(\phi)w(\phi)d\phi_1 \ldots d\phi_m. \tag{2.7}$$

Proof. Since the function $dw(\phi_t(\phi))/dt$ is continuous with respect to t, ϕ, it follows that

$$\int_0^{2\pi} \ldots \int_0^{2\pi} \dot{w}(\phi) d\phi_1 \ldots d\phi_m = \frac{dI(t)}{dt}\bigg|_{t=0}, \tag{2.8}$$

where $I(t)$ denotes the integral

$$I(t) = \int_0^{2\pi} \ldots \int_0^{2\pi} w(\phi_t(\phi)) d\phi_1 \ldots d\phi_m.$$

Consider this integral. We make a change of variables by introducing the variables ψ as follows:

$$\phi = \phi_{-t}(\psi), \quad \psi = \phi_t(\phi).$$

The Jacobian of the transformation $\phi \to \psi$ is the function

$$J(t, \psi) = \det \frac{\partial \phi_{-t}(\psi)}{\partial \psi}.$$

Since the matrix $\partial \phi_{-t}(\psi)/\partial \psi$ satisfies both the matrix equation

$$\frac{dz}{dt} = -\frac{\partial a(\phi_{-t}(\psi))}{\partial \phi} z$$

and the condition $\frac{\partial \phi_{-t}}{\partial \psi}\big|_{t=0} = E$, it follows from the Ostrogradskii-Liouville theorem that

$$J(t, \psi) = \det \frac{\partial \phi_{-t}(\psi)}{\partial \psi} =$$

$$= \exp\left\{ -\int_0^t \mathrm{tr}\, \frac{\partial a(\phi_{-t}(\psi))}{\partial \phi} dt \right\} = \exp\left\{ -\int_0^t \mu(\phi_{-t}(\psi)) d\tau \right\}.$$

Thus we have the following expression for I:

$$I = \int_{\psi_m^0}^{\psi_m^1} \ldots \int_{\psi_1^0}^{\psi_1^1} w(\psi) |J(t, \psi)| d\psi_1 \ldots d\psi_m =$$

$$= \int_{\psi_m^0}^{\psi_m^1} \ldots \int_{\psi_1^0}^{\psi_1^1} w(\psi) \exp\left\{ -\int_0^t \mu(\phi_{-\tau}(\psi)) d\tau \right\} d\psi_1 \ldots d\psi_m,$$

where ψ_j^0, ψ_j^1 ($j = 1, \ldots, m$) are the new limits of integration.

Taking into account the fact that $\phi_t(\phi) - \phi$ is periodic with respect to ϕ_ν with period 2π for $\nu = 1, \ldots, m$, we obtain via the change of variable formulae the equalities

$$\psi_j^1 = 2\pi + \psi_j^0, \quad j = 1, \ldots, m.$$

Since the functions under the integral sign are periodic, the above equalities yield the following expression for the integral I:

$$I = \int_0^{2\pi} \cdots \int_0^{2\pi} w(\psi) \exp\left\{ - \int_0^t \mu(\phi_{-\tau}(\psi)) d\tau \right\} d\psi_1 \ldots d\psi_m. \qquad (2.9)$$

Differentiating equality (2.9) with respect to t we find that

$$\dot{I} = \frac{dI}{dt}\bigg|_{t=0} = - \int_0^{2\pi} \cdots \int_0^{2\pi} w(\psi)\mu(\psi) d\psi_1 \ldots d\psi_m.$$

This relation together with (2.8) leads to the identity (2.7).

We now define the action of the operators L and L^* on functions $u \in C'(\mathcal{T}_m)$ by setting

$$
\begin{aligned}
Lu(\phi) &= \dot{u}(\phi) + b(\phi)u(\phi), \\
L^*u(\phi) &= -\dot{u}(\phi) + [b^*(\phi) - \mu(\phi)]u(\phi).
\end{aligned}
\qquad (2.10)
$$

It is clear that L and L^* defined by relation (2.10) are the same as the operators given by equalities (2.2) and (2.3) when $u \in C'(\mathcal{T}_m) \cap H^1(\mathcal{T}_m)$.

Lemma 2 enables us to prove that these operators are adjoint in the space $C'(\mathcal{T}_m)$. In fact we have the following statement.

Lemma 3. *For any functions $u, v \in C'(\mathcal{T}_m)$ equality (2.4) holds.*

To prove the lemma we use the identity

$$\left\langle \frac{du(\phi_t(\phi))}{dt} + b(\phi_t(\phi))u(\phi_t(\phi)), v(\phi_t(\phi)) \right\rangle =$$

$$= \frac{d}{dt}\langle u(\phi_t(\phi)), v(\phi_t(\phi))\rangle - \left\langle u(\phi_t(\phi)), \frac{dv(\phi_t(\phi))}{dt} \right\rangle + \langle u(\phi_t(\phi)), b^*(\phi_t(\phi))v(\phi_t(\phi))\rangle$$

to obtain

$$(Lu, v)_0 = \int_0^{2\pi} \cdots \int_0^{2\pi} \dot{w}(\phi) d\phi_1 \ldots d\phi_m - (u, \dot{v})_0 + (u, b^*v)_0,$$

where $w(\phi) = \langle u(\phi), v(\phi) \rangle$. Since the function w as well as the functions c and v belong to the space $C'(\mathcal{T}_m)$, according to Lemma 2 we have

$$\int_0^{2\pi} \ldots \int_0^{2\pi} \dot{w}(\phi) d\phi_1 \ldots d\phi_m = -\int_0^{2\pi} \ldots \int_0^{2\pi} \mu(\phi) w(\phi) d\phi_1 \ldots d\phi_m =$$

$$= -(\mu(\phi) u(\phi), v(\phi))_0.$$

But then

$$(Lu, v)_0 = (u, -\dot{v} + b^* v - \mu v)_0 = (u, L^* v)_0,$$

as required.

Lemma 4. *Suppose that the system of equations* (1.1) *has an invariant torus*

$$x = u(\phi), \quad \phi \in \mathcal{T}_m. \tag{2.11}$$

Then $u \in C'(\mathcal{T}_m)$ *and* $Lu(\phi) = f(\phi)$.

Indeed, if equality (2.11) defines an invariant torus of system (1.1) then $u \in C(\mathcal{T}_m)$ and

$$\frac{du(\phi_t(\phi_0))}{dt} = P(\phi_t(\phi_0)) u(\phi_t(\phi_0)) + f(\phi_t(\phi_0)) \tag{2.12}$$

for all $t \in \mathbf{R}$, $\phi_0 \in \mathcal{T}_m$. Because the right hand side of identity (2.12) is continuous and periodic in ϕ, the function u belongs to the space $C'(\mathcal{T}_m)$ and the equality obtained from (2.12) for $t = 0$ implies the identity $Lu(\phi_0) = f(\phi_0)$ for all $\phi_0 \in \mathcal{T}_m$.

Using Lemmas 3 and 4 we formulate necessary conditions for the existence of an invariant torus of system (1.1) in the form of the following statement.

Theorem 1. *For system of equations* (1.1) *to have an invariant torus it is necessary and sufficient that the equality*

$$(f, v)_0 = 0 \tag{2.13}$$

holds for any function v that defines the invariant torus $x = v(\phi)$ ($\phi \in \mathcal{T}_m$) of the adjoint system of equations (2.6).

Proof. Let

$$x = u(\phi), \quad \phi \in \mathcal{T}_m$$

be an invariant torus of the system of equations (1.1). By Lemma 4, $u \in C'(\mathcal{T}_m)$ and $Lu(\phi) = f(\phi)$.

Because the function v belongs to the space $C'(\mathcal{T}_m)$ and satisfies the identity $L^* v(\phi) = 0$ by virtue of the same lemma, we obtain from Lemma 3 the chain of equalities:

$$(f, v)_0 = (Lu, v)_0 = (u, L^* v)_0 = (u, 0)_0 = 0,$$

which leads to relation (2.13).

The example of the system of equations

$$\frac{d\phi}{dt} = \omega, \quad \frac{dx}{dt} = f(\phi), \tag{2.14}$$

where $\omega = (\omega_1, \ldots, \omega_m)$ are constants, shows that the necessary condition for existence of an invariant torus for system (1.1) given in Theorem 1 is not, in general, a sufficient condition.

Indeed, for the homogeneous system of equations corresponding to (2.14) the adjoint system has the form

$$\frac{d\phi}{dt} = -\omega, \quad \frac{dx}{dt} = 0.$$

Invariant tori for this system are given by the equations

$$x = c, \quad c = \text{const.}$$

The necessary condition in Theorem 1 is that the function $f(\phi)$ should have the mean value equal to zero:

$$\frac{1}{(2\pi)^m} \int_0^{2\pi} \cdots \int_0^{2\pi} f(\phi) d\phi_1 \ldots d\phi_m = 0. \tag{2.15}$$

The system of equations (2.14) has invariant tori only if the first integral of the function $f(\omega t)$ is quasi-periodic. But for this, as follows from Borel's theorem, it is not sufficient that relation (2.15) holds.

3.3. Necessary conditions for the existence of an invariant torus of a linear system with arbitrary non-homogeneity in $C(\mathcal{T}_m)$

An invariant torus $x = u(\phi)$, $\phi \in \mathcal{T}_m$, of the system of equations (1.1) will be called *degenerate* if $u(\phi) \not\equiv 0$ and

$$\lim_{|t| \to +\infty} u(\phi_t(\phi_0)) = 0 \tag{3.1}$$

for any point $\phi_0 \in T_m$.

An invariant torus $x = u(\phi), \phi \in T_m$, for which $u(\phi) \not\equiv 0$ and there exists a point $\phi_0 \in T_m$ such that (3.1) does not hold, is called a *non-degenerate* invariant torus of the system of equations (1.1). The trivial invariant torus of the homogeneous system of equations (3.1) will not be considered either as degenerate or as non-degenerate.

Theorem 1. *For the system of equations* (1.1) *to have an invariant torus for an arbitrary function $f \in C(T_m)$ it is necessary that the homogeneous system* (2.6) *does not have invariant non-degenerate tori and that adjoint system of equations* (2.1) *should not have invariant tori other than the trivial one,* $x = 0, \phi \in T_m$.

Proof. Suppose that for an arbitrary function f in $C(T_m)$ the system of equations (1.1) has an invariant torus (1.2). By Theorem 1 of §3.2, the relation

$$(f, v)_0 = 0, \tag{3.2}$$

holds, where the function $v \in C(T_m)$ defines an invariant torus of the adjoint system of equations (2.6). If $v \not\equiv 0$ then for $f = v$ the relation (3.2) is contradictory: $\|v\|_0^2 = 0$ for $v \not\equiv 0$. This contradiction shows that the adjoint system of equations cannot have an invariant torus other than the trivial one.

Suppose that the homogeneous system (2.1) has a non-degenerate invariant torus

$$x = u_0(\phi), \quad \phi \in T_m. \tag{3.3}$$

Let

$$x = u_j(\phi), \quad \phi \in T_m, \quad j = 1, \ldots, n,$$

be invariant tori of the system of equations (1.1) for

$$f = u_{j-1}, \quad j = 1, \ldots, n.$$

The following relation holds for these tori:

$$u_j(\phi_t(\phi)) = \Omega_0^t(\phi)\left[u_j(\phi) + u_{j-1}(\phi)t + \ldots + u_0(\phi)\frac{t^j}{j!}\right], \tag{3.4}$$

where $\Omega_0^t(\phi)$ denotes the *matricent* of the linear system of equations

$$\frac{dx}{dt} = P(\phi_t(\phi))x, \tag{3.5}$$

that is, the fundamental matrix of solutions of system (3.5) which is equal to the identity matrix E when $t = 0$.

Indeed, since the torus (3.3) is invariant, it follows that

$$u_0(\phi_t(\phi)) = \Omega_0^t(\phi)u_0(\phi), \ t \in \mathbf{R}.$$

But then

$$u_1(\phi_t(\phi)) = \Omega_0^t(\phi)\left[u_1(\phi) + \int_0^t \Omega_\tau^0(\phi)u_0(\phi_\tau(\phi))d\tau\right] = \Omega_0^t(\phi)[u_1(\phi) + u_0(\phi)t],$$

$$u_2(\phi_t(\phi)) = \Omega_0^t(\phi)\left[u_2(\phi) + \int_0^t \Omega_\tau^0(\phi)u_1(\phi_\tau(\phi))d\tau\right] =$$

$$= \Omega_0^t(\phi)\left[u_2(\phi) + u_1(\phi)t + u_0(\phi)\frac{t^2}{2!}\right].$$

Similarly one can prove (3.4) for $j = 3, \ldots, n$.

Using relations (3.4), we express the functions $\Omega_0^t(\phi)u_j(\phi)$ in terms of the $u_j(\phi_t(\phi))$. We set

$$u = \text{column}(u_0(\phi_t(\phi)), \ u_1(\phi_t(\phi)), \ldots, u_n(\phi_t(\phi))),$$

$$\bar{u} = \text{column}(\Omega_0^t(\phi)u_0(\phi), \ \Omega_0^t(\phi)u_1(\phi), \ldots, \Omega_0^t(\phi)u_n(\phi))$$

and rewrite relations (3.4) as the matrix equality

$$u = T\bar{u},$$

where the matrix T has the form

$$T = \begin{bmatrix} E & 0 & 0 & 0 \\ tE & E & 0 & 0 \\ \dfrac{t^2E}{2!} & tE & E & 0 \\ \cdots\cdots\cdots\cdots\cdots\cdots\cdots\cdots\cdots\cdots \\ \dfrac{t^n}{N!}E & \dfrac{t^{n-1}}{(n-1)!}E & \dfrac{t^{n-2}}{(n-2)!}E & E \end{bmatrix} = e^{Zt} = \sum_{k=0}^{\infty} \frac{Z^k t^k}{k!},$$

where E is the n-dimensional identity matrix and

$$Z = \begin{bmatrix} 0 & 0 & \ldots & 0 & 0 \\ E & 0 & \ldots & 0 & 0 \\ 0 & E & \ldots & 0 & 0 \\ \cdots\cdots\cdots\cdots\cdots\cdots\cdots \\ 0 & 0 & \ldots & E & 0 \end{bmatrix}.$$

Then

$$\bar{u} = T^{-1}u = e^{-Zt}u,$$

which is equivalent to the system of equalities

$$\Omega_0^t(\phi)u_j(\phi) = u_j(\phi_t(\phi)) - tu_{j-1}(\phi_t(\phi)) + \frac{t^2}{2!}u_{j-2}(\phi_t(\phi)) + \dots$$

$$\dots + (-1)^j\frac{t^j}{j!}u_0(\phi_t(\phi)), \quad j = 1, \dots, n. \tag{3.6}$$

We use this system to prove that the torus (3.3) is degenerate. To do this, taking into consideration the linear dependence of the vectors

$$u_0(\phi), u_1(\phi), \dots, u_n(\phi),$$

we divide T_m into two subsets N_0 and $N_1 = T_m \backslash N_0$ by attributing to N_0 all the points $\phi \in T_m$ for which the vectors

$$u_0(\phi), u_1(\phi), \dots, u_{n-1}(\phi) \tag{3.7}$$

are linearly independent. For a fixed point ϕ_0 in N_0 one can find constants $\alpha_0, \alpha_1, \dots, \alpha_{n-1}$ such that the vector $u_n(\phi_0)$ decomposes with respect to the system of vectors (3.7) as:

$$u_n(\phi_0) = \alpha_0 u_0(\phi_0) + \alpha_1 u_1(\phi_0) + \dots + \alpha_{n-1}u_{n-1}(\phi_0).$$

Using relations (3.6) we obtain the representation

$$\Omega_0^t(\phi_0)u_n(\phi_0) = \alpha_0 u_0(\phi_t(\phi_0)) + \alpha_1[u_1(\phi_t(\phi_0) - tu_0(\phi_t(\phi_0))] + \dots$$

$$\dots + \alpha_{n-1}\left[u_{n-1}(\phi_t(\phi_0)) - tu_{n-2}(\phi_t(\phi_0)) + \dots + (-1)^{n-1}\frac{t^{n-1}}{(n-1)!}u_0(\phi_t(\phi_0))\right],$$

which together with equality (3.6) for $j = n$, $\phi = \phi_0$ leads to the identity

$$\left[\alpha_0 - t\alpha_1 + \dots + (-1)^{n-1}\frac{t^{n-1}}{(n-1)!}\alpha_{n-1}\right]u_0(\phi_t(\phi_0)) +$$

$$+ \left[\alpha_1 - t\alpha_2 + \dots + (-1)^{n-2}\frac{t^{n-2}}{(n-2)!}\alpha_{n-1}\right]u_1(\phi_t(\phi_0)) + \dots$$

$$\dots + \alpha_{n-1}u_{n-1}(\phi_t(\phi_0)) = (-1)^n\frac{t^n}{n!}u_0(\phi_t(\phi_0)) +$$

$$+ (-1)^{n-1}\frac{t^{n-1}}{(n-1)!}u_1(\phi_t(\phi_0)) + \dots + (-1)tu_{n-1}(\phi_t(\phi_0)) + u_n(\phi_t(\phi_0)). \tag{3.8}$$

Letting $|t|$ tend to $+\infty$ in (3.8) we obtain the limit relation

$$\lim_{|t| \to +\infty} \|u_0(\phi_t(\phi_0))\| = 0. \tag{3.9}$$

Thus, equality (3.9) holds for points $\phi_0 \in N_0$.

Let $\phi_0 \in N_1$. This means that the functions (3.7) are linearly independent for $\phi = \phi_0$. We choose those points $\phi \in N_1$ for which the system of vectors

$$u_0(\phi), u_1(\phi), \ldots, u_{n-2}(\phi) \tag{3.10}$$

is linearly independent. For each such point $\phi = \phi_0$ the vector $u_{n-1}(\phi_0)$ has a decomposition in the vectors (3.10) similar to the decomposition of the vector $u_n(\phi_0)$ in the vectors (3.7) for the point $\phi_0 \in N_0$. This leads to an identity analogous to (3.8) and proves limit relation (3.9) for those points of N_1 for which vectors (3.10) are linearly independent. Continuing this reasoning we see that relation (3.9) holds for all those points $\phi_0 \in T_m$ for which at least one of the systems of vectors

$$u_0(\phi_0), u_1(\phi_0), \ldots, u_j(\phi_0), \quad j = 1, \ldots, n-1, \tag{3.11}$$

is linearly independent.

Let ϕ_0 be a point for which all systems of vectors (3.11) are linearly dependent. Then there exist constants α_0, α_1 such that $\alpha_0^2 + \alpha_1^2 \neq 0$ and

$$\alpha_0 u_0(\phi_0) + \alpha_1 u_1(\phi_0) = 0.$$

If $\alpha_1 \neq 0$ then $u_1(\phi_0) = -\frac{\alpha_0}{\alpha_1} u_0(\phi_0)$ and from equalities (3.6) for $j = 1$ we obtain

$$-\frac{\alpha_0}{\alpha_1} u_0(\phi_t(\phi_0)) = u_1(\phi_t(\phi_0)) - t u_0(\phi_t(\phi_0)).$$

On passing to the limit in the last equality as $|t| \to +\infty$ we obtain relation (3.9). This proves limit relation (3.9) for all points ϕ_0 except those for which $u_0(\phi_0) = 0$.

Let N be the set of such points. It follows from the existence and uniqueness of a solution of the Cauchy problem for system of equations (3.5) that a solution of this system $x(t)$ satisfies the identity $x(t) = 0$ for $t \in \mathbf{R}$ whenever $x(0) = 0$. Thus if $\phi_0 \in N$, then $x = u(\phi_t(\phi_0)) = 0$ for all $t \in \mathbf{R}$. This proves limit relation (3.9) for all points $\phi_0 \in N$.

Consequently, (3.9) holds for all points $\phi_0 \in T_m$. Therefore the torus (3.3) turns out to be degenerate, which contradicts our assumption. This contradiction completes the proof of the theorem.

The following example of the system of equations

$$\frac{d\phi}{dt} = 0, \quad \frac{dx}{dt} = P(\phi)x + f(\phi), \quad \det P(\phi) = \text{const} \qquad (3.12)$$

shows that the necessary conditions for existence of an invariant torus given by Theorem 1 may also be sufficient. Indeed, an invariant torus of system (3.12) for an arbitrary function $f \in C(T_m)$ exists if and only if

$$\det P(\phi) \neq 0, \quad \phi \in T_m. \qquad (3.13)$$

The necessary conditions given by Theorem 1 also reduce to the requirement that (3.13) should hold. Consequently, they are the same as the sufficient condition for the existence of an invariant torus of system (3.12) for an arbitrary function $f \in C(T_m)$.

The example of the system of equations

$$\frac{d\phi}{dt} = \sin \phi, \quad \frac{dx}{dt} = (\cos \phi)x + f(\phi) \qquad (3.14)$$

shows the extent to which the necessary conditions are "best possible" by proving that it is impossible to replace them by the requirement that the homogeneous system have no invariant tori. Indeed, the homogeneous system of equations corresponding to the system of equations (3.14) has an invariant torus given by the equation

$$x = \sin \phi, \quad \phi \in T_1. \qquad (3.15)$$

Since

$$\sin \phi_t(\phi) = \frac{2ce^t}{1 + c^2 e^{2t}},$$

where

$$c = \begin{cases} 0 & \phi = 0, \ \phi = \pi, \\ \tan(\phi/2), & 0 < \phi < \pi, \ \pi < \phi < 2\pi, \end{cases}$$

the torus (3.15) is degenerate:

$$\lim_{|t| \to +\infty} |\sin \phi_t(\phi)| = 0$$

for every $\phi \in T_1$. Consequently, the homogeneous system corresponding to system (3.14) has a degenerate invariant torus.

We show that in spite of this, the non-homogeneous system of equations (3.14) has an invariant torus for any function $f \in C(T_1)$. To do this, we consider the function u defined for $\phi \in [0, 2\pi)$ by the relation

$$
u(\phi) = \begin{cases}
-f(0), & \phi = 0, \\[2mm]
\sin \phi \displaystyle\int_{\pi/2}^{\psi} \frac{f(\psi)}{\sin^2 \psi}, & 0 < \phi < \pi, \\[3mm]
f(\pi), & \phi = \pi, \\[2mm]
\sin \phi \displaystyle\int_{3\pi/2}^{\phi} \frac{f(\psi)}{\sin^2 \psi} d\psi, & \pi < \phi < 2\pi,
\end{cases}
$$

and extended to all ϕ by periodicity. Since

$$
\lim_{\phi \to +\infty} \sin \phi \int_{\pi/2}^{\phi} \frac{f(\psi)}{\sin^2 \psi} d\psi = \lim_{\phi \to +0} \frac{\int_{\pi/2}^{\phi} (f(\psi)/\sin^2 \psi) d\psi}{\sin^{-1} \phi} =
$$

$$
= \lim_{\phi \to +0} \frac{f(\phi)/\sin^2 \phi}{-\cos \phi / \sin^2 \phi} = -f(0),
$$

$$
\lim_{\phi \to \pi \pm 0} \sin \phi \int_{k\pi/2}^{\phi} \frac{f(\psi)}{\sin^2 \psi} d\psi = \lim_{\phi \to \pi \pm 0} \frac{\int_{k\pi/2}^{\phi} (f(\psi)/\sin^2 \psi) d\psi}{\sin^{-1} \phi} =
$$

$$
= \lim_{\phi \to \pi \pm 0} \frac{f(\phi)}{-\cos \phi} = f(\pi), \quad k = 2 \pm 1,
$$

$$
\lim_{\phi \to 2\pi - 0} \sin \phi \int_{3\pi/2}^{\phi} \frac{f(\psi)}{\sin^2 \psi} d\psi = \lim_{\phi \to 2\pi - 0} \frac{f(\phi)}{-\cos \phi} = -f(2\pi) = -f(0),
$$

the function u is continuous for $\phi \in (-\infty, \infty)$. Consequently, u belongs to the space $C(T_1)$.

Now we have

$$
u(\phi_t(0)) \equiv u(0), \quad u(\phi_t(\pi)) \equiv u(\pi),
$$

so that $x = u(\phi_t(0))$ and $x = u(\phi_t(\pi))$ are solutions of the equation

$$
\frac{dx}{dt} = (\cos \phi_t(\phi))x + f(\phi_t(\phi))
$$

for $\phi = 0$ and $\phi = \pi$. The function $x = u(\phi_t(\phi))$ also satisfies this equation for $\phi \in (0, \pi)$ and $\phi \in (\pi, 2\pi)$ since, by (3.16),

$$\frac{du(\phi_t(\phi))}{dt} = \frac{du}{d\phi}\frac{\partial\phi}{\partial t} = \cos\phi_t(\phi)\sin\phi_t(\phi)\times$$

$$\times \int_{\frac{\pi}{2}\vee\frac{3\pi}{2}}^{\phi_t(\phi)} \frac{f(\psi)}{\sin^2\psi}d\psi + \sin\phi_t(\phi)\frac{f(\phi_t(\phi))}{\sin^2\phi_t(\phi)}\sin\phi_t(\phi) \equiv$$

$$\equiv \cos\phi_t(\phi)u(\phi_t(\phi)) + f(\phi_t(\phi)).$$

From what has been said it follows that the equation

$$x = u(\phi), \quad \phi \in T_1,$$

defines an invariant torus of the system of equations (3.17) for an arbitrary function $f \in C(T_1)$.

Finally we note that the set N of zeros of the function u defining an invariant torus of the homogeneous system of equations (1.4) is an invariant set of the system

$$\frac{d\phi}{dt} = a(\phi). \tag{3.17}$$

Since N is closed and in view of the relation

$$\lim_{|t|\to+\infty} \rho(\phi_t(\phi), N) = 0, \quad \phi \in T_m,$$

which follows from (3.1), the set N contains all non-wandering points of system (3.17). Thus the homogeneous system of equations (1.4) can have non-degenerate invariant tori only if the set of wandering points of system (3.17) is not empty in T_m. In particular, it follows that the homogeneous quasi-periodic system of equations-system (1.4) cannot have degenerate invariant tori when $a = \omega$, $\omega = $ const.

3.4. The Green's function. Sufficient conditions for the existence of an invariant torus

As before, let $\Omega_\tau^t(\phi)$ denote the fundamental matrix of the solutions of the system of equations

$$\frac{dx}{dt} = P(\phi_t(\phi))x, \tag{4.1}$$

which takes the value of the identity matrix E for $t = \tau$.

Because the functions a, P belong to the space $C^r(\mathcal{T}_m)$ it immediately follows that $\phi_t(\phi) - \phi$ and $\Omega_\tau^t(\phi)$ also belong to the space $C^r(\mathcal{T}_m)$ for all $t, \tau \in \mathbf{R}$ and any $r \geq 1$. We claim that

$$\Omega_\tau^t(\phi_\theta(\phi)) = \Omega_{\tau+\theta}^{t+\theta}(\phi) \tag{4.2}$$

for any $t, \tau, \theta \in \mathbf{R}$. To prove this, we replace ϕ by $\phi_\theta(\phi)$ in the identity defining $\Omega_\tau^t(\phi)$:

$$\frac{d\Omega_\tau^t(\phi)}{dt} = P(\phi_t(\phi))\Omega_\tau^t(\phi)$$

and taking into account the group property of the function $\phi_t(\phi)$ expressed by the relation

$$\phi_t(\phi_\theta(\phi)) = \phi_{t+\theta}(\phi),$$

we obtain

$$\frac{d\Omega_\tau^t(\phi_\theta(\phi))}{dt} = P(\phi_{t+\theta}(\phi))\Omega_\tau^t(\phi_\theta(\phi)).$$

The last identity holds for arbitrary $t, \tau, \theta \in \mathbf{R}$, $\phi \in \mathcal{T}_m$ and means that $\Omega_\tau^t(\phi_\theta(\phi))$ is the fundamental matrix of the solutions of the system of equations

$$\frac{dx}{dt} = P(\phi_{t+\theta}(\phi))x, \tag{4.3}$$

which is equal to the identity matrix E for $t = \tau$. The matrix $\Omega_{\tau+\theta}^{t+\theta}(\phi)$ also has the same property. This is possible only if the matrices $\Omega_\tau^t(\phi_\theta(\phi))$ and $\Omega_{\tau+\theta}^{t+\theta}(\phi)$ are the same, and this proves the required equality (4.2).

Let $C(\phi)$ be a matrix belonging to the space $C(\mathcal{T}_m)$. We set

$$G_0(\tau, \phi) = \begin{cases} \Omega_\tau^0(\phi)C(\phi_\tau(\phi)), & \tau \leq 0, \\ -\Omega_\tau^0(\phi)[E - C(\phi_\tau(\phi))], & \tau > 0 \end{cases} \tag{4.4}$$

and call $G_0(\tau, \phi)$ a *Green's function* for the system of equations

$$\frac{d\phi}{dt} = a(\phi) \qquad \frac{dx}{dt} = P(\phi)x \tag{4.5}$$

in the case when the integral $\int_{-\infty}^{\infty} \|G_0(\tau, \phi)\| d\tau$ is uniformly bounded with respect to ϕ:

$$\int_{-\infty}^{\infty} \|G_0(\tau, \phi)\| d\tau \leq K < \infty. \tag{4.6}$$

We give the simplest properties of the Green's function (4.4). It follows from its definition that $G_0(\tau, \phi) \in C(T_m)$ for each τ and

$$G_0(0, \phi) - G_0(+0, \phi) = E.$$

Let $G_t(\tau, \phi)$ be the matrix defined by the relation

$$G_t(\tau, \phi) = \begin{cases} \Omega_\tau^t(\phi) C(\phi_\tau(\phi)), & t \geq \tau, \\ -\Omega_\tau^t(\phi)[E - C(\phi_\tau(\phi))], & t < \tau. \end{cases} \tag{4.7}$$

Then

$$G_0(\tau, \phi_t(\phi)) = G_t(\tau + t, \phi). \tag{4.8}$$

This follows immediately from the property of the matrix $\Omega_\tau^t(\phi)$ expressed by equality (4.2)

$$G_0(\tau, \phi_t(\phi)) = \begin{cases} \Omega_\tau^0(\phi_t(\phi)) C(\phi_\tau(\phi_t(\phi))), & \tau \leq 0 \\ -\Omega_\tau^0(\phi_t(\phi))[E - C(\phi_\tau(\phi_t(\phi)))], & \tau > 0 \end{cases} =$$

$$= \begin{cases} \Omega_{\tau+t}^t(\phi) C(\phi_{\tau+t}(\phi)), & \tau + t \leq t \\ -\Omega_{\tau+t}^t(\phi)[E - C(\phi_{\tau+t}(\phi))], & \tau + t > t \end{cases} = G_t(\tau + t, \phi).$$

It follows from equality (4.8) that the matrix

$$G_0(-t, \phi_t(\phi)) = G_t(0, \phi) = \begin{cases} \Omega_0^t(\phi) C(\phi), & t \geq 0, \\ -\Omega_0^t(\phi)[E - C(\phi)], & t < 0 \end{cases}$$

consists of solutions of the homogeneous system (4.1) considered for $t \geq 0$ and $t < 0$ respectively.

Let $f \in C(T_m)$. Consider the integral

$$\int_{-\infty}^{\infty} G_0(\tau, \phi) f(\phi_\tau(\phi)) d\tau. \tag{4.9}$$

This integral is majorized by a convergent integral, as follows from the estimate

$$\left\| \int_{-\infty}^{\infty} G_0(\tau, \phi) f(\phi_\tau(\phi)) d\tau \right\| \leq \int_{-\infty}^{\infty} \|G_0(\tau, \phi)\| d\tau \cdot |f|_0.$$

We set

$$u(\phi) = \int_{-\infty}^{\infty} G_0(\tau, \phi) f(\phi_\tau(\phi)) d\tau$$

and show that $u \in C(T_m)$.

Indeed, let

$$u_i(\phi) = \int_{-i}^{i} G_0(\tau, \phi) f(\phi_\tau(\phi)) d\tau, \quad i = 1, 2, \ldots.$$

The sequence of functions $u_i(\phi)$ obviously belongs to the space $C(\mathcal{T}_m)$ and satisfies the estimate

$$\|u(\phi) - u_i(\phi)\| \le \left[\int_{-\infty}^{-i} \|G_0(\tau, \phi)\| d\tau + \int_{i}^{\infty} \|G_0(\tau, \phi)\| d\tau \right] |f|_0.$$

The uniform convergence of integral (4.6) with respect to ϕ ensures that the limit relation

$$\lim_{i \to \infty} u_i(\phi) = u(\phi)$$

is uniform with respect to $\phi \in \mathcal{T}_m$. This means that the sequence $u_i(\phi)$ ($i = 1, 2, \ldots$) converges with respect to the $C(\mathcal{T}_m)$ norm. Since $C(\mathcal{T}_m)$ is a complete space, $u \in C(\mathcal{T}_m)$.

Integral (4.9) defines an operator T on the function space $C(\mathcal{T}_m)$:

$$Tf(\phi) = \int_{-\infty}^{\infty} G_0(\tau, \phi) f(\phi_\tau(\phi)) d\tau. \tag{4.10}$$

We show that the set

$$x = u(\phi) = Tf(\phi), \quad \phi \in \mathcal{T}_m, \tag{4.11}$$

defines an invariant torus of the system of equations

$$\frac{d\phi}{dt} = a(\phi), \quad \frac{dx}{dt} = P(\phi)x + f(\phi). \tag{4.12}$$

For this purpose we consider the function

$$x(t, \phi) = u(\phi_t(\phi)).$$

According to equalities (4.8) and (4.10) we have the representation

$$u(\phi_t(\phi)) = \int_{-\infty}^{\infty} G_0(\tau, \phi_t(\phi)) f(\phi_\tau(\phi_t(\phi))) d\tau =$$

$$= \int_{-\infty}^{\infty} G_t(\tau + t, \phi) f(\phi_{\tau+t}(\phi)) d\tau = \int_{-\infty}^{\infty} G_t(\tau, \phi) f(\phi_\tau(\phi)) d\tau$$

for $u(\phi_t(\phi))$. Differentiating it with respect to t, we find that

$$\frac{du(\phi_t(\phi))}{dt} = C(\phi_t(\phi))f(\phi_t(\phi)) + (E - C(\phi_t(\phi)))f(\phi_t(\phi)) +$$

$$+P(\phi_t(\phi))\int_{-\infty}^{\infty} G_t(\tau,\phi)f(\phi_\tau(\phi))d\tau = P(\phi_t(\phi))u(\phi_t(\phi)) + f(\phi_t(\phi))$$

for any $t \in \mathbf{R}$. This shows that $u \in C'(\mathcal{T}_m)$ and the set (4.11) is an invariant torus of the system of equations (4.12).

Thus, if there exists a Green's function for system of equations (4.5), then there is an invariant torus of the system of equations (4.12) for an arbitrary function $f \in C(\mathcal{T}_m)$.

It should be noted that in the above argument we have used the fact that the functions a, P belong to the space $C^r(\mathcal{T}_m)$ only to prove that the matrix $\Omega_\tau^t(\phi)$ is continuous with respect to ϕ. Thus, everything said remains true for $a \in C_{\mathrm{Lip}}(\mathcal{T}_m)$, $P \in C(\mathcal{T}_m)$. This leads to the following statement which gives sufficient conditions for the existence of an invariant torus of system (4.12).

Theorem 1. *Suppose that the right hand side of the system of equations* (4.5) *satisfies the conditions* $a \in C_{\mathrm{Lip}}^r(\mathcal{T}_m)$, $P \in C^r(\mathcal{T}_m)$, $r \geq 0$, *and that the system itself has a Green's function* $G_0(\tau,\phi)$. *Then for any* $f \in C^r(\mathcal{T}_m)$ *the system of equations* (4.12) *has an invariant torus defined by relation* (4.11) *and satisfying the estimate*

$$|u|_0 = |Tf|_0 \leq K|f|_0$$

$$K = \max_{\phi \in \mathcal{T}_m} \int_{-\infty}^{\infty} \|G_0(\tau,\phi)\|d\tau.$$

We now give an example showing that the invariant torus of Theorem 1 may only be continuous even in the case when the functions a, P and f are analytic with respect to ϕ.

Consider the system of equations from [86]:

$$\frac{d\phi}{dt} = -\sin\phi,$$

$$\frac{dx}{dt} = -x + \sin\phi.$$

By definition (4.4), the Green's function $G_0(\tau,\phi)$ with the matrix $C(\phi) \equiv 1$ has the form:

$$G_0(\tau,\phi) = \begin{cases} e^\tau, & \tau \leq 0 \\ 0, & \tau > 0. \end{cases}$$

The invariant torus (4.11) is defined by the expression

$$x = u(\phi) = \int_{-\infty}^{0} e^{\tau} \sin \phi_{\tau}(\phi) d\tau, \quad \phi \in T_1,$$

where

$$\sin \phi_{\tau}(\phi) = \begin{cases} 0, & \phi = k\pi, \ k = 0, \pm 1, \pm 2, \ldots, \\ \dfrac{2e^{\tau} \tan(\phi/2)}{e^{2\tau} + \tan^2(\phi/2)}, & \phi \neq k\pi. \end{cases}$$

Calculation of the function u leads to the expression

$$u(\phi) = \begin{cases} 0, & \phi = k\pi \\ \tan(\phi/2) \ln \sin^2(\phi/2), & \phi \neq k\pi \end{cases}$$

which shows that the function u does not have a finite derivative for $\phi = 0$:

$$\lim_{\phi \to 0} \frac{du(\phi)}{d\phi} = \lim_{\phi \to 0} \left(\frac{\ln \sin^2(\phi/2)}{2 \cos^2(\phi/2)} + 1 \right) = -\infty.$$

3.5. Conditions for the existence of an exponentially stable invariant torus

Suppose that the matrix $\Omega_0^t(\phi)$ for the system of equations (4.1) satisfies the inequality:

$$\|\Omega_0^t(\phi)\| \leq Ke^{-\gamma t}, \quad t \geq 0 \tag{5.1}$$

for all $\phi \in T_m$ and some positive K and γ independent of ϕ. We show that in this case the function

$$G_0(\tau, \phi) = \begin{cases} \Omega_\tau^0(\phi), & \tau \leq 0, \\ 0, & \tau > 0 \end{cases} \tag{5.2}$$

is a Green's function for the system (4.5) ((5.2) is obtained from (4.4) by setting $C(\phi) \equiv E$ in (4.4)).

Indeed, it follows from inequality (5.1) that

$$\|\Omega_\tau^0(\phi)\| = \|\Omega_0^{-\tau}(\phi_\tau(\phi))\| \leq Ke^{\gamma t}, \quad \tau \leq 0.$$

Thus, we have for function (5.2) the inequality

$$\|G_0(\tau, \phi)\| \leq Ke^{-\gamma|\tau|}, \quad \tau \in \mathbf{R},$$

which implies estimate (4.6) for the integral of $\|G_0(\tau, \phi)\|$. The function (5.2) is therefore a Green's function for the system of equations (4.5).

The set

$$x = u(\phi) = Tf(\phi), \quad \phi \in T_m, \tag{5.3}$$

where T is the operator (4.10), is an invariant torus of system (4.12). We show that this torus is exponentially stable.

We denote the general solution of the system of equations (4.1) by $x = x(t, \phi, x_0) = \Omega_0^t(\phi)x_0$. Using the properties of the matrix $\Omega_0^t(\phi)$ for this solution we obtain the estimate

$$\|x(t, \phi, x_0)\| = \|\Omega_0^t(\phi)x_0\| = \|\Omega_\tau^t(\phi)\Omega_0^\tau(\phi)x_0\| =$$

$$= \|\Omega_\tau^{t-\tau+\tau}(\phi)x(\tau, \phi, x_0)\| \le \|\Omega_0^{t-\tau}(\phi_\tau(\phi))\| \, \|x(\tau, \phi, x_0)\| \le$$

$$\le Ke^{-\gamma(t-\tau)}\|x(\tau, \phi, x_0)\|, \tag{5.4}$$

which holds for all $t \ge \tau$ and arbitrary $\phi \in T_m$. The estimate (5.4) is sufficient for the invariant torus (5.3) to be exponentially stable.

Consequently, if the matrix $\Omega_0^t(\phi)$ satisfies inequality (5.1) then the system of equations (4.12) has an exponentially stable invariant torus which is given by the relation

$$x = Tf(\phi) = \int_{-\infty}^0 \Omega_\tau^0(\phi)f(\phi_\tau(\phi))d\tau, \quad \phi \in T_m, \tag{5.5}$$

and satisfies the estimate

$$|Tf|_0 \le K\gamma^{-1}|f|_0. \tag{5.6}$$

We introduce an index which enables us to check whether inequality (5.1) holds for the matrix $\Omega_0^t(\phi)$ of the system of equations (4.1). We denote by $\mathfrak{R}_0$ the set of $n \times n$ positive-definite matrices $S(\phi)$ belonging to the space $C'(T_m)$. As before we denote by $\dot{S}(\phi)$ the matrix

$$\lim_{t \to 0} \frac{dS(\phi_t(\phi))}{dt} = \dot{S}(\phi).$$

We set

$$\inf_{s \in \mathfrak{R}_0} \max_{\|x\|=1} \frac{\langle (S(\phi)P(\phi) + \frac{1}{2}\dot{S}(\phi))x, x \rangle}{\langle S(\phi)x, x \rangle} \le -\beta(\phi)$$

and define the number β_0 as

$$\beta_0 = \inf_{\phi \in \mathcal{T}_m} \beta(\phi). \tag{5.7}$$

Lemma 1. *For any $\mu > 0$ one can find $K = K(\mu) > 0$ such that the solution $x(t, \phi, x_0) = \Omega_0^t(\phi)x_0$ of the system of equations (4.1) satisfies the inequality*

$$\|x(t, \phi, x_0)\| \leq K e^{-(\beta_0 - \mu)(t-\tau)} \|x(\tau, \phi, x_0)\|$$

for all $t \geq \tau$ and arbitrary $\phi \in \mathcal{T}_m$.

It follows from the lemma that inequality (5.1) always holds when

$$\beta_0 > 0. \tag{5.8}$$

We now turn to the proof of the lemma. If the infimum is attained on the set $\mathfrak{R}_0$, then there exists a matrix $S(\phi) \in \mathfrak{R}_0$ such that

$$\max_{\|x\|=1} \frac{\langle (S(\phi)P(\phi) + \frac{1}{2}\dot{S}(\phi))x, x \rangle}{\langle S(\phi)x, x \rangle} \leq -\beta(\phi).$$

But then

$$\left\langle S(\phi_t(\phi)) \frac{dx(t, \phi, x_0)}{dt},\ x(t, \phi, x_0) \right\rangle = \langle S(\phi_t(\phi))P(\phi_t(\phi))x(t, \phi, x_0),\ x(t, \phi, x_0) \rangle,$$

which leads to the inequality

$$\frac{d}{dt}\left\langle S(\phi_t(\phi))x(t, \phi, x_0), x(t, \phi, x_0) \right\rangle \leq$$

$$\leq -2\beta(\phi_t(\phi))\langle S(\phi_t(\phi))x(t, \phi, x_0), x(t, \phi, x_0) \rangle.$$

Taking the integral of the last inequality, we obtain

$$\langle S(\phi_t(\phi))x(t, \phi, x_0), x(t, \phi, x_0) \rangle \leq$$

$$\leq \exp\left\{-2 \int_\tau^t \beta(\phi_s(\phi))ds\right\} \langle S(\phi_\tau(\phi))x(\tau, \phi, x_0), x(\tau, \phi, x_0) \rangle \tag{5.9}$$

for all $t \geq \tau$.

Since the matrix $S(\phi)$ is periodic and positive definite, we have the estimate:

$$K_0\langle x, x \rangle \leq \langle S(\phi)x, x \rangle \leq K^0\langle x, x \rangle$$

for some positive K_0, K^0 which are independent of $\phi \in T_m$, $x \in E^n$.

Therefore inequality (5.9) leads to the estimate

$$K_0\|x(t,\phi,x_0)\|^2 \leq \exp\left\{-2\int_\tau^t \beta(\phi_s(\phi))ds\right\}K^0\|x(\tau,\phi,x_0)\|^2 \qquad (5.10)$$

for $t \geq \tau$, whence the required estimate follows:

$$\|x(t,\phi,x_0)\| \leq Ke^{-\beta_0(t-\tau)}\|x(\tau,\phi,x_0)\|, \quad t \geq \tau,$$

where $K^2 = K^0/K_0$ does not depend on ϕ, x_0, τ.

If the infimum is not attained on the set $\mathfrak{R}_0$, then there exists a sequence of matrices $S_\nu(\phi)$, $\nu = 1, 2, \ldots, S_\nu(\phi) \in \mathfrak{R}_0$, such that

$$\lim_{\nu\to\infty} \max_{\|x\|=1} \frac{\langle (S_\nu(\phi)P(\phi) + \frac{1}{2}\dot{S}_\nu(\phi))x, x\rangle}{\langle S_\nu(\phi)x, x\rangle} \leq -\beta(\phi).$$

Then for an arbitrary $\mu > 0$ there exists a matrix $S_\mu(\phi) \in \mathfrak{R}_0$ such that

$$\max_{\|x\|=1} \frac{\langle (S_\mu(\phi)P(\phi) + \frac{1}{2}\dot{S}_\mu(\phi))x, x\rangle}{\langle S_\mu(\phi)x, x\rangle} \leq -\beta(\phi) + \mu.$$

By replacing $S_\mu(\phi)$ by $S(\phi)$ and $\beta(\phi) - \mu$ by $\beta(\phi)$ in the above reasoning, we obtain the required estimate for $x(t, \phi, x_0)$. This completes the proof of Lemma 1.

Corollary. *If inequality (5.8) holds then there exists an exponentially stable invariant torus of system (4.12) for an arbitrary function $f \in C^r(T_m)$.*

We prove the converse. Suppose that the invariant torus $x = 0$, $\phi \in T_m$ of the system of equations (4.5) is exponentially stable. Then

$$\|x(t,\phi,x_0)\| \leq Ke^{-\gamma(t-\tau)}\|x(\tau,\phi,x_0)\|, \quad t \geq \tau, \qquad (5.11)$$

for arbitrary $\phi \in T_m$, $x_0 \in E^n$, $\tau \in \mathbf{R}$ and some positive K, γ independent of ϕ, x_0, τ.

Since

$$x(t, \phi, x_0) = \Omega_0^t(\phi)x_0$$

estimate (5.1) for the matrix $\Omega_0^t(\phi)$ follows from (5.11). We use this estimate to prove inequality (5.8). Denote by $S(\phi)$ the matrix defined by the expression

$$S(\phi) = \int_0^\infty (\Omega_0^\tau(\phi))^*\Omega_0^\tau(\phi)d\tau. \qquad (5.12)$$

We show that $S \in \mathfrak{R}_0$ and that

$$\max_{\|x\|=1} \frac{\langle (S(\phi)P(\phi) + \frac{1}{2}\dot{S}(\phi))x, x\rangle}{\langle S(\phi)x, x\rangle} \leq -\beta,$$

where β is a positive constant.

Since $\|(\Omega_0^\tau(\phi))^*\| = \|\Omega_0^\tau(\phi)\|$, inequality (5.1) ensures that the integral (5.12) converges uniformly, so that the matrix $S(\phi)$ belongs to the space $C(T_m)$. Because the integrand matrix in formula (5.12) is symmetric, the matrix $S(\phi)$ is symmetric.

Now we have

$$S(\phi_t(\phi)) = \int_0^\infty (\Omega_t^{t+\tau}(\phi))^* \Omega_t^{t+\tau}(\phi) d\tau =$$

$$= (\Omega_t^0(\phi))^* \int_t^\infty (\Omega_0^\tau(\phi))^* \Omega_0^\tau(\phi) d\tau \Omega_t^0(\phi), \qquad (5.13)$$

which shows that the function $S(\phi_t(\phi))$ is differentiable with respect to t for $t \in \mathbf{R}$. This means that $S \in C'(T_m)$. We now show that the matrix $S(\phi)$ is positive definite. Using the inequality

$$\langle S(\phi)x, x\rangle = \frac{\langle S(\phi)x, x\rangle}{\langle x, x\rangle}\|x\|^2 \geq \min_{\phi \in T_m, \|\xi\|=1}\langle S(\phi)\xi, \xi\rangle\|x\|^2 = \langle S(\phi_0)\xi_0, \xi_0\rangle\|x\|^2,$$

where ϕ_0, ξ_0 are fixed values in the set T_m, $\|\xi\| = 1$, we write the estimate

$$\langle S(\phi)x, x\rangle \geq \langle S(\phi_0)\xi_0, \xi_0\rangle\|x\|^2 = \left\langle \int_0^\infty (\Omega_0^\tau(\phi_0))^* \Omega_0^\tau(\phi_0) d\tau \xi_0, \xi_0 \right\rangle\|x\|^2 =$$

$$= \int_0^\infty \langle \Omega_0^\tau(\phi_0)\xi_0, \Omega_0^\tau(\phi_0)\xi_0\rangle d\tau\|x\|^2 =$$

$$= \int_0^\infty \|\Omega_0^\tau(\phi_0)\xi_0\|^2 d\tau\|x\|^2 = \lambda_1\|x\|^2, \qquad (5.14)$$

where we have set

$$\lambda_1 = \int_0^\infty \|\Omega_0^\tau(\phi_0)\xi_0\|^2 d\tau.$$

Since $\Omega_0^\tau(\phi_0)\xi_0$ is a solution of the homogeneous system of equations (4.1) taking the value $\xi_0 \neq 0$ for $\tau = 0$, it follows that

$$\|\Omega_0^\tau(\phi_0)\xi_0\|^2 > 0$$

for all $\tau \in \mathbf{R}$. Thus $\lambda_1 > 0$ and this shows that the matrix $S(\phi)$ is positive definite. This proves that $S \in \mathfrak{R}_0$.

We now find $\dot{S}(\phi)$. Taking the derivative of (5.13) with respect to t we obtain

$$\frac{dS(\phi_t(\phi))}{dt} = -P^*(\phi_t(\phi))S(\phi_t(\phi)) - E - S(\phi_t(\phi))P(\phi_t(\phi)),$$

therefore $\dot{S}(\phi)$ is defined by the expression:

$$\dot{S}(\phi) = -P^*(\phi)S(\phi) - S(\phi)P(\phi) - E.$$

Now the equalities

$$\langle (S(\phi)P(\phi) + \tfrac{1}{2}\dot{S}(\phi))x, x \rangle = \tfrac{1}{2}\langle (S(\phi)P(\phi) + P^*(\phi)S(\phi) -$$
$$-P^*(\phi)S(\phi) - S(\phi)P(\phi) - E)x, x \rangle = -\tfrac{1}{2}\|x\|^2$$

together with inequalities (5.14) lead to the estimate:

$$\max_{\|x\|=1} \frac{\langle (S(\phi)P(\phi) + \tfrac{1}{2}\dot{S}(\phi))x, x \rangle}{\langle S(\phi)x, x \rangle} = \max_{\|x\|=1} \frac{-\|x\|^2}{2\langle S(\phi)x, x \rangle} =$$
$$= -\frac{1}{2} \frac{1}{\max_{\|x\|=1} \langle S(\phi)x, x \rangle} = -\frac{1}{2\lambda_2},$$

where $\lambda_2 \geq \lambda_1 > 0$. Consequently

$$\max_{\|x\|=1} \frac{\langle (S(\phi)P(\phi) + \tfrac{1}{2}\dot{S}(\phi))x, x \rangle}{\langle S(\phi)x, x \rangle} \leq -\beta, \tag{5.15}$$

where $\beta = 1/(2\lambda_2)$ is a positive constant. Inequality (5.15) leads to the estimate

$$\beta_0 \geq \beta \geq 0$$

for the index (5.7), as required.

Summing up, we have the following statement.

Theorem 1. *Suppose that the right hand side of the system of equations* (4.5) *satisfies the conditions:* $a \in C_{\mathrm{Lip}}^r(\mathcal{T}_m)$, $P \in C^r(\mathcal{T}_m)$, $r \geq 0$. *Then the system of equations* (4.12) *has an exponentially stable invariant torus for any* $f \in C^r(\mathcal{T}_m)$ *if and only if the index* (5.7) *is positive.*

In practice, to see whether β_0 is positive one chooses S in $\mathfrak{R}_0$ so that the derivative along the solutions of the system of equations (4.5) of the form

$$V = \langle S(\phi)x, x \rangle$$

is negative definite. Since this derivative has the form

$$\dot{V} = 2\langle (S(\phi)P(\phi) + \tfrac{1}{2}\dot{S}(\phi))x, x \rangle = \langle \hat{S}(\phi)x, x \rangle,$$

where $\hat{S}(\phi)$ is the symmetric matrix given by

$$\hat{S}(\phi) = S(\phi)P(\phi) + P^*(\phi)S(\phi) + \dot{S}(\phi),$$

it follows that $\dot{V}$ is a negative definite form when $\hat{S}(\phi)$ is a positive definite matrix. Using Sylvester's criterion [39] to determine whether a symmetric matrix is positive definite we conclude that β_0 is positive whenever there exists a symmetric matrix $S(\phi) \in C'(\mathcal{T}_m)$ such that its principal minors and the principal minors of the matrix $-\hat{S}(\phi)$ corresponding to it are positive for all $\phi \in \mathcal{T}_m$.

These conditions hold, in particular, when the principal minors of the symmetric constant matrix S and the matrix $-\hat{S}(\phi) = -(SP(\phi)+P^*(\phi)S)$ are positive for all $\phi \in \mathcal{T}_m$.

3.6. Uniqueness conditions for the Green's function and the properties of this function

It follows from the necessary conditions for the existence of an invariant torus of the system of equations (4.12) for an arbitrary function $f \in C^r(\mathcal{T}_m)$ and Theorem 4.1 that a necessary condition for the existence of a Green's function for the homogeneous system of equations (4.5) is that there be no non-degenerate invariant tori of this system. The conditions under which the system of equations (4.5) that has degenerate invariant tori would have a Green's function are not clear at the present time. However, the peculiar properties of the latter due to the existence of degenerate invariant tori of system (4.5) have been studied to some extent. We consider some of them. First we show that the absence of degenerate tori of system (4.5) ensures that the Green's function for this system is unique.

Theorem 1. *Let $a \in C_{\mathrm{Lip}}(\mathcal{T}_m)$, $P \in C(\mathcal{T}_m)$ and suppose that the system of equations (4.5) has a Green's function $G_0(\tau, Q)$. Then the torus $x = 0$, $\phi \in \mathcal{T}_m$ is a unique invariant torus of system (4.5) if and only if this system does not have any Green's functions other than $G_0(\tau, \phi)$.*

Proof. Suppose that the system of equations (4.5) has a Green's function $G_1(\tau, \phi)$ besides the function $G_0(\tau, \phi)$. We set

$$u_0(\phi) = \int_{-\infty}^{\infty} G_1(\tau, \phi) f(\phi_\tau(\phi)) d\tau - \int_{-\infty}^{\infty} G_0(\tau, \phi) f(\phi_\tau(\phi)) d\tau, \qquad (6.1)$$

where f is an arbitrary function in $C(\mathcal{T}_m)$.

It is clear that the set

$$x = u_0(\phi), \quad \phi \in \mathcal{T}_m,$$

defines an invariant torus of the system of equations (4.5). We prove that

$$u_0 \not\equiv 0 \qquad (6.2)$$

for some function $f \in C(\mathcal{T}_m)$.

It follows from the definition of the Green's function and formula (6.1) that

$$u_0(\phi) = \int_{-\infty}^{\infty} \Omega_\tau^0(\phi) R(\phi_\tau(\phi)) f(\phi_\tau(\phi)) d\tau,$$

where R is a non-zero matrix in $C(\mathcal{T}_m)$. To prove inequality (6.2) it is sufficient to show that an operator T_0 on the function space $C(\mathcal{T}_m)$ defined by the relation

$$T_0 f(\phi) = \int_{-\infty}^{\infty} G_1(\tau, \phi) f(\phi_\tau(\phi)) d\tau - \int_{-\infty}^{\infty} G_0(\tau, \phi) f(\phi_\tau(\phi)) d\tau =$$

$$= \int_{-\infty}^{\infty} \Omega_\tau^0(\phi) R(\phi_\tau(\phi)) f(\phi_\tau(\phi)) d\tau \qquad (6.3)$$

is not the zero operator on this space.

Suppose this is not the case, that is,

$$T_0 f \equiv 0 \quad \forall f \in C(\mathcal{T}_m). \qquad (6.4)$$

We show that (6.4) implies the equality

$$R(\phi) = 0 \quad \forall \phi \in \mathcal{T}_m, \qquad (6.5)$$

thus contradicting the assumption that $G_0(\tau, \phi) \neq G_1(\tau, \phi)$ which implies that $R(\phi) \not\equiv 0$, $\phi \in \mathcal{T}_m$.

Turning to the proof of equality (6.5), we first consider the values of the matrix R at the points of trajectories of system (4.5) that are periodic on $\mathcal{T}_m$. Let ϕ_0 be such a point, $\phi = \phi_{t+T}(\phi_0) = \phi_t(\phi_0) \bmod 2\pi$, $t \in \mathbf{R}$, the corresponding trajectory, T its period. It follows from the Floquet-Lyapunov theory that

$$\Omega_0^t(\phi_0) = \Phi(t)e^{At},$$

where Φ is a periodic non-singular matrix with period T and A is a constant matrix. But then

$$\Omega_\tau^0(\phi_0) = (\Omega_0^\tau(\phi_0))^{-1} = e^{-A\tau}\Phi^{-1}(\tau),$$

so that

$$T_0 f(\phi_0) = \int_{-\infty}^{\infty} e^{-A\tau}\Phi^{-1}(\tau)R(\phi_\tau(\phi_0))f(\phi_\tau(\phi_0))d\tau =$$

$$= \sum_{p=-\infty}^{+\infty} \int_{pT}^{(p+1)T} e^{-A\tau}\Phi^{-1}(\tau)R(\phi_\tau(\phi_0))f(\phi_\tau(\phi_0))d\tau =$$

$$= \sum_{p=-\infty}^{+\infty} \int_0^T e^{-ApT}e^{-A\tau}\Phi^{-1}(\tau)R(\phi_\tau(\phi_0))f(\phi_\tau(\phi_0))d\tau =$$

$$= \sum_{p=-\infty}^{+\infty} (e^{-AT})^p B_f,$$

where

$$B_f = \int_0^T e^{-A\tau}\Phi^{-1}(\tau)R(\phi_\tau(\phi_0))f(\phi_\tau(\phi_0))d\tau.$$

Since the series $\sum_{p=-\infty}^{+\infty}(e^{-AT})^p$ is divergent for any matrix A, it follows that $T_0 f(\phi_0) = 0$ only if

$$B_f = 0. \tag{6.6}$$

Consider equality (6.6). We distinguish the cases when ϕ_0 is a stationary point of the dynamical system on $\mathcal{T}_m$ and when ϕ_0 is not such a point. In the first case equality (6.6) holds for any $T > 0$. This is possible only if

$$e^{-A\tau}\Phi^{-1}(\tau)R(\phi_\tau(\phi_0))f(\phi_\tau(\phi_0)) \equiv 0.$$

But then

$$R(\phi_\tau(\phi_0))f(\phi_\tau(\phi_0)) \equiv 0,$$

and, taking into account the identity $\phi_\tau(\phi_0) \equiv \phi_0$ and the fact that the function f is arbitrary, we see that this can hold only if

$$R(\phi_0) = 0. \tag{6.7}$$

In the second case there exists a number $T > 0$, the real period of the periodic trajectory on the torus, such that

$$\phi_{t+T}(\phi_0) = \phi_t(\phi_0) \bmod 2\pi, \quad \phi_{t+\tau}(\phi_0) \neq \phi_t(\phi_0) \bmod 2\pi$$

for any $0 < \tau < T$. Then there is one-to-one correspondence between a point $\phi_t(\phi_0)$ on the torus T_m and a point $\psi_t = \nu t$, $\nu = 2\pi/T$, on the circle T_1. In view of this correspondence, for any function $F \in C(T_1)$ we can find a function $f \in C(T_m)$ such that

$$f(\phi_t(\phi_0)) = F(\nu t), \ t \in \mathbf{R}.$$

In this case equality (6.6) becomes

$$\int_0^T e^{-A\tau} \Phi^{-1}(\tau)R(\phi_\tau(\phi_0))F(\nu\tau)d\tau = 0. \tag{6.8}$$

Since equality (6.8) holds for an arbitrary function F in $C(T_1)$, this is possible only if

$$e^{-A\tau}\Phi^{-1}(\tau)R(\phi_\tau(\phi_0)) \equiv 0.$$

Because the matrices $e^{-A\tau}$ and $\Phi^{-1}(\tau)$ are non-singular, this leads to the identity

$$R(\phi_\tau(\phi_0)) \equiv 0, \quad \tau \in [0,T]. \tag{6.9}$$

Indeed, we set

$$F(\nu t) = \{0, \ldots, F_j(\nu t), \ldots, 0\},$$

$$r_{ij}(t) = \{e^{-A\tau}\Phi^{-1}(\tau)R(\phi_\tau(\phi_0))\}_{ij} \quad (i,j = 1,\ldots,n)$$

and rewrite equality (6.8) for the chosen function $F(\nu t)$. We have:

$$\int_0^T r_{ij}(\tau)F_j(\nu\tau)d\tau = 0. \tag{6.10}$$

In (6.10) we replace F_j by the function in $C(\mathcal{T}_1)$ defined for $\psi \in [0, 2\pi]$ by the equality

$$F_j(\psi) = \bar{r}_{ij}(\psi/\nu)f_0(\psi),$$

where $\bar{r}_{ij}$ is the complex conjugate of r_{ij}, and f_0 is a function in $C(\mathcal{T}_1)$ equal to zero for $\psi = 0$ and positive for $\psi = (0, 2\pi)$:

$$f_0(0) = f_0(2\pi) = 0, \quad f_0(\psi) > 0 \text{ when } \psi \in (0, 2\pi).$$

As a result we obtain the equality

$$\int_0^T |r_{ij}(\tau)|^2 f_0(\nu\tau)d\tau = 0.$$

This is possible only if

$$|r_{ij}(\tau)|^2 f_0(\nu\tau) = 0, \quad \tau \in (0, T),$$

which is equivalent to

$$r_{ij}(\tau) = 0, \quad \tau \in [0, T]. \tag{6.11}$$

Since r_{ij} is an arbitrary element of the matrix $e^{-A\tau}\Phi^{-1}(\tau)R(\phi_\tau(\phi_0))$, identity (6.11) is equivalent to (6.9). This proves (6.7) for any point ϕ_0 that belongs to a trajectory of system (4.5) that is periodic on $\mathcal{T}_m$.

We now consider the values of the matrix $R(\phi)$ at the other points of the torus $\mathcal{T}_m$. Let ϕ_0 be such a point, and $\phi = \phi_t(\phi_0)$, $t \in \mathbf{R}$, the trajectory corresponding to it. Because the integral

$$\int_{-\infty}^{\infty} \|G_i(\tau, \phi)\|d\tau, \ i = 0, 1,$$

is uniformly bounded, the integral

$$\int_{-\infty}^{\infty} \|\Omega_\tau^0(\phi)R(\phi_\tau(\phi))\|d\tau$$

is also uniformly bounded. But then one can find a sequence of positive numbers ϵ_n, monotonically decreasing to zero $\epsilon_{n+1} < \epsilon_n$, $\lim_{n\to\infty} \epsilon_n = 0$ and a sequence of positive numbers t_n monotonically increasing to $+\infty$, $t_{n+1} > t_n$, $\lim_{n\to\infty} t_n = +\infty$ such that the following inequality holds:

$$\int_{-\infty}^{-t_n} \|\Omega_\tau^0(\phi_0)R(\phi_\tau(\phi_0))\|d\tau + \int_{t_n}^{+\infty} \|\Omega_\tau^0(\phi_0)R(\phi_\tau(\phi_0))\|d\tau < \epsilon_n. \tag{6.12}$$

But since $T_0 f(\phi_0) = 0$, it follows that

$$\left\| \int_{-t_n}^{t_n} \Omega_\tau^0(\phi_0) R(\phi_\tau(\phi_0)) f(\phi_\tau(\phi_0)) d\tau \right\| \leq$$

$$\leq \left\| \int_{-\infty}^{\infty} \Omega_\tau^0(\phi_0) R(\phi_\tau(\phi_0)) f(\phi_\tau(\phi_0)) d\tau \right\| +$$

$$+ \int_{-\infty}^{-t_n} \| \Omega_\tau^0(\phi_0) R(\phi_\tau(\phi_0)) \| d\tau |f|_0 +$$

$$+ \int_{t_n}^{+\infty} \| \Omega_\tau^0(\phi_0) R(\phi_\tau(\phi_0)) \| d\tau |f|_0 \leq \epsilon_n |f|_0.$$

The arc $\phi = \phi_t(\phi_0)$, $-t_n \leq t \leq t_n$ of the trajectory on T_m does not intersect itself, therefore for any continuous function $F(t)$, defined for $t \in \mathbf{R}$, one can find a function $f_n \in C(T_m)$ related to F as follows:

$$f_n(\phi_t(\phi_0)) = F(t), \quad -t_n \leq t \leq t_n.$$

Suppose that $R(\phi_t(\phi_0)) \not\equiv 0$ for $t \in \mathbf{R}$. Then $\Omega_t^0(\phi_0) \times R(\phi_t(\phi_0)) \not\equiv 0$ for $t \in \mathbf{R}$. Let $r_{ij}(t) = \{\Omega_t^0(\phi_0) R(\phi_t(\phi_0))\}_{ij}$ be an element of the matrix $\Omega_t^0(\phi_0) R(\phi_t(\phi_0))$ satisfying the inequality $r_{ij}(t) \not\equiv 0$, $t \in \mathbf{R}$.

We set

$$\hat{r}_{ij}(t) = \begin{cases} \bar{r}_{ij}(t), & |r_{ij}(t)| \leq 1, \\ \operatorname{sign} \bar{r}_{ij}(t), & |r_{ij}(t)| > 1. \end{cases}$$

It is obvious that the function $\hat{r}_{ij}(t)$ is continuous for $t \in \mathbf{R}$ and $\max_{t \in \mathbf{R}} |\hat{r}_{ij}(t)| \leq 1$.

We choose scalar functions $f_n \in C(T_m)$ such that the conditions

$$f_n(\phi_t(\phi_0)) = \hat{r}_{ij}(t) \text{ for } -t_n \leq t \leq t_n, \quad |f_n|_0 \leq 1,$$

hold and consider inequality (6.12) for the functions $f_n(\phi) e_j$, where $e_j = \{0, \ldots, 1, \ldots, 0\}$ is the jth unit vector. We have:

$$\left\| \int_{-t_n}^{t_n} \Omega_\tau^0(\phi_0) R(\phi_\tau(\phi_0)) f_n(\phi_\tau(\phi_0)) e_j d\tau \right\| =$$

$$= \left| \int_{-t_n}^{t_n} r_{ij}(\tau) \hat{r}_{ij}(\tau) d\tau \right| = \int_{-t_n}^{t_n} r_{ij}(\tau) \hat{r}_{ij}(\tau) d\tau \leq \epsilon_n. \qquad (6.13)$$

Since $r_{ij}(t) \hat{r}_{ij}(t) \not\equiv 0$ for $t \in \mathbf{R}$, it follows that

$$\int_{-t_N}^{t_N} r_{ij}(\tau) \hat{r}_{ij}(\tau) d\tau = \delta_1 > 0$$

for some sufficiently large N. But then, since the function $r_{ij}(t)\hat{r}_{ij}(t)$ is non-negative, it follows that

$$\int_{-t_n}^{t_n} r_{ij}(\tau)\hat{r}_{ij}(\tau)d\tau \geq \delta_1 \quad \forall n \geq N,$$

which contradicts inequality (6.13) and the condition

$$\lim_{n \to \infty} \epsilon_n = 0.$$

This contradiction derived from the assumption that

$$r_{ij}(t) \not\equiv 0, \quad t \in \mathbf{R},$$

proves that

$$\Omega_t^0(\phi_0)R(\phi_t(\phi_0)) \equiv 0, \quad t \in \mathbf{R}. \tag{6.14}$$

For any point $\phi_0 \in T_m$ not belonging to a trajectory of system (4.5) that is periodic on T_m, relation (6.7) follows from equality (6.14).

The above arguments in combination prove equality (6.5). But this, as has been noted, contradicts the assumption that the Green's function is not unique. Thus if the Green's function is not unique it will necessarily imply that there exists an invariant torus of system (4.5) different from the trivial one $x = 0$, $\phi \in T_m$. The "only if" part of Theorem 1 is proved. We now prove the reverse implication. To do this, we suppose that uniqueness of the Green's function does not imply the uniqueness of an invariant torus of system (4.5). First we establish two important equalities that follow from the assumption that the Green's function is unique.

Lemma 1. *Suppose that the function $G_0(\tau, \phi)$ is the only Green's function for the system of equations (4.5). Then the matrix $C(\phi) = G_0(0, \phi)$ satisfies the equalities*

$$C(\phi_t(\phi)) = \Omega_0^t(\phi)C(\phi)\Omega_t^0(\phi) \quad \forall t \in \mathbf{R}, \ \phi \in T_m \tag{6.15}$$

$$C^2(\phi) = C(\phi) \quad \forall \phi \in T_m. \tag{6.16}$$

Indeed, suppose that the first relation does not hold, that is, for some $t_1 \neq 0$

$$C(\phi_{t_1}(\phi)) \neq \Omega_0^{t_1}(\phi)C(\phi)\Omega_{t_1}^0(\phi), \quad \phi \in T_m.$$

We set

$$R(\phi) = \Omega^0_{t_1}(\phi)C(\phi_{t_1}(\phi)) - C(\phi)\Omega^0_{t_1}(\phi)$$

and consider the function

$$G_1(\tau,\phi) = G_0(\tau,\phi) + \Omega^0_\tau(\phi)R(\phi_\tau(\phi)).$$

For this function we have the chain of inequalities

$$\int_{-\infty}^{\infty} \|G_1(\tau,\phi)\|d\tau \leq \int_{-\infty}^{\infty} \|G_0(\tau,\phi)\|d\tau + \int_{-\infty}^{\infty} \|\Omega^0_\tau(\phi)R(\phi_\tau(\phi))d\tau \leq$$

$$\leq K + \int_{-\infty}^{\infty} \|\Omega^0_{\tau+t_1}(\phi)C(\phi_{\tau+t_1}(\phi)) - \Omega^0_\tau(\phi)C(\phi_\tau(\phi))\Omega^0_{t_1}(\phi_\tau(\phi))\|d\tau \leq$$

$$\leq K + \int_{-\infty}^{-t_1} \|\Omega^0_{\tau+t_1}(\phi)C(\phi_{\tau+t_1}(\phi))\|d\tau + \int_{-t_1}^{\infty} \|\Omega^0_{\tau+t_1}(\phi)[E - C(\phi_{\tau+t_1}(\phi))]\|d\tau +$$

$$+ \int_{-\infty}^{-t_1} \|\Omega^0_\tau(\phi)C(\phi_\tau(\phi))\|d\tau|\Omega^0_{t_1}|_0 + \int_{-t_1}^{\infty} \|\Omega^0_{\tau+t_1}(\phi) - \Omega^0_\tau(\phi)C(\phi_\tau(\phi)) \times$$

$$\times \Omega^0_{t_1}(\phi_\tau(\phi))\|d\tau \leq 2K + \left[\int_{-\infty}^{-t_1} \|\Omega^0_\tau(\phi)C(\phi_\tau(\phi))\|d\tau + \right.$$

$$\left. + \int_{-t_1}^{\infty} \|\Omega^0_\tau(\phi)[E - C(\phi_\tau(\phi))]\|d\tau \right] |\Omega^0_{t_1}|_0 \leq$$

$$\leq 2K + K|\Omega^0_{t_1}|_0 + \left[\left| \int_{-t_1}^{0} \|\Omega^0_\tau(\phi)C(\phi_\tau(\phi))\|d\tau \right| + \right.$$

$$\left. + \left| \int_0^{-t_1} \|\Omega^0_\tau(\phi)[E - C(\phi_\tau(\phi))]\|d\tau \right| \right] |\Omega^0_{t_1}|_0 \leq$$

$$\leq 2K + (K + \overline{K})|\Omega^0_{t_1}|_0 = K_1,$$

from which it follows that

$$\int_{-\infty}^{\infty} \|G_1(\tau,\phi)\|d\tau \leq K_1$$

uniformly with respect to $\phi \in T_m$. But then the function $G_1(\tau,\phi)$ satisfies all the conditions necessary for it to be a Green's function for the system of equations (4.5). This contradicts the assumption that the Green's function is unique. The contradiction proves equality (6.15).

Now let

$$C^2(\phi) \not\equiv C(\phi), \quad \phi \in \mathcal{T}_m.$$

Then the function

$$G_1(\tau, \phi) = G_0(\tau, \phi) + \Omega_\tau^0(\phi)(C^2(\phi_\tau(\phi)) - C(\phi_\tau(\phi)))$$

satisfies the chain of inequalities

$$\int_{-\infty}^{\infty} \|G_1(\tau, \phi)\| d\tau \le K + \int_{-\infty}^{0} \|\Omega_\tau^0(\phi)C(\phi_\tau(\phi))\| d\tau |C - E|_0 +$$

$$+ \int_{0}^{\infty} \|\Omega_\tau^0(\phi)[E - C(\phi_\tau(\phi))] d\tau |C|_0 \le K[1 + \max(|C - E|_0, |C|_0)] = \tilde{K}_1,$$

which shows that it is also a Green's function for the system of equations (4.5).
The latter contradicts the assumption that the Green's function is unique and
thus proves equality (6.16).

We now use equality (6.15) to complete the proof of Theorem 1. Suppose
that the Green's function $G_0(\tau, \phi)$ is unique, but that the torus

$$x = u_0(\phi), \quad \phi \in \mathcal{T}_m, \tag{6.17}$$

is an invariant torus of system (4.5) different from the trivial one. Then

$$u_0(\phi) \not\equiv 0, \quad \phi \in \mathcal{T}_m. \tag{6.18}$$

Consider the function $u_1(\phi) \in C(\mathcal{T}_m)$ defining the invariant torus

$$x = u_1(\phi), \quad \phi \in \mathcal{T}_m,$$

of the system of equations

$$\frac{d\phi}{dt} = a(\phi), \quad \frac{dx}{dt} = P(\phi)x + u_0(\phi). \tag{6.19}$$

Since the torus (6.17) of the homogeneous system of equations correspond-
ing to the system (6.19) is invariant, it follows that $u_0(\phi_t(\phi)) = \Omega_0^t(\phi)u_0(\phi)$.
Therefore

$$u_1(\phi) = \int_{-\infty}^{+\infty} G_0(\tau, \phi)u_0(\phi_\tau(\phi))d\tau = \int_{-\infty}^{\infty} G_0(\tau, \phi)\Omega_0^\tau(\phi)u_0(\phi)d\tau =$$

$$= \int_{-\infty}^{0} \Omega_\tau^0(\phi)C(\phi_\tau(\phi))\Omega_0^\tau(\phi)u_0(\phi)d\tau -$$

$$- \int_{0}^{\infty} \Omega_\tau^0(\phi)[E - C(\phi_\tau(\phi))]\Omega_0^\tau(\phi)u_0(\phi)d\tau, \tag{6.20}$$

which together with (6.15) leads to the relation

$$u_1(\phi) = \int_{-\infty}^{0} C(\phi)u_0(\phi)d\tau - \int_{0}^{\infty} (E - C(\phi))u_0(\phi)d\tau \qquad (6.21)$$

This is possible only if

$$C(\phi)u_0(\phi) = 0, \quad C(\phi)u_0(\phi) = u_0(\phi)$$

simultaneously. But then $u_0(\phi) = 0$, $\forall \phi \in \mathcal{T}_m$, which contradicts inequality (6.18).

The main properties of the Green's function that follow from the assumption that it is unique are defined by relations (6.15) and (6.16). The first means that the matrix $C(\phi)$ belongs to the space $C'(\mathcal{T}_m)$ and is a solution of the matrix equation

$$C = P(\phi)C - CP(\phi),$$

while the second means that the matrix $C(\phi)$ is a projection.

We shall show that equality (6.15) is a characteristic property for a Green's function to be unique, namely, if the system of equations (4.5) has more than one Green's function then none of them satisfies equality (6.15). We formulate this in the form of the following statement.

Lemma 2. *Let $a \in C_{\mathrm{Lip}}(\mathcal{T}_m)$, $P \in C(\mathcal{T}_m)$ and suppose that the system of equations (4.5) has an invariant torus different from the trivial one $x = 0$, $\phi \in \mathcal{T}_m$. Then none of the Green's functions for system (4.5) satisfies relation (6.15).*

Turning to the proof of the lemma we suppose that the torus (6.17) is an invariant torus of system (4.5) different from a trivial one. Let $G_0(\tau, \phi)$ be a Green's function for system (4.5) satisfying relation (6.15). Then the system of equations (6.19) has an invariant torus defined by the function (6.20) which, according to equality (6.21), can only be equal to zero:

$$u_1(\phi) \equiv 0, \quad \phi \in \mathcal{T}_m.$$

But then $\dot{u}_1(\phi) \equiv 0$, $\phi \in \mathcal{T}_m$, and it follows from the second equation of (6.19) that $u_0(\phi) \equiv 0$, $\phi \in \mathcal{T}_m$. But this is a contradiction since the torus (6.17) is trivial. The contradiction proves that the function $G_0(\tau, \phi)$ cannot satisfy relation (6.15).

Finally we note that the condition that there be no invariant tori of system (4.5) is sufficient for the Green's function of this system to be unique. Therefore the quasi-periodic system of equations, system (4.5) with $a(\phi) = \omega = $ const can have only one Green's function such that relations (6.15), (6.16) hold.

3.7. Separatrix manifolds. Decomposition of a linear system

Consider the system of equations (4.5):

$$\frac{d\phi}{dt} = a(\phi), \quad \frac{dx}{dt} = P(\phi)x \qquad (7.1)$$

under the assumption that $a \in C_{\text{Lip}}(\mathcal{T}_m)$, $P \in C(\mathcal{T}_m)$ and that this system does not have an invariant torus other than the trivial one $x = 0$, $\phi \in \mathcal{T}_m$. Suppose that there exists a Green's function $G_0(\tau, \phi)$ for the system of equations (7.1).

These assumptions introduce certain properties of the behaviour of trajectories of system (7.1). We now discuss this behaviour.

First we note that the matrix $C(\phi) = G_0(0, \phi)$ satisfying equality (6.16) is a projection for all $\phi \in \mathcal{T}_m$. It follows from this that all the eigenvalues of the matrix $C(\phi)$ equal 1 or 0. Since the matrix $C(\phi)$ is continuous in ϕ, it follows that its rank does not depend on ϕ and is equal to the multiplicity of 1 considered as an eigenvalue of the matrix $C(\phi)$. Thus the Jordan form of the matrix $C(\phi)$ is the matrix $E_1 = \text{diag}\{E, 0\}$, where E is the $r \times r$ identity matrix. We denote by $S(\phi)$ the matrix converting $C(\phi)$ to Jordan form:

$$C(\phi) = S(\phi)E_1 S^{-1}(\phi) \quad \forall \phi \in \mathcal{T}_m.$$

Let M^+ be a set of the phase space of system (7.1) defined as the set of points (ϕ, x) satisfying the equation

$$C(\phi)x = x, \qquad (7.2)$$

and M^- the analogous set of points (ϕ, x) satisfying the equation

$$C(\phi)x = 0. \qquad (7.3)$$

We now discuss the geometric structure of the set M^+. For each fixed ϕ, equation (7.2) defines in E^n a plane E_ϕ^+ spanned by the vectors $S_1(\phi), \ldots, S_r(\phi)$

which are the first r columns of the matrix $S(\phi)$. Since the matrix $C(\phi)$ is periodic, it follows that

$$E_\phi^+ = E_{\phi \bmod 2\pi}^+.$$

Because the matrix $S(\phi)$ is non-singular for all $\phi \in T_m$ it follows that E_ϕ^+ is an r-dimensional plane. We define on M^+ a neighbourhood system by taking a δ-neighbourhood of a point $P_0 = \{\phi_0, x_0\} \in M^+$ to be the intersection of the set M^+ with a δ-neighbourhood of the point P_0 in $T_m \times E^n$,

$$U_\delta(P_0/M^+) = M^+ \cap [U_\delta(\phi_0) \times U_\delta(x_0)].$$

Since the Jordan form of the matrix $C(\phi)$ does not depend on ϕ and is C-continuous with respect to ϕ, it follows from a result in [18] that the matrix $S(\phi)$ converting $C(\phi)$ to Jordan form E_1 can be taken to be continuous with respect to ϕ for $\phi \in U_\delta(\phi_0)$, where $U_\delta(\phi_0)$ is understood as a δ-neighbourhood of the point ϕ_0 on the torus T_m.

Having chosen $S(\phi)$ to be continuous, we make a change of variables in the system of equations (7.2) by setting

$$x = S(\phi)y. \tag{7.4}$$

As a result we obtain the system of equations

$$E_1 y = y, \tag{7.5}$$

which is equivalent to the system

$$y_j = 0, \; j = r + 1, \ldots, n.$$

Hence it follows that the last $n - r$ coordinates of the vector $S^{-1}(\phi)x$ are equal to zero for any point $P = \{\phi, x\} \in U_\delta(P_0/M^+)$. In view of all this, the change of variables (7.4) defines a continuous one-to-one mapping S of the neighbourhood $U_\delta(P_0/M^+)$ onto a neighbourhood of the point $S(P_0) \in T_m \times E^r$, where E^r is the r-dimensional Euclidean space defined in E^n by the system (7.5).

Let $SU_\delta(P_0/M^+)$ be the image of the neighbourhhood $U_\delta(P_0/M^+)$ under the map S, and Q a point in $SU_\delta(P_0/M^+)$. We define its coordinates (ϕ, c) using the coordinates (ϕ, x) of the inverse image $P = S^{-1}Q$ by setting:

$$\begin{bmatrix} c \\ 0 \end{bmatrix} = S^{-1}(\phi)x, \quad c = (c_1, \ldots, c_r).$$

We take a point $P_1 = \{\phi_1, x_1\} \in M^+$ close to P_0 so that a neighbourhood $U_\delta(P_1/M^+)$ of it has a non-empty intersection with $U_\delta(P_0/M^+)$:

$$U_\delta(P_0/M^+) \bigcap U_\delta(P_1/M^+) = D \neq \varnothing.$$

Let $S_1(\phi)$ be a continuous matrix in ϕ for $\phi \in U_\delta(\phi_1)$ converting $C(\phi)$ to Jordan form, and S_1 the map from the neighbourhood $U_\delta(P_1/M)$ into $\mathcal{T}_m \times E^r$ induced by this matrix. Every point $P = \{\phi, x\}$ in the set D has the coordinates (ϕ, c) defined by the map S and the coordinates (ϕ, c_1) defined by the map S_1. To find out the relation between the coordinates c and c_1 we prove the following lemma.

Lemma 1. *Let $S(\phi)$ and $S_1(\phi)$ be matrices transforming the projection matrix $C(\phi)$ to the Jordan form E_1. Then*

$$S_1^{-1}(\phi)S(\phi) = \mathrm{diag}\{D_1(\phi), D_2(\phi)\}, \tag{7.6}$$

where $D_1(\phi)$ and $D_2(\phi)$ are non-singular matrices of dimensions r and $n - r$ respectively.

To prove the lemma we split the matrix $S_1^{-1}(\phi)S(\phi)$ into blocks

$$S_1^{-1}S = \begin{bmatrix} D_1 & D_{12} \\ D_{21} & D_2 \end{bmatrix}$$

and write down the equality relating S_1, C and E_1:

$$C = S_1 E_1 S_1^{-1} = S E_1 S^{-1}.$$

According to this equality,

$$E_1 S_1^{-1} S = S_1^{-1} S E_1,$$

so that (using the form of the matrix E_1) we have the relation

$$\begin{bmatrix} D_1 & D_{12} \\ 0 & 0 \end{bmatrix} = \begin{bmatrix} D_1 & 0 \\ D_{21} & 0 \end{bmatrix}.$$

This is possible only if $D_{12} = 0$ and $D_{21} = 0$. Consequently the matrix $S_1^{-1}(\phi)S(\phi)$ has a quasi-diagonal form. Since $S_1^{-1}(\phi)S(\phi)$ is non-singular, the diagonal blocks of $\mathrm{diag}\{D_1(\phi), D_2(\phi)\}$ are also non-singular. This completes the proof of the lemma.

Using the lemma we express the relation between the coordinates c_1 and c. By definition,

$$\begin{bmatrix} c \\ 0 \end{bmatrix} = S^{-1}(\phi)x, \qquad \begin{bmatrix} c_1 \\ 0 \end{bmatrix} = S_1^{-1}(\phi)x.$$

Thus it follows from equality (7.6) that

$$\begin{bmatrix} c_1 \\ 0 \end{bmatrix} = S_1^{-1}(\phi)S(\phi)\begin{bmatrix} c \\ 0 \end{bmatrix} = \begin{bmatrix} D_1(\phi)c \\ 0 \end{bmatrix}.$$

This is equivalent to the equality

$$c_1 = D_1(\phi)c. \tag{7.7}$$

Since the matrices $S_1^{-1}(\phi)$ and $S(\phi)$ are continuous for $\phi \in \mathcal{D}$, the matrix $D_1(\phi)$ is also continuous for $\phi \in \mathcal{D}$, where $\mathcal{D} = U_\delta(\phi_0) \cap U_\delta(\phi_1)$.

Relation (7.7) shows that the local coordinates of the point P considered as a point of $U_\delta(P/M^+)$ can be expressed using a continuous and invertible change of variables in terms of the local coordinates of this point considered as a point of $U_\delta(P_0/M^+)$. This suffices for M^+ to be a *topological manifold* of dimension $m + r$.

To emphasize that a small neighbourhood of any point of the manifold M^+ is homeomorphic to some neighbourhood of the manifold $T_m \times E^r$ we shall say that the manifold M^+ is *locally homeomorphic* to the product $T_m \times E^r$.

The set M^- has a similar structure. To prove this we only need to write equation (7.3) using the matrix $C_1(\phi) = E - C(\phi)$ in the form (7.2) and apply the above reasoning to it.

We now show that the manifolds M^+ and M^- are invariant sets of the system of equations (7.1). Suppose that $(\phi_0, x_0) \in M^+$. Let $\phi_t(\phi_0), x_t(\phi_0, x_0) = \Omega_0^t(\phi_0)x_0$, $t \in \mathbf{R}$ be a trajectory of system (7.1) that passes through the point (ϕ_0, x_0) at $t = 0$. The condition $(\phi_0, x_0) \in M^+$ is equivalent to

$$C(\phi_0)x_0 = x_0.$$

Thus, to prove that the manifold M^+ is invariant it is sufficient to prove the equality analogous to the one given above for any point $(\phi_t(\phi_0), x_t(\phi_0, x_0))$ for any $t \in \mathbf{R}$.

Taking into consideration relation (6.15) we have

$$C(\phi_t(\phi_0))x_t(\phi_0, x_0) = C(\phi_t(\phi_0))\Omega_0^t(\phi_0)x_0 =$$
$$= \Omega_0^t(\phi_0)C(\phi_0)x_0 = x_t(\phi_0, x_0) \quad \forall t \in \mathbf{R}.$$

This means that $(\phi_t(\phi_0), x_t(\phi_0, x_0)) \in M^+$ for any $t \in \mathbf{R}$. Thus we have proved that the set M^+ is invariant.

The proof that M^- is invariant is similar.

As a result we have the following statement.

Theorem 1. *Under the assumptions made above on the system of equations* (7.1), *the sets* M^+ *and* M^- *are invariant manifolds of this system that are locally homeomorphic to the products* $T_m \times E^r$ *and* $T_m \times E^{n-r}$ *respectively.*

We give some further properties of the manifolds M^+ and M^-.

First we note that

$$M^+ \bigcap M^- = T_m, \quad M^+ \oplus M^- = T_m \times E^n. \tag{7.8}$$

The first relation states that the intersection of the manifolds M^+ and M^- is the torus

$$x = 0, \ \phi \in T_m, \tag{7.9}$$

while the second asserts that their direct sum is the whole phase space of the dynamical system (7.1).

The first relation of (7.8) follows because the inclusion $(\phi, x) \in M^+ \cap M^-$ necessarily implies equality (7.9) under the assumption that (7.2) and (7.3) hold simultaneously. The second follows since $\forall \phi \in T_m$ the planes E_ϕ^+ and E_ϕ^- by defining the decomposition of E^n into the direct sum of subspaces $E^r = E_\phi^+$ and $E^{n-r} = E_\phi^-$ uniquely define the decomposition of $T_m \times E^n$ into a direct sum of M^+ and M^-.

Let us find out the properties as $|t| \to +\infty$ of the motions of system (7.1) that start on the manifolds M^+ and M^- respectively.

Suppose that the Green's function $G_0(\tau, \phi)$ is damped with respect to τ as $|\tau| \to +\infty$, in other words,

$$\|G_0(\tau, \phi)\| \le d(|\tau|), \ \tau \in \mathbf{R}, \ \phi \in T_m, \tag{7.10}$$

where $d(t)$ is a positive function that is monotonically decreasing to zero as $t \to +\infty$. The motions that start on the manifold M^+ at $t = 0$ are damped as $t \to +\infty$ and the motions that start on the manifold M^- are damped as $t \to -\infty$ and have the same damping character as the function $G_0(\tau, \phi)$:

$$\|x_t(\phi, x)\| \le d(t - \tau)\|x_\tau(\phi, x)\|, \quad t \ge \tau, \ (\phi, x) \in M^+,$$
$$\|x_t(\phi, x)\| \le d(\tau - t)\|x_\tau(\phi, x)\|, \quad t \le \tau, \ (\phi, x) \in M^-. \tag{7.11}$$

Indeed, suppose that inequality (7.10) holds and $(\phi, x) \in M^+$. The chain of relations

$$\|x_t(\phi, x)\| = \|\Omega_0^t(\phi)x\| = \|\Omega_0^t(\phi)C(\phi)x\| = \|\Omega_\tau^t(\phi)\Omega_0^\tau(\phi)C(\phi)x\| =$$

$$= \|\Omega_\tau^t(\phi)C(\phi_\tau(\phi))\Omega_0^\tau(\phi)x\| = \|\Omega_\tau^t(\phi)C(\phi_\tau(\phi))x_\tau(\phi, x)\| =$$

$$= \|\Omega_{\tau-t}^0(\phi_t(\phi))C(\phi_{\tau-t}(\phi_t(\phi)))x_\tau(\phi, x)\| \le$$

$$\le \|G_0(\tau - t, \phi_t(\phi))\| \, \|x_\tau(\phi, x)\| \le d(t - \tau)\|x_\tau(\phi, x)\|,$$

which hold for $t \ge \tau$, proves the first inequality of (7.11).

The similar chain of relations, which hold for $t \le \tau$ and $(\phi, x) \in M^-$, proves the second inequality of (7.11).

Now let $(\phi_0, x_0) \notin M^+ \cup M^-$. We claim that a motion $\phi_t(\phi_0)$, $x_t(\phi_0, x_0)$ that starts at the point (ϕ_0, x_0) leaves any neighbourhood of torus (7.9) both for $t > 0$ and for $t < 0$.

Indeed, let

$$x_0' = C(\phi_0)x_0, \quad x_0'' = C_1(\phi_0)x_0,$$

where $C_1(\phi) = E - C(\phi)$. Then $(\phi_0, x_0') \in M^+$, $(\phi_0, x_0'') \in M^-$, $\|x_0'\| \ne 0$, $\|x_0''\| \ne 0$.

In fact, since the matrices $C(\phi)$ and $C_1(\phi)$ are projections, it follows that

$$C(\phi_0)x_0' = C^2(\phi_0)x_0 = C(\phi_0)x_0 = x_0',$$

$$C(\phi_0)x_0'' = C(\phi_0)C_1(\phi_0)x_0 = 0,$$

which proves that the point (ϕ_0, x_0') belongs to the manifold M^+ and the point (ϕ_0, x_0') to the manifold M^-. Now we have the decomposition

$$x_0 = x_0' + x_0'', \tag{7.12}$$

which proves the inequalities $\|x_0'\| \ne 0$, $\|x_0''\| \ne 0$ since we assumed that $(\phi_0, x_0) \notin M^+$ and $(\phi_0, x_0) \notin M^-$.

Consider the function $x_t(\phi_0, x_0)$. According to decomposition (7.12), we have

$$x_t(\phi_0, x_0) = \Omega_0^t(\phi_0)x_0' + \Omega_0^t(\phi_0)x_0'', \quad t \in \mathbf{R}.$$

For $t > 0$ this leads to the estimate

$$\|x_t(\phi_0, x_0)\| \ge \|\Omega_0^t(\phi_0)x_0''\| - \|\Omega_0^t(\phi_0)x_0'\| \ge \|x_0''\|/d(t) - d(t)\|x_0'\|,$$

which proves that $\lim\limits_{t \to +\infty} \|x_t(\phi_0, x_0)\| = +\infty$.

For $t < 0$ we have the similar estimate

$$\|x_t(\phi_0, x_0)\| \geq \|\Omega_0^t(\phi_0)x_0'\| - \|\Omega_0^t(\phi_0)x_0''\| \geq \|x_0'\|/d(t) - d(t)\|x_0''\|,$$

which proves that $\lim\limits_{t \to -\infty} \|x_t(\phi_0, x_0)\| = +\infty$.

Combining the obtained estimates we have the needed relation

$$\lim\limits_{|t| \to +\infty} \|x_t(\phi_0, x_0)\| = +\infty.$$

Since the manifolds M^+ and M^- are invariant and in view of the behaviour of the trajectories of system (7.1), which start on these manifolds and outside of them, the manifolds M^+ and M^- are called *separatrix manifolds* of the system of equations (7.1). The above discussion can be summarized as follows.

Theorem 2. *Suppose that the system of equations* (7.1) *satisfies the conditions:*

1) $a \in C_{\mathrm{Lip}}(\mathcal{T}_m)$, $P \in C(\mathcal{T}_m)$;

2) *the trivial torus* $x = 0$, $\phi \in \mathcal{T}_m$ *is a unique invariant torus of this system;*

3) *there is a Green's function* $G_0(\tau, \phi)$ *for this system satisfying inequality* (7.10).

Then equations (7.2) *and* (7.3) *define separatrix manifolds of system* (7.1) *on which the damping properties of the motions are determined by relations* (7.11) *as* $|t| \to +\infty$.

We now consider the question of the C^S-block decomposability of system (7.1) by relating this problem to the question of global homeomorphism between the manifolds M^+, M^- and the products $\mathcal{T}_m \times E^r$, $\mathcal{T}_m \times E^{n-r}$ respectively.

Suppose that such a homeomorphism exists, so that the matrix $S(\phi)$ defining it belongs to the space $C^S(\mathcal{T}_m) \cap C'(\mathcal{T}_m)$.

We subdivide a vector y into two blocks y_1 and y_2 of dimensions r and $n - r$ respectively: $y = (y_1; y_2)$ and write the equation for the manifold M^+ in terms of the variables $y = S^{-1}(\phi)x$. Using the change of variables formulae, the equation for M^+ takes the form

$$y_2 = 0, \ \phi \in \mathcal{T}_m. \tag{7.13}$$

Since the change of coordinates from x to y is made by means of a "global" change of variables, the invariance of the manifold M^+ is not violated.

The transformed system of equations is the system

$$\frac{d\phi}{dt} = a(\phi), \quad \frac{dy}{dt} = Q(\phi)y, \tag{7.14}$$

where the matrix $Q(\phi)$ is defined by the expression

$$Q(\phi) = S^{-1}(\phi)P(\phi)S(\phi) - S^{-1}(\phi)\dot{S}(\phi), \tag{7.15}$$

and thus the coordinate plane (7.13) is invariant.

But this is possible only if the block $Q_{21}(\phi)$ of the block decomposition of the matrix $Q(\phi)$, which corresponds to the decomposition of y into y_1 and y_2:

$$Q(\phi) = \begin{bmatrix} Q_1(\phi) & Q_{12}(\phi) \\ Q_{21}(\phi) & Q_2(\phi) \end{bmatrix},$$

is zero:

$$Q_{21}(\phi) = 0 \quad \forall \phi \in T_m.$$

The manifold M^- is given in terms of the variables (y_1, y_2) by the equation:

$$y_1 = 0, \quad \phi \in T_m. \tag{7.16}$$

But then the plane (7.16) is also invariant with respect to system (7.14). This necessarily implies the equality

$$Q_{12}(\phi) = 0 \quad \forall \phi \in T_m.$$

From what has been said one can see that the transformed system of equations, system (7.14) has the block diagonal form:

$$\frac{d\phi}{dt} = a(\phi), \quad \frac{dy_1}{dt} = Q_1(\phi)y_1, \quad \frac{dy_2}{dt} = Q_2(\phi)y_2.$$

The properties of the motions of system (7.1) do not change under the transformation to the form (7.14); therefore the solution $\phi_t(\phi_0), y_1(t, \phi_0, y_1^0)$, $y_2(t, \phi_0, y_2^0)$ of system (7.14) that takes the value ϕ_0, y_1^0, y_2^0 at $t = 0$ satisfies the inequalities

$$\|y_1(t, \phi_0, y_1^0)\| \leq K d(t - \tau)\|y_1(\tau, \phi_0, y_1^0)\|, \quad t \geq \tau,$$
$$\|y_2(t, \phi_0, y_2^0)\| \leq K d(\tau - t)\|y_2(\tau, \phi_0, y_2^0)\|, \quad t \leq \tau,$$

for any $\phi_0 \in \mathcal{T}_m$, $y_1^0 \in E^r$, $y_2^0 \in E^{n-r}$, $\tau \in \mathbf{R}$, where K is a positive constant. As a result we have the following statement.

Theorem 3. *Suppose that conditions 1 and 2 of Theorem 2 hold and that there exists a Green's function $G_0(\tau, \phi)$ for system (7.1) such that the matrix $C(\phi) = G_0(0, \phi)$ is reduced to Jordan form by the matrix*

$$S(\phi) \in C^s(\mathcal{T}_m) \cap C'(\mathcal{T}_m).$$

Then the change of variables

$$x = S(\phi)y$$

splits system (7.1) into $C^S(\mathcal{T}_m)$ blocks thus reducing it to system (7.14).

The problem whether the system of equations (7.1) can be split into $C^s(\mathcal{T}_m)$-block form is reduced by Theorem 3 to the algebraic problem of the membership of the eigenvectors of the projection $C(\phi)$ to the space $C^S(\mathcal{T}_m)$. We shall return to a discussion of this point later. For the moment we merely note that this problem is completely solved only for $m = 1$ and for a matrix $C(\phi)$ that has rank $r = n - 1$ [118].

3.8. Sufficient conditions for exponential dichotomy
of an invariant torus

To find out the conditions under which the trivial torus $x = 0$, $\phi \in \mathcal{T}_m$, of system (7.1) is exponentially dichotomous it is convenient to use the machinery of sign-changing Lyapunov functions that are quadratic forms of the variable x. This leads to the following result which gives necessary conditions for the invariant torus to be exponentially dichotomous [117].

Theorem 1. *Let $a \in C_{\mathrm{Lip}}(\mathcal{T}_m)$, $P \in C(\mathcal{T}_m)$ and suppose that there exists a non-singular symmetric matrix $S(\phi) \in C'(\mathcal{T}_m)$ such that the matrix*

$$\hat{S}(\phi) = \dot{S}(\phi) + S(\phi)P(\phi) + P^*(\phi)S(\phi)$$

is negative-definite for all $\phi \in \mathcal{T}_m$. Then the trivial torus $x = 0$, $\phi \in \mathcal{T}_m$ of system (7.1) is exponentially dichotomous.

To prove the theorem, we assume that the matrix $S(\phi)$ has r positive and $n - r$ negative eigenvalues.

Let

$$S_t = (\Omega_0^t(\phi))^* S(\phi_t(\phi))\Omega_0^t(\phi),$$

and let K_t be the set of points x of E^n satisfying the inequality

$$\langle S_t x, x \rangle \geq 0, \tag{8.1}$$

where ϕ and t are fixed numbers, $\phi \in T_m$, $t \geq 0$.

We set

$$K^+ = \bigcap_{t \geq 0} K_t$$

and prove that K^+ is an r-dimensional linear subspace of E^n, which plays the role of the space E^+ in the definition of the dichotomy of the trivial torus for system (7.1).

Using the expressions for matrices S_t and $\hat{S}(\phi)$ we find that

$$\frac{dS_t}{dt} = (\Omega_0^t(\phi))^* \hat{S}(\phi_t(\phi))\Omega_0^t(\phi).$$

Thus for every $x \in E^n$, $x \neq 0$,

$$\frac{d\langle S_t x, x \rangle}{dt} = \langle \hat{S}(\phi_t(\phi))\Omega_0^t(\phi)x, \Omega_0^t(\phi)x \rangle =$$

$$= \langle \hat{S}(\phi_t(\phi))x_t(\phi, x), x_t(\phi, x) \rangle,$$

and, since the matrix $\hat{S}(\phi)$ is negative definite, this leads to the inequality

$$\langle S_t x, x \rangle < \langle S_\tau x, x \rangle \quad \forall t > \tau, \ \forall \tau \in \mathbf{R}. \tag{8.2}$$

It follows from inequality (8.2) that the set K_t is contained in K_τ for $t > \tau : K_t \subset K_\tau$ and that the intersection of the boundary ∂K_t of K_t with the boundary ∂K_τ of K_τ contains only the point zero: $\partial K_t \cap \partial K_\tau = \{0\}$.

Thus, K^+ is the intersection of sets K_t that are totally ordered by inclusion. We show that K^+ has r linearly independent vectors.

Since the matrix S_t is continuous with respect to t and non-singular, it has the same number of positive eigenvalues as the matrix $S_0 = S(\phi)$. Being symmetric, it is orthogonally similar to its Jordan form $\mathfrak{F}_t$ which can be represented in the form

$$\mathfrak{F}_t = \text{diag}\{D_1(t), -D_2(t)\},$$

where $D_1(t) = \text{diag}\{\lambda_1(t), \ldots, \lambda_r(t)\}$, $D_2(t) = \text{diag}\{\lambda_{r+1}(t), \ldots, \lambda_n(t)\}$, and the $\lambda_j(t)$ are positive numbers.

But then

$$S_t = O^*(t)\mathfrak{F}_t O(t) = O_1^*(t)\text{diag}\{E_1, -E_2\}O_1(t) \qquad (8.3)$$

where $O(t)$ is an orthogonal matrix, and E_1 and E_2 are the $(r \times r)$ and $(n - r) \times (n - r)$ identity matrices respectively,

$$O_1(t) = O(t)\text{diag}\{D_1^{1/2}(t), D_2^{1/2}(t)\}.$$

Under the linear coordinate transformation according to the formulae

$$x = By, \quad \det B \neq 0,$$

the form $\langle S_t x, x \rangle$ will be evidently transformed into the form $\langle \widetilde{S}_t y, y \rangle$ with matrix

$$\widetilde{S}_t = B^* S_t B.$$

By making an appropriate choice of the coordinate system in E^n we can transform the quadratic form with matrix $S_0 = S(\phi)$ to the form with matrix $\widetilde{S}_0 = \text{diag}\{E_1, -E_2\}$. Without loss of generality we can suppose that the coordinate system x has already been chosen in such a way that the quadratic form $\langle S(\phi)x, x \rangle$ can be written as $x_1^2 + \ldots + x_r^2 - x_{r+1}^2 - \ldots - x_n^2$. That is, we can suppose that

$$S(\phi) = \text{diag}\{E_1, -E_2\}.$$

We now consider the intersection of the set K_t and the plane P_0 which is given by the equations

$$x_1 = 0, \ldots, \ x_j = 1, \ldots, \ x_r = 0. \qquad (8.4)$$

The set $K_0 \cap P_o$ consists of the points satisfying the inequality

$$x_{r+1}^2 + \ldots + x_n^2 \leq 1,$$

and so it is a closed ball III_0 in an $(n - r)$-dimensional subspace of E^n which is the plane P_0.

The set $K_t \cap P_0$ consists of the points satisfying simultaneously inequality (8.1) and equalities (8.4). We claim that the set $K_t \cap P_0$ is not empty. To see this we write inequality (8.1) defining K_t in the coordinate system y related to x by the formula

$$y = O(t)x,$$

where $O(t)$ is the matrix from the representation of S_t in the form (8.3).

We have

$$\langle S_t x, x \rangle = \langle \mathfrak{F}_t O(t)x, O(t)x \rangle = \langle \mathfrak{F}_t y, y \rangle =$$
$$= \lambda_1(t)y_1^2 + \ldots + \lambda_r(t)y_r^2 - \lambda_{r+1}(t)y_{r+1}^2 - \ldots - \lambda_n(t)y_n^2 \geq 0,$$

or

$$\lambda_1(t)y_1^2 + \ldots + \lambda_r(t)y_r^2 \geq \lambda_{r+1}(t)y_{r+1}^2 + \ldots + \lambda_n(t)y_n^2. \qquad (8.5)$$

Since the functions $\lambda_j(t)$ are positive, the values

$$y_1 = c_1, \ldots, \ y_r = c_r, \ y_{r+1} = \ldots = y_n = 0,$$

where $c_1, \ldots, c_r$ are arbitrary constants, satisfy inequality (8.5).

Thus the set K_t contains the subspace $E^r(t)$ defined by the equation

$$x = O(t) \begin{bmatrix} c \\ 0 \end{bmatrix}, \ c = \text{column}(c_1, \ldots, c_r)$$

or by the equation

$$x = \begin{bmatrix} O_1(t) \\ O_{21}(t) \end{bmatrix} c,$$

where O_1 and O_{21} are the $r \times r$ and $n \times r$ matrices respectively formed by the first r columns of the matrix $O(t)$.

Using the inclusion $K_t \subset K_0$ we find that the linear subspace $E^r(t)$ is contained in the set

$$x_1^2 + \ldots + x_r^2 \geq x_{r+1}^2 + \ldots + x_n^2.$$

But then the matrix $O_1(t)$ is non-singular:

$$\det O_1(t) \neq 0,$$

since otherwise, a point of $E^r(t)$ defined by a non-zero c with $O_1(t)c = 0$ (which, in view of the relation $O(t)\begin{bmatrix} c \\ 0 \end{bmatrix} \neq 0$, entails the inequality $O_{21}(t)c \neq 0$) does not belong to the set K_0:

$$0 \geq \sum_{j=1}^{n-r} (O_{21}(t)c)_j^2,$$

which is impossible.

Because the matrix $O_1(t)$ is non-singular, there exist a vector $O_1^{-1}(t)e_j$, where $e_j = (0, \ldots, 1, \ldots, 0)$ is the r-dimensional unit vector, and a point

$$x = \begin{bmatrix} O_1(t) \\ O_{21}(t) \end{bmatrix} O_1^{-1}(t)e_j = \begin{bmatrix} e_j \\ O_{21}(t)O_1^{-1}(t)e_j \end{bmatrix},$$

belonging to the intersection of the planes P_0 and $E^r(t)$ so that it also belongs to the set $K_t \cap P_0$. Thus the set $K_t \cap P_0$ is non-empty. We show that $K_t \cap P_0$ either consists of a single point or it is homeomorphic to a closed ball in an $(n-r)$-dimensional subspace of E^n, namely, the plane P_0.

To do this we write inequality (8.2) for $\tau = 0$:

$$x_1^2 + \ldots + x_r^2 - x_{r+1}^2 - \ldots - x_n^2 \geq \langle S_t x, x \rangle, \quad t > 0. \tag{8.6}$$

If we write S_t in the block form

$$S_t = \begin{bmatrix} S_1 & S_{12} \\ S_{21} & S_2 \end{bmatrix},$$

and regard S_1 as an $r \times r$ matrix, then using inequality (8.6) we see that the matrix $(S_2 + S_2^*)/2$ is negative-definite:

$$\langle S_2 \bar{x}_2, \bar{x}_2 \rangle < -\|\bar{x}_2\|^2$$

for each vector $\bar{x}_2 = (x_{r+1}, \ldots, x_n) \neq 0$.

But then the matrix $(S_2 + S_2^*)/2$ can be represented in the form

$$(S_2 + S_2^*)/2 = -O_2^* O_2,$$

where O_2 is a non-singular matrix and the change of variables

$$y_2 = O_2 \bar{x}_2, \quad y_2 = (y_{r+1}, \ldots, y_n)$$

reduces the quadratic form $\langle S_2 \bar{x}_2, \bar{x}_2 \rangle$ to the form

$$\langle S_2 \bar{x}_2, \bar{x}_2 \rangle = -\langle O_2 \bar{x}_2, O_2 \bar{x}_2 \rangle = -\|y_2\|^2.$$

Setting $x = (\bar{x}_1, \bar{x}_2)$, $\bar{x}_1 = (x_1, \ldots, x_r)$, we rewrite the inequality defining K_t in the form

$$\langle S_1 \bar{x}_1, \bar{x}_1 \rangle + \langle (S_{12}^* + S_{21}) \bar{x}_1, \bar{x}_2 \rangle + \langle S_2 \bar{x}_2, \bar{x}_2 \rangle \geq 0. \tag{8.7}$$

Thus, the set $K_t \cap P_0$ consists of the points satisfying the inequality

$$\langle S_1 e_j, e_j \rangle + \langle (S_{12}^* + S_{21}) e_j, \bar{x}_2 \rangle + \langle S_2 \bar{x}_2, \bar{x}_2 \rangle \geq 0. \qquad (8.8)$$

In the coordinate system on P_0 chosen according to (8.7) inequality (8.8) becomes

$$\|y_2\|^2 - \langle (S_{12}^* + S_{21}) e_j, \ O_2^{-1} y_2 \rangle - \langle S_1 e_j, e_j \rangle \leq 0,$$

or

$$(y_{r+1} - \beta_1)^2 + \ldots + (y_n - \beta_{n-r})^2 \leq f, \qquad (8.9)$$

where $\beta_1, \ldots, \beta_{n-r}$, f are functions of t depending on e_j. Since the set $K_t \cap P_0$ is non-empty, there is at least one value of y_2 satisfying inequality (8.9), and this is possible only if $f \geq 0$.

Let $f = 0$ for some $t = t_0$. Then the set $K_{t_0} \cap P_0$ consists of the single point

$$\bar{x}_1 = e_j, \quad y_{r+\nu} = \beta_\nu, \quad \beta_\nu = \beta_\nu(t_0), \quad \nu = 1, \ldots, n - r. \qquad (8.10)$$

Since $K_t \subset K_{t_0}$ when $t > t_0$, it follows that $(K_t \cap P_0) \subset (K_{t_0} \cap P_0)$ when $t > t_0$, so that the set $K_t \cap P_0$ contains the point (8.10) and only this point when $t > t_0$.

If $f > 0$, then the set (8.9) is a closed ball in the space P_0, so that $K_t \cap P_0$ is homeomorphic to a closed ball $Ш_t$ of an $(n - r)$-dimensional subspace of E^n, namely, the plane P_0.

Since $Ш_t \subset Ш_\tau$ when $t > \tau$, the set $\bigcap\limits_{t>0} Ш_t$ consists either of a single point (8.10) or is the intersection of a family of closed non-empty "balls" that is totally ordered by inclusion and therefore contains at least one point

$$\bar{x}_1 = e_j, \quad \bar{x}_1 = x_2^0.$$

In both cases the set $K^+ \cap P_0$ contains at least one point. By changing the unit vector e_j and setting $j = 1, 2, \ldots, r$ we obtain r points belonging to the set K^+.

Let

$$x_1, \ldots, x_r \qquad (8.11)$$

be these points. Since $x_j = (e_j, x_{2,j}^0 \ (j = 1, \ldots, r)$, it follows that the vectors joining the origin and points (8.11) are linearly independent. Moreover, since the segments joining the origin and points (8.11) are formed by the points of the form μx_j, $0 \le \mu \le 1$, they belong to the set K^+.

Let x be a point of K^+. We prove that the function

$$x_t = \Omega_0^t(\phi)x$$

satisfies the inequality

$$\|x_t\| \le Ke^{-\gamma(t-\tau)}\|x_\tau\| \quad \forall t > \tau, \quad \forall \tau \in \mathbf{R}, \tag{8.12}$$

where K and γ are positive constants not dependent on the choice of the point ϕ.

It follows from the inclusion $x \in K^+$ that

$$\langle S_t x, x \rangle \ge 0 \quad \forall t > 0.$$

This inequality together with inequality (8.2) implies that

$$\langle S_t x, x \rangle \ge 0 \quad \forall t \in \mathbf{R}.$$

To obtain estimate (8.12) we set

$$V_\epsilon(t) = \langle S_t x, x \rangle + \epsilon \langle x_t, x_t \rangle$$

considering ϵ to be a small positive constant. For an arbitrary function $V_\epsilon(t)$ we then have the estimate

$$\frac{dV_\epsilon(t)}{dt} = \langle \hat{S}(\phi_t(\phi))x_t, x_t \rangle + \epsilon\langle(P(\phi_t(\phi)) + P^*(\phi_t(\phi)))x_t, x_t \rangle \le$$

$$\le -\gamma_1\|x_t\|^2 + \epsilon M\|x_t\|^2 = -(\gamma_1 - \epsilon M)\|x_t\|^2 = -\gamma_2\|x_t\|^2,$$

where

$$M = \max_{\phi \in T_m, \|x\|=1} |\langle(P)\phi) + P^*(\phi))x, x \rangle|, \quad \gamma_2 = \gamma_1 - \epsilon M > 0.$$

Since also

$$V_\epsilon(t) \le \langle S_t x, x \rangle + \epsilon\|x_t\|^2 =$$

$$= \langle S(\phi_t(\phi))x_t, x_t \rangle + \epsilon\|x_t\|^2 \le (M_1 + \epsilon)\|x_t\|^2,$$

$$\|x_t\|^2 \le \frac{\langle S_t x, x \rangle + \epsilon\|x_t\|^2}{\epsilon} = \frac{V_\epsilon(t)}{\epsilon}, \tag{8.13}$$

where $M_1 = \max\limits_{\phi \in T_m, \|x\|=1} |\langle S(\phi)x, x\rangle|$, we have

$$\frac{dV_\epsilon(t)}{dt} \leq -\gamma_2 \|x_t\|^2 \leq -\frac{\gamma_2}{M_1 + \epsilon} V_\epsilon(t) = -2\gamma V_\epsilon(t).$$

Upon integrating this inequality we obtain

$$V_\epsilon(t) \leq V_\epsilon(\tau)e^{-2\gamma(t-\tau)} \quad \forall t > \tau$$

or, by using inequalities (8.13),

$$\|x_t\|^2 \leq \frac{V_\epsilon(t)}{\epsilon} \leq \left(\frac{V_\epsilon(\tau)}{\epsilon}\right)e^{-2\gamma(t-\tau)} \leq \left(1 + \frac{M_1}{\epsilon}\right)e^{-2\gamma(t-\tau)} \times$$

$$\times \|x_\tau\|^2 \leq K^2 e^{-2\gamma(t-\tau)}\|x_\tau\|^2 \quad \forall t > \tau, \forall \tau \in \mathbf{R}.$$

Since M_1 and γ are positive constants and do not depend on the choice of ϕ, the last inequality is the required inequality (8.12).

We denote the subspace of E^n spanned by the vectors (8.11) by E^r. We claim that $E^r \subset K^+$.

For if not, then let

$$y = \sum_{j=1}^r \mu_j x_j$$

be a linear combination of vectors (8.11) not belonging to K^+. It follows from $y \notin K^+$ that one can find $t_0 > 0$ such that

$$\langle S_t y, y\rangle < 0 \quad \forall t \geq t_0.$$

But then using inequality (8.2) we find that

$$\langle -S_t y, y\rangle > \langle -S_{t_0} y, y\rangle = \text{const} > 0 \quad \forall t > t_0. \tag{8.14}$$

On the other hand,

$$\langle S_t y, y\rangle = \sum_{j=1}^r \mu_j \langle S_t x_j, x_j\rangle = \sum_{j=1}^r \mu_j \langle S(\phi_t(\phi))x_t(x_j), x_t(x_j)\rangle,$$

where $x_t(x_j) = \Omega_0^t(\phi)x_j$ satisfies estimate (8.12) because $x_j \in K^+$ and thus

$$\lim_{t \to +\infty} x_t(x_j) = 0. \tag{8.15}$$

Because the matrix $S(\phi_t(\phi))$ is bounded for $t > 0$, by using relation (8.15) we find that

$$\lim_{t \to +\infty} \langle -S_t y, y \rangle = 0.$$

This contradicts inequality (8.14). The contradiction proves that $E^r \subset K^+$.

Since $E^r \subset K^+$, the solution $\phi_t(\phi)$, $x_t(\phi, x) = \Omega_0^t(\phi)x$ of system (7.1) with $x \in E^r$ satisfies inequality (8.12) and thus E^r can be considered as the space E^+ featuring in the definition of the exponential dichotomy of the trivial torus for system (7.1).

A similar argument for the function $-\langle S_t x, x \rangle$ establishes without difficulty that there exists an $(n-r)$-dimensional space E^{n-r} in the intersection K^- of the sets $\widetilde{K}_t$ formed by the points $x \in E^n$ for which

$$\langle S_t x, x \rangle \leq 0, \quad t \leq 0.$$

For each point $x \in K^-$ the function $x_t = \Omega_0^t(\phi)x$ satisfies the inequality

$$\|x_t\| \leq K e^{\gamma(t-\tau)} \|x_\tau\| \quad \forall t \leq \tau, \ \forall \tau \in \mathbf{R}. \tag{8.16}$$

This is true because the derivative of the function

$$V_\epsilon^1(t) = -\langle S_t x, x \rangle + \epsilon \|x_t\|^2$$

satisfies the following inequality for small positive ϵ:

$$\frac{dV_\epsilon^1(t)}{dt} = -\langle \hat{S}(\phi_t(\phi))x_t, x_t \rangle + \epsilon \langle (P(\phi_t(\phi))+$$

$$+P^*(\phi_t(\phi)))x_t, x_t \rangle \geq (\gamma_1 - \epsilon M)\|x_t\|^2 = \gamma_2 \|x_t\|^2 \geq$$

$$\geq \frac{\gamma_2}{M_1 + \epsilon} V_\epsilon^1(t) = 2\gamma V_\epsilon^1(t) \quad \forall t \in \mathbf{R},$$

which on integration yields the estimate

$$V_\epsilon^1(\tau) \geq e^{2\gamma(\tau-t)} V_\epsilon^1(t) \quad \forall r \leq \tau, \ \forall \tau \in \mathbf{R},$$

which in turn leads to inequality (8.16):

$$\|x_t\|^2 \leq \frac{V_\epsilon^1(t)}{\epsilon} \leq \left(\frac{V_\epsilon^1(\tau)}{\epsilon}\right) e^{2\gamma(t-\tau)} \leq$$

$$\leq \left(1 + \frac{M_1}{\epsilon}\right) e^{2\gamma(t-\tau)} \|x_\tau\|^2 = K^2 e^{2\gamma(t-\tau)} \|x_\tau\|^2 \quad \forall t \leq \tau, \ \forall \tau \in \mathbf{R}.$$

Because of inequality (8.16), the space E^{n-r} can be considered relative to the solutions of system (7.1) as the space E^- featuring in the definition of the exponential dichotomy of the trivial torus for system (7.1).

Since $\langle S(\phi)x, x \rangle > 0$ for points $x \in E^r$, $x \neq 0$ and $\langle S(\phi)x, x \rangle < 0$ for points $x \in E^{n-r}$, $x \neq 0$, it follows that the sets E^r and E^{n-r} intersect in a single point

$$E^r \cap E^{n-r} = \{0\}.$$

The equalities $K^+ = E^r$, $K^- = E^{n-r}$, $E^n = K^+ \oplus K^-$ now follow from the equality $E^n = E^r \oplus E^{n-r}$, inequalities (8.12), (8.16) and the relation $\lim\limits_{|t| \to +\infty} \|\Omega_0^t(\phi)x\| = +\infty$, $x \notin E^r \cup E^{n-r}$ which can be obtained by applying an argument similar to that used to obtain the relation $\lim\limits_{|t| \to +\infty} \|x_t(\phi_0, x_0)\| = +\infty$ for $(\phi_0, x_0) \notin M^+ \cup M^-$ in §3.7.

Since the choice of $\phi \in T_m$ was arbitrary and K, γ do not depend on ϕ, the invariant torus $x = 0$, $\phi \in T_m$ of system (7.1) is exponentially dichotomous. This completes the proof of the theorem.

Remark. It follows from the proof of the theorem that the manifold $M^+\big|_{\phi=\text{const}}$ lies in the "cone" in the space E^n defined by the inequality

$$\langle S(\phi)x, x \rangle > 0, \tag{8.17}$$

for every $\phi \in T_m$, while the manifold $M^-\big|_{\phi=\text{const}}$ lies in the "cone" in the space E^n, defined by the inequality

$$\langle S(\phi)x, x \rangle < 0.$$

3.9. Necessary conditions for an invariant torus to be exponentially dichotomous

We show that the conditions given in the preceding section as sufficient are also necessary for exponential dichotomy of the trivial torus $x = 0$, $\phi \in T_m$, of system (7.1). Thus, we shall make the assumption that the torus $x = 0$, $\phi \in T_m$ is an exponentially dichotomous invariant torus of system (7.1). For each $\phi \in T_m$, the space E^n can be decomposed into a direct sum of the spaces $E^r(\phi)$ and $E^{n-r}(\phi) : E^n = E^r(\phi) \oplus E^{n-r}(\phi)$ such that the solution $x_t(\phi, x_0) = \Omega_0^t(\phi)x_0$ of the system of equations

$$\frac{dx}{dt} = P(\phi_t(\phi))x, \tag{9.1}$$

taking the value $x_0 \in E^r(\phi)$ at $t = 0$, satisfies the estimate

$$\|x_t(\phi, x_0)\| \leq K e^{-\gamma(t-\tau)} \|x_\tau(\phi, x_0)\| \quad (t \geq \tau, \ t, \tau \in \mathbf{R}),$$

while the solution taking the value $x_0 \in E^{n-r}(\phi)$ satisfies the estimate

$$\|x_t(\phi, x_0)\| \leq K e^{\gamma(t-\tau)} \|x_\tau(\phi, x_0)\| \quad (t \leq \tau, \ t, \tau \in \mathbf{R}),$$

where K, γ are positive constants independent of $\phi \in T_m$.

The above property of solutions of system (9.1) characterizes it as an exponentially dichotomous system on the entire axis $\mathbf{R} = (-\infty, \infty)$.

Denote the projections corresponding to the decomposition of E^n into the direct sum of the spaces $E^r(\phi)$ and $E^{n-r}(\phi)$ by $C(\phi)$ and $C_1(\phi) = E - C(\phi)$. Then the function

$$G_t(\tau, \phi) = \begin{cases} \Omega_0^t(\phi) C(\phi) \Omega_\tau^0(\phi), & t \geq \tau, \\ -\Omega_0^t(\phi) C_1(\phi) \Omega_\tau^0(\phi), & t < \tau \end{cases} \tag{9.2}$$

satisfies the estimate [42]

$$\|G_t(\tau, \phi)\| \leq K_1 e^{-\gamma|t-\tau|}, \quad t, \tau \in \mathbf{R}. \tag{9.3}$$

We use this function to determine the properties of the matrix $C(\phi)$.

Lemma 1. *Let* $a \in C_{\mathrm{Lip}}(T_m)$, $P \in C_{\mathrm{Lip}}(T_m)$ *and suppose that the function* (9.2) *satisfies* (9.3). *Then the matrix* $C(\phi)$ *belongs to the space* $C(T_m)$.

To prove this, we consider the matrix

$$Z_t(\phi, \bar{\phi}) = \begin{cases} \Omega_0^t(\phi) C(\phi) - \Omega_0^t(\bar{\phi}) C(\bar{\phi}), & t \geq 0, \\ -\Omega_0^t(\phi) C_1(\phi) + \Omega_0^t(\bar{\phi}) C_1(\bar{\phi}), & t < 0, \end{cases} \tag{9.4}$$

which equals the matrix $C(\phi) - C(\bar{\phi})$ at $t = 0$. It is continuous with respect to t, satisfies the inequality

$$\|Z_t(\phi, \bar{\phi})\| \leq 2K_1 e^{-\gamma|t|}, \quad t \in \mathbf{R} \tag{9.5}$$

and is a solution of the differential equation

$$\frac{dZ}{dt} = P(\phi_t(\bar{\phi})) Z + [P(\phi_t(\phi)) - P(\phi_t(\bar{\phi}))] G_t(0, \phi). \tag{9.6}$$

Because system (9.1) is exponentially dichotomous, a solution of system (9.6) that is bounded on **R** is unique and can be represented in the form of the integral

$$Z_t(\phi, \bar{\phi}) = \int_{-\infty}^{\infty} G_t(\tau, \bar{\phi})[P(\phi_\tau(\phi)) - P(\phi_\tau(\bar{\phi}))]G_\tau(0, \phi)d\tau. \tag{9.7}$$

Hence it follows that

$$C(\phi) - C(\bar{\phi}) = \int_{-\infty}^{\infty} G_0(\tau, \bar{\phi})[P(\phi_\tau(\phi)) - P(\phi_\tau(\bar{\phi}))]G_\tau(0, \phi)d\tau.$$

We find an estimate for the integrand in formula (9.7). Let

$$\|a(\phi) - a(\bar{\phi})\| \le \alpha\|\phi - \bar{\phi}\|, \quad \|P(\phi) - P(\bar{\phi})\| \le \beta\|\phi - \bar{\phi}\|$$

for any $\phi, \bar{\phi} \in T_m$ and some positive α and β.

One can easily derive from system of equations (7.1) the following estimates

$$\|\phi_t(\phi) - \phi_t(\bar{\phi})\| \le e^{\alpha|t|}\|\phi - \bar{\phi}\|,$$

$$\|\phi_t(\phi) - \phi_t(\bar{\phi})\| \le \|\phi - \bar{\phi}\| + 2|a|_0|t|,$$

as a result of which we obtain

$$\|\phi_t(\phi) - \phi_t(\bar{\phi})\| \le e^{\alpha|t|/(\nu+1)}(\|\phi - \bar{\phi}\| + 2|a|_0|t|)^{\nu/(\nu+1)}\|\phi - \bar{\phi}\|^{1/(\nu+1)}$$

for all $t \in \mathbf{R}$ and arbitrary $\nu \ge 0$.

Applying inequality (9.3) to the function $G_t(\tau, \phi)$, this leads to the estimate

$$\|Z_t(\phi, \bar{\phi})\| \le \beta K_1^2\|\phi - \bar{\phi}\|^{1/(\nu+1)} \int_{-\infty}^{\infty} \exp\Big\{-\gamma(|t - \tau| + |\tau|)+$$

$$+\frac{a}{\nu + 1}|\tau|\Big\}(2|a|_0|\tau| + \|\phi - \bar{\phi}\|)^{\nu/(\nu+1)}d\tau, \tag{9.8}$$

which implies that the integral (9.7) is convergent for a sufficiently large ν: $\nu + 1 > \alpha/\gamma$ and that the inequality for $Z_0(\phi, \bar{\phi}) = C(\phi) - C(\bar{\phi})$

$$\|C(\phi) - C(\bar{\phi})\| \le \beta K_1^2 M\|\phi - \bar{\phi}\|^{1/(\nu+1)} \tag{9.8}$$

holds. Here M is a positive constant equal to the value of the integral for $t = 0$ in inequality (9.8) and $\|\phi - \bar{\phi}\| = 2\pi\sqrt{m}$. Inequality (9.9) proves that the matrix $C(\phi)$ is continuous with respect to ϕ for $\phi \in T_m$. It follows from the definition of the matrix $C(\phi)$ that it is periodic with respect to ϕ. Hence, $C(\phi) \in C(T_m)$, as required.

We now consider the function

$$G_0(\tau, \phi) = \begin{cases} \Omega_\tau^0(\phi)C(\phi_\tau(\phi)), & \tau \le 0, \\ -\Omega_\tau^0(\phi)C_1(\phi_\tau(\phi)), & \tau > 0, \end{cases} \tag{9.10}$$

where $C(\phi)$ is taken to be the matrix in formula (9.2). It can be seen from formulae (9.2) and (9.10) that

$$G_0(\tau, \phi) = G_t(0, \phi_\tau(\phi))\big|_{t=-\tau},$$

and so the inequality

$$\|G_0(\tau, \phi)\| \le K_1 e^{-\gamma|\tau|}, \quad \tau \in \mathbf{R}, \tag{9.11}$$

holds.

Because $C(\phi) \in C(T_m)$, the function (9.10) is a Green's function for system (7.1). Because system (9.1) does not have bounded solutions on $\mathbf{R}$ other than the trivial one, this function is unique, and the equality

$$G_0(\tau, \phi) = \begin{cases} C(\phi)\Omega_\tau^0(\phi), & \tau \le 0 \\ -C_1(\phi)\Omega_\tau^0(\phi), & \tau > 0 \end{cases} = G_t(\tau, \phi)\big|_{t=0}$$

holds, which proves that $G_0(\tau, \phi)$ can be obtained from function (9.2) for $t = 0$. This proves that the notations of (9.2) and (9.10) agree.

Thus, the exponential dichotomy of the trivial torus $x = 0$, $\phi \in T_m$, of system (7.1) implies that a Green's function for this system exists and is unique, and is exponentially damped as $|\tau| \to +\infty$, in accordance with (9.11).

We shall use properties of the functions $G_t(\tau, \phi)$ and $G_0(\tau, \phi)$ to prove the main statement of this section.

Theorem 1. *Let $a \in C_{\text{Lip}}(T_m)$, $P \in C_{\text{Lip}}(T_m)$ and suppose that the trivial torus of system of equations (7.1) is exponentially dichotomous. Then there exists a non-singular symmetric matrix $S(\phi) \in C'(T_m)$ such that the matrix*

$$\hat{S}(\phi) = \dot{S}(\phi) + S(\phi)P(\phi) + P^*(\phi)S(\phi)$$

is negative-definite for all $\phi \in T_m$.

To prove the theorem we define the matrix

$$S(\phi) = S_1(\phi) - S_2(\phi), \tag{9.12}$$

by setting

$$S_1(\phi) = \int_0^\infty C^*(\phi)(\Omega_0^\tau(\phi))^* \Omega_0^\tau(\phi) C(\phi) d\tau,$$

$$S_2(\phi) = \int_{-\infty}^0 C_1^*(\phi)(\Omega_0^\tau(\phi))^* \Omega_0^\tau(\phi) C_1(\phi) d\tau. \tag{9.13}$$

Inequality (9.3) ensures that the integrals (9.4) converge uniformly. And this leads to the inclusions $S_i(\phi) \in C(\mathcal{T}_m)$ for $i = 1, 2$ because the matrices $C(\phi)$ and $\Omega_0^\tau(\phi)$ belong to the space $C(\mathcal{T}_m)$ for every fixed $\tau \in \mathbf{R}$.

We show that $S_i(\phi) \in C'(\mathcal{T}_m)$. To do this we use the fact that the matrix $C(\phi)$ satisfies equality (6.15) (this follows because the Green's function $G_0(\tau, \phi)$ is unique) to transform the expression for $S_1(\phi_t(\phi))$ in the following way:

$$S_1(\phi_t(\phi)) = \int_0^\infty C^*(\phi_t(\phi))(\Omega_t^{t+\tau}(\phi))^* \Omega_t^{t+\tau}(\phi) C(\phi_t(\phi)) d\tau =$$

$$= \int_0^\infty (\Omega_0^{t+\tau}(\phi) C(\phi) \Omega_t^0(\phi))^* \Omega_0^{t+\tau}(\phi) C(\phi) \Omega_t^0(\phi) d\tau =$$

$$= \int_t^\infty (\Omega_t^0(\phi))^* C^*(\phi)(\Omega_0^\tau(\phi))^* \Omega_0^\tau(\phi) C(\phi) \Omega_t^0(\phi) d\tau. \tag{9.14}$$

It follows from formula (9.14) that the function $S_1(\phi_t(\phi))$ is continuously differentiable with respect to t, therefore S_1 belongs to the space $C'(\mathcal{T}_m)$. For the derivative of $S_1(\phi_t(\phi))$ we have the expression

$$\frac{dS_1(\phi_t(\phi))}{dt} = -C^*(\phi_t(\phi)) C(\phi_t(\phi)) - P^*(\phi_t(\phi)) S_1(\phi_t(\phi)) - S_1(\phi_t(\phi)) P(\phi_t(\phi)),$$

from which it follows that

$$\dot{S}_1(\phi) = -C^*(\phi) C(\phi) - P^*(\phi) S_1(\phi) - S_1(\phi) P(\phi). \tag{9.15}$$

One can similarly transform $S_2(\phi_t(\phi))$ to the form

$$S_2(\phi_t(\phi)) = \int_{-\infty}^t (\Omega_t^0(\phi))^* C_1^*(\phi)(\Omega_0^\tau(\phi))^* \Omega_0^\tau(\phi) C_1(\phi) \Omega_t^0(\phi) d\tau,$$

from which it follows that $S_2 \in C'(T_m)$ and

$$\dot{S}_2(\phi) = C_1^*(\phi)C_1(\phi) - P^*(\phi)S_2(\phi) - S_2(\phi)P(\phi). \tag{9.16}$$

Equalities (9.15), (9.16) imply that the matrix $\hat{S}(\phi)$ is negative definite. Indeed, since

$$
\begin{aligned}
\hat{S} &= S + SP + P^*S = \dot{S}_1 - \dot{S}_2 + SP + P^*S = \\
&= -C^*C - P^*S_1 - S_1P - C_1^*C_1 + P^*S_2 + S_2P + SP + P^*S = \\
&= -C^*C - C_1^*C_1 - P^*(S_1 - S_2) - (S_1 - S_2)P + P^*S + SP = \\
&= -C^*C - C_1^*C_1,
\end{aligned}
$$

it follows that

$$\langle \hat{S}x, x \rangle = -\langle Cx, Cx \rangle - \langle C_1x, C_1x \rangle \leq -(\|Cx\|^2 + \|C_1x\|^2) \leq -\tfrac{1}{2}\|x\|^2,$$

which means that the matrix $\hat{S}(\phi)$ is negative definite for all $\phi \in T_m$.

To complete the proof we need to show that the matrix $S(\phi)$ is non-singular for $\phi \in T_m$. Taking into consideration the fact that $C^2(\phi) = C(\phi)$ and $\|C(\phi)\| = \|\Omega_r^0(\phi)\Omega_0^\tau(\phi)C(\phi)\| \leq \|\Omega_r^0(\phi)\| \times \|\Omega_0^\tau(\phi)C(\phi))\| \leq \|\Omega_r^0(\phi)\|Ke^{-\gamma\tau}$ for $\tau \geq 0$, we have the following estimates for $\langle S_1(\phi)x, x \rangle$:

$$\langle S_1(\phi)x, x \rangle = \int_0^\infty \|\Omega_0^\tau(\phi)C(\phi)x\|^2 d\tau \leq$$

$$\leq \int_0^\infty \|\Omega_0^\tau(\phi)C(\phi)\|^2 d\tau \|C(\phi)x\|^2 \leq$$

$$\leq \frac{K^2}{2\gamma}\|C(\phi)x\|^2 \leq \lambda_2\|C(\phi)x\|^2,$$

$$\langle S_1(\phi)x, x \rangle \geq \int_0^\infty \frac{d\tau}{\|\Omega_\tau^0(\phi)\|^2}\|C(\phi)x\|^2 \geq$$

$$\geq \int_0^1 \frac{d\tau}{\|\Omega_\tau^0(\phi)\|^2}\|C(\phi)x\|^2 \geq \lambda_1\|C(\phi)x\|^2.$$

By combining these estimates we obtain the inequality

$$\lambda_1\|C(\phi)x\|^2 \leq \langle S_1(\phi)x, x \rangle \leq \lambda_2\|C(\phi)x\|^2, \tag{9.17}$$

which holds for all $x \in E^n$, $\phi \in T_m$ and some positive λ_1 and λ_2.

In similar fashion we obtain an inequality for $\langle S_2(\phi)x, x\rangle$ in the form:

$$\lambda_3\|C_1(\phi)x\|^2 \leq \langle S_2(\phi)x, x\rangle \leq \lambda_4\|C_1(\phi)x\|^2,$$

which holds for all $x \in E^n$, $\phi \in T_m$ and some positive λ_3 and λ_4.

Let $T(\phi)$ be the matrix that reduces $C(\phi)$ to diagonal form:

$$C(\phi) = T(\phi)\mathfrak{F}T^{-1}(\phi), \quad \mathfrak{F} = \text{diag}\{E_1, 0\},$$

where E_1 is the $r \times r$ identity matrix.

By applying to inequalities (9.17) the change of variables according to the formulae

$$x = T(\phi)y$$

and taking into account the fact that

$$\|\mathfrak{F}y\| = \|T^{-1}CTy\| \leq \|T^{-1}\|\,\|CTy\| \leq \|T^{-1}\|\,\|T\|\,\|\mathfrak{F}y\|,$$

we obtain from inequalities (9.17) the new inequalities

$$\lambda_1\|T^{-1}\|^{-2}\|\mathfrak{F}y\|^2 \leq \lambda_1\|CTy\|^2 \leq$$

$$\leq \langle T^*S_1Ty, y\rangle \leq \lambda_2\|CTy\|^2 \leq \lambda_2\|T\|^2\|\mathfrak{F}y\|^2,$$

from which it follows that the matrix T^*S_1T has the block-diagonal form

$$T^*S_1T = \text{diag}\{D_1, 0\},$$

where D_1 is an $r \times r$ positive-definite matrix for any $\phi \in T_m$.

Similar reasoning for the matrix $S_2(\phi)$ leads to the inequalities

$$\lambda_3\|T^{-1}\|^{-2}\|(E - \mathfrak{F})y\|^2 \leq \langle T^*S_2Ty, y\rangle \leq \lambda_4\|T\|^2\|(E - \mathfrak{F})y\|^2,$$

from which it follows that the matrix T^*S_2T also has the block-diagonal form

$$T^*S_2T = \text{diag}\{0, D_2\},$$

where D_2 is an $(n - r) \times (n - r)$ positive-definite matrix for any $\phi \in T_m$.

The matrix $S(\phi) = S_1(\phi) - S_2(\phi)$ can then be represented as

$$S(\phi) = (T^{-1}(\phi))^*\text{diag}\{D_1(\phi), -D_2(\phi)\}T^{-1}(\phi) \tag{9.18}$$

and has exactly r positive and $n - r$ negative eigenvalues for each $\phi \in T_m$. This proves that the matrix $S(\phi)$ is non-singular.

Remark. It is clear from what has been said that if the trivial torus of system (7.1) is dichotomous, then the matrix $C(\phi)$ defining the Green's function $G_0(\phi, \tau)$ and the matrix $S(\phi)$ defined by equalities (9.12) and (9.13) are related by the property that any matrix $T(\phi)$ that reduces $C(\phi)$ to Jordan form can be used to represent $S(\phi)$ in the form (9.18) with the dimensions of the blocks $D_1(\phi)$ and $D_2(\phi)$ equal to r and $n - r$ respectively.

We now combine the results of this and the previous two sections in the form of a single assertion on the exponential dichotomy of the trivial torus of system (7.1), making it a separate theorem.

Theorem 2. *Let $a \in C_{\mathrm{Lip}}(T_m)$ and $P \in C_{\mathrm{Lip}}(T_m)$. The trivial torus of system of equations (7.1) is exponentially dichotomous if and only if there exists a non-singular symmetric matrix $S(\phi) \in C'(T_m)$ for which the matrix $\hat{S}(\phi) = \dot{S}(\phi) + S(\phi)P(\phi) + P^*(\phi)S(\phi)$ is negative-definite for all $\phi \in T_m$ or if the Green's function (9.10), (9.11) for system (7.1) exists and is unique.*

It follows from Theorem 2 that the necessary and sufficient conditions given there are equivalent for the trivial torus of system (7.1) to be exponentially dichotomous.

3.10. Conditions for the $C'(T_m)$-block decomposability of an exponentially dichotomous system

We shall be considering system (7.1) under the assumption that its trivial torus is exponentially dichotomous. We shall give conditions under which such a system is $C'(T_m)$-block decomposable. If the system in question is $C'(T_m)$-block decomposable, then we shall require that the separatrix manifolds be "straightened", that is, they can be turned into the subspaces E^r and E^{n-r} for any $\phi \in T_m$.

Theorem 1. *For the exponentially dichotomous system of equations (7.1) to be $C'(T_m)$-block decomposable it is necessary and sufficient that there exist a non-singular symmetric matrix $S(\phi) \in C'(T_m)$ for which the matrix $\hat{S}(\phi) = \dot{S}(\phi) + S(\phi)P(\phi) + P^*(\phi)S(\phi)$ is negative definite for all $\phi \in T_m$ and can be* *represented as*

$$S(\phi) = Q(\phi)\mathrm{diag}\{D_1(\phi), -D_2(\phi)\}Q^*(\phi), \qquad (10.1)$$

where $Q(\phi)$ is a non-singular matrix in $C'(T_m)$ and $D_1(\phi)$, $D_2(\phi)$ are symmetric positive-definite matrices for all $\phi \in T_m$.

Necessity. Suppose that the system of equations (7.1) is exponentially dichotomous and $C'(T_m)$-block decomposable. Suppose that the change of variables

$$x = T(\phi)y, \ T \in C'(T_m),$$

reduces system (7.1) to the split form

$$\frac{d\phi}{dt} = a(\phi), \quad \frac{dy_1}{dt} = Q_1(\phi)y_1, \quad \frac{dy_2}{dt} = Q_2(\phi)y_2, \qquad (10.2)$$

where $y = (y_1, y_2)$, $Q_1(\phi)$ is an $r \times r$ matrix and $Q_2(\phi)$ an $(n-r) \times (n-r)$ matrix. Here $Q_i(\phi)$ $(i = 1,2)$ are such that

$$\|\Omega_\tau^t(Q_1)\| \leq Ke^{-\gamma(t-\tau)}, \quad t \geq \tau,$$
$$\|\Omega_\tau^t(Q_2)\| \leq Ke^{\gamma(t-\tau)}, \quad t \leq \tau, \qquad (10.3)$$

where K and γ are positive constants independent of ϕ and r is an arbitrary number in **R**. Because of inequalities (10.3), the system (10.2) has a Green's function in the form

$$\overline{G}_0(\tau, \phi) = \begin{cases} \Omega_\tau^0(\phi)\mathfrak{F}, & \tau \leq 0, \\ -\Omega_\tau^0(\phi)(E - \mathfrak{F}), & \tau > 0, \end{cases}$$

where $\Omega_\tau^0(\phi) = \mathrm{diag}\{\Omega_\tau^0(Q_1), \Omega_\tau^0(Q_2)\}$, $\mathfrak{F} = \mathrm{diag}\{E_1, 0\}$, and E_1 is the $r \times r$ identity matrix.

The initial system of equations (7.1) has a Green's function

$$G_0(\tau, \phi) = T(\phi_\tau(\phi))\overline{G}_0(\tau, \phi)T^{-1}(\phi_\tau(\phi)),$$

the projection $C(\phi) = G_0(0, \phi)$ of which admits the representation

$$C(\phi) = T(\phi)\mathfrak{F}T^{-1}(\phi),$$

where $\mathfrak{F}$ is its Jordan form.

According to the Remark of the previous section, the matrix $S(\phi)$ constructed in the proof of Theorem 1 of §3.9 admits the representation (9.18):

$$S(\phi) = (T^{-1}(\phi))^{*}\mathrm{diag}\{D_1(\phi), -D_2(\phi)\}T^{-1}(\phi)$$

with blocks $D_1(\phi)$ and $D_2(\phi)$ of dimensions r and $n - r$ respectively that are positive-definite for all $\phi \in T_m$. By setting $Q(\phi) = (T^{-1}(\phi))^{*}$ we obtain for $S(\phi)$ representation (10.1).

Sufficiency. Suppose that there exists a matrix $S(\phi)$ that satisfies the conditions of Theorem 1. Then by Theorem 2 of §3.9 there exists a unique Green's function $G_0(\tau, \phi)$ for system (7.1). Thus, the matrix $C(\phi) = G_0(0, \phi)$ is a projection for any $\phi \in T_m$ and belongs to the space $C'(T_m)$. This is sufficient to reduce the $C'(T_m)$-block decomposability problem to one of converting the matrix $C(\phi)$ to Jordan form $\mathfrak{F} = \mathrm{diag}\{E_1, 0\}$ using a matrix $T(\phi)$ in the space $C'(T_m)$.

Since $C(\phi)$ is a projection, the columns of the matrix $T(\phi)$ reducing $C(\phi)$ to Jordan form are composed of linearly independent solutions of the equations

$$C(\phi)x = 0, \tag{10.4}$$

$$C(\phi)x = x \tag{10.5}$$

defining the eigenvectors of the matrix $C(\phi)$.

Equation (10.4) is the same as equation (7.3) which defines the separatrix manifold M^- and equation (10.5) is the same as equation (7.2) which defines the separatrix manifold M^+ of system (7.1). According to the Remark in §3.8, the non-zero solutions of equation (10.4) satisfy inequality

$$\langle S(\phi)x, x \rangle < 0, \tag{10.6}$$

while the non-zero solutions of equation (10.5) satisfy the inequality

$$\langle S(\phi)x, x \rangle > 0. \tag{10.7}$$

Finally, a maximal family of linearly independent solutions of equations (10.4) and (10.5) forms a basis in E^n for any $\phi \in T_m$.

We now show that a maximal family of linearly independent solutions of equations (10.4) and (10.5) can be chosen so that they belong to the space $C'(T_m)$.

To prove this we make a change of variables in equations (10.4) and (10.5) by setting

$$y = Q^*(\phi)x. \tag{10.8}$$

Multiplying the result by $Q^*(\phi)$ we have, instead of equations (10.4) and (10.5), the equations

$$A(\phi)y = 0, \tag{10.9}$$

$$(A(\phi) - E)y = 0, \tag{10.10}$$

where $A(\phi) = Q^*(\phi)C(\phi)(Q^*(\phi))^{-1}$ is a matrix in $C'(\mathcal{T}_m)$ with rank r and E is the $n \times n$ identity matrix.

After the transformation (10.8), inequalities (10.6) and (10.7) become

$$\begin{aligned}
\langle D_1(\phi)y_1, y_1 \rangle < \langle D_2(\phi)y_2, y_2 \rangle, \\
\langle D_1(\phi)y_1, y_1 \rangle > \langle D_2(\phi)y_2, y_2 \rangle,
\end{aligned} \tag{10.11}$$

where $(y_1, y_2) = y$.

Let

$$A(\phi) = \begin{bmatrix} A_1(\phi) & A_{12}(\phi) \\ A_{21}(\phi) & A_2(\phi) \end{bmatrix}$$

be the decomposition of $A(\phi)$ into blocks corresponding to the decomposition of y into y_1 and y_2. We write system (10.9) in the form of two equations

$$\begin{aligned}
A_1(\phi)y_1 + A_{12}(\phi)y_2 = 0, \\
A_{21}(\phi)y_1 + A_2(\phi)y_2 = 0.
\end{aligned} \tag{10.12}$$

We claim that the rank of the matrix

$$\begin{bmatrix} A_1(\phi \\ A_{21}(\phi) \end{bmatrix} \tag{10.13}$$

is equal to r for any $\phi \in \mathcal{T}_m$. For suppose, on the contrary, that the rank of matrix (10.13) is equal to $r_1 < r$ for some value $\phi \in \mathcal{T}_m$. Then system (10.12) has a non-zero solution of the form

$$y_1 = y_1^0(\phi), \quad y_2 = 0.$$

This solution satisfies inequality (10.11); therefore

$$\langle D_1(\phi)y_1^0(\phi), \, y_1^0(\phi) \rangle < 0,$$

which is impossible since the matrix $D_1(\phi)$ is positive-definite.

Thus the rank of matrix (10.13) is equal to r. An orthogonalization process enables us to represent matrix (10.13) in the form

$$\begin{bmatrix} A_1(\phi) \\ A_{21}(\phi) \end{bmatrix} = \mathcal{E}(\phi)T_1(\phi) = \begin{bmatrix} \mathcal{E}_1(\phi) & T_1(\phi) \\ \mathcal{E}_{21}(\phi) & T_1(\phi) \end{bmatrix},$$

where $\mathcal{E}(\phi)$ is a matrix with orthogonal columns, $T_1(\phi)$ is an $r \times r$ non-singular matrix, and $\mathcal{E}(\phi)$ and $T_1(\phi)$ belong to the space $C'(\mathcal{T}_m)$.

We transform equation (10.12) by setting

$$T_1(\phi)y_1 = z_1.$$

As a result we obtain the equations

$$\mathcal{E}_1(\phi)z_1 + A_{12}(\phi)y_2 = 0,$$
$$\mathcal{E}_{21}(\phi)z_1 + A_2(\phi)y_2 = 0. \tag{10.14}$$

Since the rank of the coefficient matrix of system (10.14) is equal to that of $A(\phi)$ and hence to that of the matrix $\mathcal{E}(\phi)$, the columns of $\begin{bmatrix} A_{12}(\phi) \\ A_2(\phi) \end{bmatrix}$ are linear combinations of the columns of $\mathcal{E}(\phi)$. Because the columns of $\mathcal{E}(\phi)$ are orthogonal, we can write the coefficients of the decomposition of columns of the matrix $\begin{bmatrix} A_{12}(\phi) \\ A_2(\phi) \end{bmatrix}$ with respect to columns of $\mathcal{E}(\phi)$ explicitly and we see that these coefficients are also elements of the space $C'(\mathcal{T}_m)$. The representation of the matrix $\begin{bmatrix} A_{12}(\phi) \\ A_2(\phi) \end{bmatrix}$ in the form

$$\begin{bmatrix} A_{12}(\phi) \\ A_2(\phi) \end{bmatrix} = \mathcal{E}(\phi)R_1(\phi),$$

where $R_1(\phi)$ is an $r \times (n - r)$ matrix belonging to $C'(\mathcal{T}_m)$, follows from the above remarks. We now write system (10.14) in the form

$$\mathcal{E}(\phi)[z_1 + R_1(\phi)y_2] = 0,$$

and, on multiplying the left hand side of this equality by the matrix

$$(\mathcal{E}^*(\phi)\mathcal{E}(\phi))^{-1}\mathcal{E}^*(\phi),$$

we obtain the following system of equations equivalent to (10.14):

$$z_1 + R_1(\phi)y_2 = 0.$$

It is clear from this that the system of functions

$$z_1 = -R_1(\phi)e_j, \quad y_2 = e_j, \quad j = 1, \ldots, n - r,$$

in the space $C'(T_m)$, where e_j is the $(n - r)$-dimensional unit vector, forms a system of linearly independent solutions. A system of linearly independent solutions of equation (10.4) is formed by the system of functions in $C'(T_m)$ defined by the equalities

$$x_{r+j}(\phi) = (Q^*(\phi))^{-1}y_j(\phi), \quad j = 1, \ldots, n - r, \tag{10.15}$$

where $y_j(\phi)$ is the n-dimensional vector

$$y_j(\phi) = (-T_1^{-1}(\phi)R_1(\phi)e_j, e_j).$$

This proves that the system of equations (10.4) has $n - r$ linearly independent solutions in $C'(T_m)$.

Applying similar reasoning to equation (10.10) we see that the rank of the matrix

$$\begin{bmatrix} A_{12}(\phi) \\ A_2(\phi) - E_2 \end{bmatrix},$$

where E_2 is the $(n - r) \times (n - r)$ identity matrix, is equal to $n - r$ for any $\phi \in T_m$.

This is sufficient to transform system of equations (10.10) to an equivalent system of equations

$$R_2(\phi)y_1 + z_2 = 0$$

and to prove that equation (10.5) has r linearly independent solutions in the space $C'(T_m)$. These solutions are the functions

$$x_\nu(\phi) = (Q^*(\phi))^{-1}y_\nu(\phi), \quad \nu = 1, \ldots, r,$$

where $y_\nu(\phi)$ is an n-dimensional vector of the form

$$y_\nu(\phi) = (e_\nu, -T_2^{-1}(\phi)R_2(\phi)e_\nu), \tag{10.16}$$

e_ν is the r-dimensional unit vector.

This proves that the system of equations (10.5) has r linearly independent solutions in $C'(T_m)$.

The matrix whose columns are formed by the solutions (10.15), (10.16) transforms the matrix $C(\phi)$ to the Jordan form $\mathfrak{F}$. It belongs to the space $C'(\mathcal{T}_m)$ and, by Theorem 3 of §3.7, splits the system (7.1) into $C'(\mathcal{T}_m)$-blocks. The theorem is proved.

Remark 1. It can be seen from the proof of the theorem that the matrix $T(\phi)$ that decomposes system (7.1) into $C'(\mathcal{T}_m)$-blocks has the same smoothness as the matrices $Q(\phi)$ and $C(\phi) = G_0(0, \phi)$. Consequently, this system is $C^S(\mathcal{T}_m)$-block decomposable when $Q(\phi) \in C^S(\mathcal{T}_m)$ and $C(\phi) \in C^S(\mathcal{T}_m)$.

Remark 2. It follows from formulae (10.15) and (10.16) that when the matrix $S(\phi)$ has the block diagonal form

$$S(\phi) = \mathrm{diag}\{D_1(\phi), -D_2(\phi)\}, \tag{10.17}$$

the matrix $T(\phi)$ that splits system (7.1) into $C'(\mathcal{T}_m)$-blocks has the form

$$T(\phi) = \begin{bmatrix} E_1 & T_1(\phi) \\ T_2(\phi) & E_2 \end{bmatrix},$$

where E_1 and E_2 are the identity matrices of dimensions r and $n-r$ respectively. In this case the matrices $T_1(\phi)$ and $T_2(\phi)$ are solutions of the matrix Riccati equations

$$\begin{aligned} \dot{T}_1 &= P_1 T_1 - T_1 P_2 - T_1 P_{21} T_1 + P_{12}, \\ \dot{T}_2 &= P_2 T_2 - T_2 P_1 - T_2 P_{12} T_2 + P_{21}, \end{aligned} \tag{10.18}$$

and the split system of equations (10.2) has the form

$$\frac{d\phi}{dt} = a(\phi), \quad \frac{dy_1}{dt} = [P_1(\phi) + P_{12}(\phi)T_2(\phi)]y_1,$$

$$\frac{dy_2}{dt} = [P_2(\phi) + P_{21}(\phi)T_1(\phi)]y_2,$$

where P_1, P_{12}, P_{21}, P_2 are the blocks of the matrix $P = \begin{bmatrix} P_1 & P_{12} \\ P_{21} & P_2 \end{bmatrix}$ of appropriate dimensions.

The requirements imposed on the matrix (10.17) by the conditions of Theorem 1 of §3.8 turn out to be sufficient for the solvability of the Riccati equations (10.18) in $C'(\mathcal{T}_m)$.

Remark 3. It can be seen from the proof of the theorem that if in representation (10.1) the matrix $Q(\phi)$ has a period equal to a multiple of 2π, then the matrix reducing system (7.1) to split form is periodic with the same period as $Q(\phi)$.

The problem on the representation of matrices $S(\phi)$ in $C'(T_m)$ in the form of (10.1) has been little studied. The following statement completely solves it only for two-dimensional matrices.

Lemma 1. *Let $S(\phi)$ be a two-dimensional matrix in $C'(T_m)$ with eigenvalues $\lambda_1(\phi)$ and $\lambda_2(\phi)$ of different signs. Then there exists a non-singular matrix $Q(\phi)$ that is periodic and continuous with respect to ϕ with period 4π, and continuously differentiable with respect to t for $\phi = \phi_t(\phi)$ such that*

$$S(\phi) = Q(\phi)\operatorname{diag}\{d_1(\phi), -d_2(\phi)\}Q^*(\phi), \qquad (10.19)$$

where $d_1(\phi)$ and $d_2(\phi)$ are positive functions for all $\phi \in T_m$.

To prove the lemma, we first note that if representation (10.19) is possible with matrix $Q(\phi)$ satisfying the conditions of the lemma, then it is possible with $d_1(\phi) = \lambda_1(\phi)$, $d_2(\phi) = -\lambda_2(\phi)$ if we assume that $\lambda_1(\phi)$ and $\lambda_2(\phi)$ are the positive and negative eigenvalues of the matrix $S(\phi)$. But then

$$\det S(\phi) = [\det Q(\phi)]^2 d_1(\phi)(-d_2(\phi)) =$$
$$= [\det Q(\phi)]^2 \lambda_1(\phi)\lambda_2(\phi) = [\det Q(\phi)]^2 \det S(\phi),$$

and consequently

$$[\det Q(\phi)]^2 = 1.$$

If $\det Q(\phi) = -1$, then the matrix

$$Q(\phi)\operatorname{diag}\{1, -1\}$$

brings about the representation (10.19) and has determinant equal to 1. Thus if representation (10.19) is possible then the representation

$$S(\phi) = Q(\phi)\operatorname{diag}\{\lambda_1(\phi), \lambda_2(\phi)\}Q^*(\phi) \qquad (10.20)$$

is also possible with the periodic matrix $Q(\phi)$ of period 4π satisfying the conditions of the lemma and such that

$$\det Q(\phi) = 1. \qquad (10.21)$$

Thus we seek a matrix $Q(\phi)$ effecting the transformation (10.20) and satisfying condition (10.21). Let

$$S(\phi) = \begin{bmatrix} a & b \\ b & d \end{bmatrix}, \quad Q(\phi) = \begin{bmatrix} x & y \\ x_1 & y_1 \end{bmatrix}.$$

Condition (10.21) takes the form of the equation

$$xy_1 - yx_1 = 1, \tag{10.22}$$

while equality (10.20) takes the form of the system

$$ay_1 - by = \lambda_1 x, \quad -ax_1 + bx = \lambda_2 y,$$
$$by_1 - dy = \lambda_1 x_1, \quad -bx_1 + dx = \lambda_2 y_1. \tag{10.23}$$

Substituting the values of x, x_1 found from the first two equations of system (10.23) into the two remaining equations, we obtain the identities

$$(ad - b^2)y = \lambda_1 \lambda_2 y, \quad (ad - b^2)y_1 = \lambda_1 \lambda_2 y_1.$$

Consequently, in order to find $Q(\phi)$ it is sufficient to choose x and x_1 using y and y_1 according to the first equalities of system (10.23) such that relation (10.22) holds. The latter, by virtue of the choice for x, x_1, takes the form of the equality

$$ay_1^2 - 2byy_1 + dy^2 = \lambda_1. \tag{10.24}$$

The problem has been reduced to that of finding two functions $y(\phi), y_1(\phi)$ that are continuous and periodic in ϕ and which turn equality (10.24) into an identity for all $\phi \in T_m$. Since two arbitrary functions $y(\phi), y_1(\phi)$ satisfying (10.24) cannot become zero simultaneously, we can set

$$y = R\cos\psi,$$
$$y_1 = R\sin\psi \tag{10.25}$$

and reduce equality (10.24) to the form

$$R^2[a\sin^2\psi - 2b\sin\psi\cos\psi + d\cos^2\psi] = \lambda_1$$

and, after obvious transformations, to the form

$$(d - a)\cos 2\psi - 2b\sin 2\psi = \frac{2\lambda_1}{R^2} - (a + d). \tag{10.26}$$

Since the eigenvalues of the matrix $S(\phi)$ have different signs, it follows that

$$(d - a)^2 + 4b^2 \neq 0,$$

which enables us to rewrite equality (10.26) as

$$\cos(2\psi + \alpha) = \frac{2\lambda_1 - (a+d)R^2}{R^2[(d-a)^2 + 4b^2]^{1/2}},\qquad(10.27)$$

where α is determined by the equalities

$$\cos\alpha = \frac{d-a}{[(d-a)^2 + 4b^2]^{1/2}},$$

$$\sin\alpha = \frac{2b}{[(d-a)^2 + 4b^2]^{1/2}}.$$

Since $\lambda_1(\phi) > 0$, $\lambda_2(\phi) < 0$, it follows that

$$(a+d)^2 < (d-a)^2 + 4b^2.$$

Thus, by choosing $R(\phi)$ sufficiently large we can make the absolute value of the right hand side of equality (10.27) less than one. With such a choice of $R(\phi)$ the function

$$\psi = \tfrac{1}{2}\cos^{-1}\frac{2\lambda_1 - (a+d)R^2}{R^2[(d-a)^2 + 4b^2]^{1/2}} - \frac{\alpha}{2}$$

is a solution of equation (10.27).

Find out the properties of this solution considered as a function of ϕ. For a sufficiently large $R = R(\phi)$ from the space $C'(T_m)$ the function

$$\cos^{-1}\frac{2\lambda_1 - (a+d)R^2}{R^2[(d-a)^2 + 4b^2]^{1/2}} = 2\Phi(\phi)$$

belongs to the space $C'(T_m)$.

By applying the argument principle to the pair of functions

$$\frac{d(\phi) - a(\phi)}{[(d(\phi) - a(\phi))^2 + 4b^2(\phi)]^{1/2}},\qquad \frac{2b(\phi)}{[(d(\phi) - a(\phi))^2 + 4b^2(\phi)]^{1/2}}$$

we see that

$$\alpha = \alpha(\phi) = (k,\phi) + 2\Phi_1(\phi),$$

where $k = (k_1,\ldots,k_m)$ is an integer vector and Φ_1 a function in $C'(T_m)$. But then

$$\Psi = \Phi(\phi) - \Phi_1(\phi) - \frac{(k,\phi)}{2} = \Phi_0(\phi) - \frac{(k,\phi)}{2},$$

where $\Phi_0 \in C'(T_m)$.

168 CHAPTER 3

It follows from formulae (10.25) that the functions

$$y = y(\phi) = R(\phi)\cos\left(\Phi_0(\phi) - \frac{(k,\phi)}{2}\right),$$
$$y_1 = y_1(\phi) = R(\phi)\sin\left(\Phi_0(\phi) - \frac{(k,\phi)}{2}\right)$$

satisfy equality (10.24) for all $\phi \in T_m$. These functions have period 4π with respect to ϕ_ν ($\nu = 1,\ldots,m$), are continuous with respect to ϕ, and have continuous derivatives with respect to t for $\phi = \phi_t(\phi)$. Then the matrix $Q(\phi)$ has similar properties because its elements are the functions $y(\phi)$, $y_1(\phi)$ and the functions $x(\phi)$, $x_1(\phi)$ which are expressible in terms of them.

To show that period doubling in the representation of the matrix $S(\phi)$ in form (10.1) can take place, we consider this representation for the matrix

$$S(\phi) = \begin{bmatrix} -\cos\phi & \sin\phi \\ \sin\phi & \cos\phi \end{bmatrix}, \tag{10.28}$$

where ϕ is a scalar.

The eigenvalues of this matrix are $\lambda_1 = 1$ and $\lambda_2 = -1$. As eigenvectors for the matrix $S(\phi)$ we can take:

$$(\sin(\phi/2), \cos(\phi/2)) \quad \text{for } \lambda_1 = 1$$
$$(\cos(\phi/2), -\sin(\phi/2)) \quad \text{for } \lambda_2 = -1,$$

and consequently the eigenvectors for the matrix $S(\phi)$ cannot be chosen to be periodic with period 2π.

We shall call system (7.1) two-dimensional if $x = (x_1, x_2)$ is a two-dimensional vector, and $C'(T_m)$-block decomposable with period doubling if the decomposition is effected by a matrix $T(\phi)$ that is continuous with respect to ϕ, periodic with respect to ϕ with period 4π and has continuous derivative with respect to t for $\phi = \phi_t(\phi)$.

The following result is a consequence of Lemma 1 and Theorem 1.

Theorem 2. *Let $a \in C_{\text{Lip}}(T_m)$, $P \in C(T_m)$ and suppose that system (7.1) is two-dimensional and exponentially dichotomous. Then it is $C'(T_m)$-block decomposable with period doubling.*

For $n \geq 3$, period doubling is not sufficient for the $C'(T_m)$-block decomposability of an exponentially dichotomous system. The reasons for this will be discussed in the following section.

3.11. On triangulation and the relation between the $C'(\mathcal{T}_m)$-block decomposability of a linear system and the problem of the extendability of an r-frame to a periodic basis in E^n

We continue our discussion of the system of equations

$$\frac{d\phi}{dt} = a(\phi), \quad \frac{dx}{dt} = P(\phi)x, \tag{11.1}$$

where a and P are functions in $C_{\text{Lip}}(\mathcal{T}_m)$ and $C(\mathcal{T}_m)$ respectively.

We set

$$Lu = \dot{u} - P(\phi)u,$$

thus defining an operator L on the function space $C'(\mathcal{T}_m)$.

A scalar function $\lambda = \lambda(\phi)$ in $C(\mathcal{T}_m)$ is called an *eigenvalue* of the operator L if the equation

$$Lu = \lambda u \tag{11.2}$$

has a non-trivial solution $u = u(\phi) \not\equiv 0$, $\phi \in \mathcal{T}_m$, in $C'(\mathcal{T}_m)$. An eigenvalue of the operator L is called an *eigennumber* if it does not depend on ϕ. A non-trivial solution in $C'(\mathcal{T}_m)$ of equation (11.2) will be called an *eigenfunction* of L corresponding to the eigenvalue λ.

Theorem 1. *Suppose that the operator L has eigenvalues*

$$\lambda_1, \ \lambda_2, \ldots, \lambda_p$$

and let

$$u_1(\phi), u_2(\phi), \ldots, u_p(\phi) \tag{11.3}$$

be the eigenfunctions of the operator L corresponding to these eigenvalues.

Then the surface

$$x = u_1(\phi)y_1 + \ldots + u_p(\phi)y_p, \quad y_j \in \mathbf{R}, \ \phi \in \mathcal{T}_m, \tag{11.4}$$

is an invariant set of system (11.1) and on this set system (11.1) is equivalent to the following system of equations

$$\frac{d\phi}{dt} = a(\phi), \quad \frac{dy_j}{dt} = -\lambda_j(\phi)y_j, \quad j = 1, \ldots, p. \tag{11.5}$$

To prove the theorem we fix a point (ϕ_0, x_0) on the surface (11.4) by setting

$$x_0 = \sum_{j=1}^{p} u_j(\phi_0)y_j^0, \quad y_j^0 \in \mathbf{R}$$

and consider the motion $\phi_t(\phi_0)$, $x_t(\phi_0, x_0)$ of system (11.1), that starts at this point. Let

$$x(t, \phi_0, x_0) = \sum_{j=1}^{p} u_j(\phi_t(\phi_0))y_j(t, \phi_0, y_j^0),$$

where $\phi_t(\phi_0)$, $y_j(t, \phi_0, y_j^0)$ is the solution of system (11.5) with initial conditions ϕ_0, y_j^0. By differentiating the function $x(t, \phi_0, x_0)$ with respect to t, we see that

$$\frac{dx(t, \phi_0, x_0)}{dt} = \sum_{j=1}^{p} [\dot{u}_j(\phi_t(\phi_0))y_j(t, \phi_0, y_j^0) - \lambda_j(\phi_t(\phi_0)) \times$$

$$\times u_j(\phi_t(\phi_0))y_j(t, \phi_0, y_j^0)] = P(\phi_t(\phi_0)) \sum_{j=1}^{p} u_j(\phi_t(\phi_0))y_j(t, \phi_0, y_j^0) =$$

$$= P(\phi_t(\phi_0))x(t, \phi_0, x_0).$$

Hence it follows that $\phi_t(\phi_0)$, $x(t, \phi_0, x_0)$ is the solution of system (11.1) defined by the initial condition ϕ_0, x_0. But then

$$x(t, \phi_0, x_0) = x_t(\phi_0, x_0), \quad t \in \mathbf{R}.$$

This means that any motion $\phi_t(\phi_0)$, $x_t(\phi_0, x_0)$ for which (ϕ_0, x_0) belongs to the surface (11.4) is defined for any $t \in \mathbf{R}$ by formula (11.4) if ϕ, y_j are replaced there by the solution $\phi_t(\phi_0), y_j(t, \phi_0, y_j^0)$ of system of equations (11.5). This proves Theorem 1.

Note that if the vectors (11.3) form a p-frame in E^n, then the surface (11.4) is a manifold diffeomorphic to the direct product of the torus $\mathcal{T}_m$ and the plane E^p. In particular, it can be a separatrix manifold of an exponentially dichotomous system.

Theorem 2. *Under the change of variables*

$$x = U(\phi)y + V(\phi)z, \quad \det[U(\phi), V(\phi)] \neq 0, \quad \phi \in \mathcal{T}_m \qquad (11.6)$$

the system of equations (11.1) *is reduced to the block-triangular form*

$$\frac{d\phi}{dt} = a(\phi), \quad \frac{dz}{dt} = Q(\phi)z, \quad \frac{dy}{dt} = -D(\phi)y + Q_1(\phi)z, \qquad (11.7)$$

where $U(\phi) = (u_1(\phi), \dots, u_p(\phi))$, $V(\phi) = (v_1(\phi), \dots, v_{n-p}(\phi))$ *are matrices in*
$C'(T_m)$, $D(\phi) = \mathrm{diag}\{\lambda_1(\phi), \dots, \lambda_p(\phi)\} \in C(T_m)$, *if and only if the functions*
$\lambda_j(\phi)$ $(j = 1, \dots, p)$ *are eigenvalues of the operator* L *and* $u_j(\phi)$ $(j = 1, \dots, p)$
are the eigenfunctions of L *corresponding to them.*

To prove the theorem, first we suppose that the $\lambda_j(\phi)$ $(j = 1, \dots, p)$ are
eigenvalues of L and the $u_j(\phi)$ $(j = 1, \dots, p)$ are the eigenfunctions of L corresponding to them, so that

$$Lu_j(\phi) = \lambda_j(\phi)u_j(\phi), \quad j = 1, \dots, p, \quad \phi \in T_m. \qquad (11.8)$$

Equalities (11.8) are equivalent to the single matrix equality

$$LU(\phi) = U(\phi)D(\phi), \quad \phi \in T_m. \qquad (11.9)$$

Let $V(\phi)$ be a matrix satisfying the conditions of Theorem 2. We show that
the change of variables (11.6) reduces system (11.1) to the form (11.7).

To do this, we show that the motion $\phi = \phi_t(\phi_0)$, $x = x_t(\phi_0, x_0)$ of system
(11.1) is defined by formula (11.6) for any $\phi_0 \in T_m$, $x_0 \in E^n$, $t \in \mathbf{R}$, by setting
$\phi = \phi_t(\phi_0)$, $y = y_t$, $z = z_t$ and choosing y_t, z_t such that they are solutions of
some system of equations (11.7). For this purpose we set

$$\dot{U}(\phi_t(\phi))y_t + U(\phi_t(\phi))\frac{dy_t}{dt} + \dot{V}(\phi_t(\phi))z_t + V(\phi_t(\phi))\frac{dz_t}{dt} =$$
$$= P(\phi_t(\phi))U(\phi_t(\phi))y_t + P(\phi_t(\phi))V(\phi_t(\phi))z_t$$

for all $t \in \mathbf{R}$, $\phi \in T_m$. Using (11.9) we obtain the following relation for dy_t/dt
and dz_t/dt

$$U(\phi_t(\phi))\Big(\frac{dy_t}{dt} + D(\phi_t(\phi))y_t\Big) + V(\phi_t(\phi))\frac{dz_t}{dt} =$$
$$= [P(\phi_t(\phi))V(\phi_t(\phi)) - \dot{V}(\phi_t(\phi))]z_t. \qquad (11.10)$$

We denote the inverse of the matrix $\Phi(\phi) = [U(\phi), V(\phi)]$ by

$$\begin{bmatrix} U_1(\phi) \\ V_1(\phi) \end{bmatrix} = \Phi^{-1}(\phi).$$

Then the identity $\Phi^{-1}(\phi)\Phi(\phi) = E$ can be written as:

$$U_1(\phi)U(\phi) = E, \quad U_1(\phi)V(\phi) = 0,$$
$$V_1(\phi)U(\phi) = 0, \quad V_1(\phi)V(\phi) = E,$$

where E and 0 are the unit and zero matrices of appropriate dimensions.

Multiplying equality (11.10) on the left by $\Phi^{-1}(\phi_t(\phi))$ we obtain the following system of relations for dy_t/dt and dz_t/dt:

$$\frac{dy_t}{dt} = -D(\phi_t(\phi))y_t + U_1(\phi_t(\phi))[P(\phi_t(\phi))V(\phi_t(\phi)) - \dot{V}(\phi_t(\phi))]z_t,$$

$$\frac{dz_t}{dt} = V_1(\phi_t(\phi))[P(\phi_t(\phi))V(\phi_t(\phi)) - \dot{V}(\phi_t(\phi))]z_t,$$

which proves that the change of variables (11.6) reduces the initial system of equations (11.1) to the form of (11.7):

$$\frac{d\phi}{dt} = a(\phi), \quad \frac{dz}{dt} = V_1(\phi)[P(\phi)V(\phi) - \dot{V}(\phi)]z,$$

$$\frac{dy}{dt} = -D(\phi)y + U_1(\phi)[P(\phi)V(\phi) - \dot{V}(\phi)]z.$$

Conversely, suppose that the system (11.1) is reduced to system (11.7) under the change of variables (11.6). The motion $\phi_t(\phi)$, $x_t(\phi, x_0)$ is related to the solution $\phi_t(\phi), y_t, z_t$ of system of equations (11.7) by the formula

$$x_t(\phi, x_0) = U(\phi_t(\phi))y_t + V(\phi_t(\phi))z_t.$$

Differentiating it with respect to t we obtain the identity:

$$U(\phi_t(\phi))y_t + U(\phi_t(\phi))[-D(\phi_t(\phi))y_t + Q_1(\phi_t(\phi))z_t]+$$
$$+\dot{V}(\phi_t(\phi))z_t + V(\phi_t(\phi))Q(\phi_t(\phi))z_t = P(\phi_t(\phi))U(\phi_t(\phi))y_t+$$
$$+P(\phi_t(\phi))V(\phi_t(\phi))z_t, \quad t \in \mathbf{R}, \ \phi \in T_m.$$

If we consider this identity for the solution y_t, z_t with $z_t \equiv 0$, we obtain the new identity

$$[\dot{U}(\phi_t(\phi)) - P(\phi_t(\phi))U(\phi_t(\phi)) - U(\phi_t(\phi))D(\phi_t(\phi))]y_t \equiv 0,$$

from which it follows for $t = 0$ that

$$\dot{U}(\phi) - P(\phi)U(\phi) = U(\phi)D(\phi), \quad \phi \in \mathcal{T}_m$$

in view of the arbitrariness of the choice of the initial condition y_0. But then $U(\phi)$ satisfies the matrix equation (11.9) and consequently the $\lambda_j(\phi)$ $(j = 1, \ldots, p)$ are eigenvalues and the $U_j(\phi)$ $(j = 1, \ldots, p)$ are the eigenfunctions of L corresponding to them.

The following statement, which clarifies the relation between the $C'(\mathcal{T}_m)$-block decomposability of the linear system (11.1) and the problem of the extendability of an r-frame to a periodic basis in E^n, easily follows from Theorem 2.

Theorem 3. *The system of equations* (11.1)*is reduced to the block-triangular form* (11.7) *by transformation* (11.6) *with the matrix* $[U(\phi), V(\phi)] \in C'(\mathcal{T}_m)$ *if and only if the p-frame* (11.3) *can be extended to a periodic basis in* E^n.

By Theorem 3, system (11.1), the eigenfunctions of which form a frame that cannot be extended to a periodic basis in E^n, cannot be transformed to the block-diagonal form (11.7) by transformation (11.6). System (11.7) has the general form of a block-triangular system if y is a one-dimensional vector. Therefore if the operator L only has eigenvalues such that the eigenfunctions corresponding to them cannot be extended to a periodic basis in E^n, then system (11.1) cannot be transformed to a block-triangular form with 1- and $(n - 1)$-dimensional blocks by a transformation with coefficients in the space $C'(\mathcal{T}_m)$. This can be used to construct exponentially dichotomous systems that are not $C'(\mathcal{T}_m)$-block decomposable. Such examples are constructed in [35] for systems of equations with quasi-periodic coefficients. We shall say that the exponentially dichotomous system (11.1) is $C'(\mathcal{T}_m)$-*block decomposable with separatrix manifolds that can be straightened* if the separatrix manifolds of the decomposed system are its coordinate planes, and the motion on them is determined by the block subsystems of the decomposed system.

Theorem 4. *Suppose that the system of equations* (11.1) *is exponentially dichotomous. Suppose further that the eigenvalues of the operator* L *of the system contain r positive eigenvalues and that the eigenfunctions corresponding to them form an r-frame in* E^n.

Then system (11.1) *is* $C'(T_m)$-*block decomposable with separatrix manifolds that can be straightened if and only if this system of vectors can be extended to a basis in* E^n.

To prove the theorem, we denote the positive eigenvalues of L by

$$\lambda_1(\phi), \ldots, \lambda_r(\phi) \tag{11.11}$$

and the eigenfunctions of this operator corresponding to them by

$$u_1(\phi), \ldots, u_r(\phi). \tag{11.12}$$

Suppose that functions (11.12) form an r-frame of vectors that can be extended to a periodic basis in E^n. Then the transformation (11.6) with matrix $V(\phi)$ in $C'(T_m)$ which extends $U(\phi) = (u_1(\phi), \ldots, u_r(\phi))$ to a periodic basis in E^n transforms system (11.1) to the form (11.7). Under the transformations (11.6) the exponential dichotomy of system (11.1) is preserved, so that system (11.7) is exponentially dichotomous.

As follows from the properties established earlier, the number r determines the dimension of the separatrix manifold M^+ of the exponentially dichotomous system. Since the eigenvalues $\lambda_j(\phi)$ $(j = 1, \ldots, r)$ are positive, it follows from the form of system (11.7) that its separatrix manifold M^+ is given by the equation

$$z = 0, \quad \phi \in T_m. \tag{11.13}$$

Using the variables ϕ, x the manifold M^+ can be given by equation (7.2)

$$C(\phi)x = x, \quad \phi \in T_m, \tag{11.14}$$

where $C(\phi)$ is a projection in $C'(T_m)$.

By expressing x in terms of y and z using (11.6) and substituting it into (11.14) we can write the equation that defines the manifold M^+ of system (11.7) in the form:

$$C(\phi)U(\phi)y + C(\phi)V(\phi)z = U(\phi)y + V(\phi)z. \tag{11.15}$$

The points of the manifold (11.13) satisfy equation (11.15), but since y was chosen arbitrarily, this is possible only if

$$C(\phi)U(\phi) = U(\phi), \quad \phi \in T_m. \tag{11.16}$$

We write down the equation for the separatrix manifold M^- (7.3):

$$C(\phi)x = 0, \quad \phi \in T_m.$$

Under the transformation (11.6) it becomes the separatrix manifold M^- of system (11.7) with equation

$$C(\phi)[U(\phi)y + V(\phi)z] = 0, \quad \phi \in T_m,$$

which, in view of identity (11.16), takes the form

$$U(\phi)y + C(\phi)V(\phi)z = 0. \tag{11.17}$$

Since the matrix $U^*(\phi)U(\phi)$ is a Gram matrix of linear independent vectors, it is non-singular for all $\phi \in T_m$. Multiplying equation (11.17) first by $U^*(\phi)$ and then by $[U^*(\phi)U(\phi)]^{-1}$, we arrive at the equality

$$y = -[U^*(\phi)U(\phi)]^{-1}U^*(\phi)C(\phi)V(\phi)z, \quad \phi \in T_m, \tag{11.18}$$

which is satisfied by any solution of equation (11.17). For a fixed $\phi \in T_m$, equality (11.18) defines an $(n - r)$-dimensional plane in the y, z space. Since the manifold M^- is an $(n - r)$-dimensional plane in the y, z space, it follows that for a fixed $\phi \in T_m$, M^- is the plane (11.18). This proves that equality (11.18) is the equation of the separatrix manifold M^- of system (11.7).

By expressing the variables y in equations (11.7) in terms of the variables y_1 according to the formula

$$y = y_1 - [U^*(\phi)U(\phi)]^{-1}U^*(\phi)C(\phi)V(\phi)z,$$

we transform system (11.7) into a system of the same form but with a separatrix manifold with equation

$$y_1 = 0, \quad \phi \in T_m.$$

The latter is possible only if the system of equations for ϕ, z, y_1 has the form (11.7) with $y = y_1$ and $Q_1 \equiv 0$ and therefore has the form of a block diagonal system with diagonal blocks $Q(\phi)$ and $-D(\phi) = -\mathrm{diag}\{\lambda_1(\phi), \ldots, \lambda_2(\phi)\}$. The $C'(T_m)$-block decomposability of system (11.1) has been proved. From the form of the decomposed system it follows that the separatrix manifolds M^+ and M^- have been "straightened".

Suppose now that the eigenfunctions (11.12) of the operator L form an r-frame which cannot be extended to a periodic basis in E^n. Suppose that system (11.1) can nevertheless be reduced to the block-diagonal form

$$\frac{d\phi}{dt} = a(\phi), \quad \frac{dy_1}{dt} = Q_1(\phi)y_1, \quad \frac{dz_1}{dt} = Q_2(\phi)z_1 \tag{11.19}$$

with straightened separatrix manifolds under the change of variables

$$x = U_1(\phi)y_1 + V_1(\phi)z_1$$

with the matrix $\Phi(\phi) = [U_1(\phi), V_1(\phi)] \in C'(T_m)$, $\det \Phi(\phi) \neq 0$ for $\phi \in T_m$.

Suppose that the equation

$$z_1 = 0, \quad \phi \in T_m$$

defines the separatrix manifold M^+ of system (11.19). Then the surface

$$x = U_1(\phi)y_1, \quad y_1 \in E^r, \quad \phi \in T_m \tag{11.20}$$

is an invariant set of system (11.1), the motion of which is defined by the first two equations of system (11.19) and consequently they are exponentially damped as $t \to +\infty$. The surface

$$x = U(\phi)y, \quad y \in E^r, \quad \phi \in T_m \tag{11.21}$$

is also an invariant set of system (11.1) by Theorem 1. The motions on it are defined by system (11.5), and since the values $\lambda_j(\phi)$ $(j = 1, \ldots, r)$ are positive, they are also exponentially damped as $t \to +\infty$.

For any fixed $\phi \in T_m$ both surfaces (11.20) and (11.21) define r-dimensional planes in x-space. The motion of system (11.1) on surfaces (11.20) and (11.21) shows that these surfaces are one and the same separatrix manifold M^+ of system (11.1) written in the different coordinate systems (ϕ, y_1) and (ϕ, y) respectively. For a fixed $\phi \in T_m$ there are defined two Euclidian coordinate systems y_1 and y in the same Euclidian space $E^r(\phi)$ which have the same origin. Consequently, there is a non-singular $r \times r$ matrix $R(\phi)$ defined for every $\phi \in T_m$, such that the transition from the coordinate system y_1, to the coordinate system y is given by the relation

$$y_1 = R(\phi)y.$$

For equalities (11.20) and (11.21) to define the same plane for any $\phi \in T_m$, it is necessary that the matrix $R(\phi)$ satisfy the identity

$$U_1(\phi)R(\phi) = U(\phi), \quad \phi \in T_m. \tag{11.22}$$

Using the matrix $\Phi(\phi)$ we can rewrite identity (11.22) as

$$\Phi(\phi)\begin{bmatrix} R(\phi) \\ 0 \end{bmatrix} = U(\phi), \quad \phi \in T_m,$$

and by taking the inverse of the matrix $\Phi(\phi)$ we see that $R(\phi) \in C'(T_m)$.

This proves that the matrix $R(\phi)$ in identity (11.22) belongs to the space $C'(T_m)$. Then we have the representation

$$[U(\phi), V_1(\phi)] = [U_1(\phi)R(\phi), V_1(\phi)] =$$

$$= [U_1(\phi), V_1(\phi)]\begin{bmatrix} R(\phi) & 0 \\ 0 & E \end{bmatrix},$$

from which it follows that

$$\det[U(\phi), V_1(\phi)] \neq 0, \quad \phi \in T_m \qquad (11.23)$$

since the matrices $\Phi(\phi) = [U_1(\phi), V_1(\phi)]$ and $R(\phi)$ are non-singular.

Since $V_1(\phi) \in C'(T_m)$, it follows that inequality (11.23) ensures that the r-frame $U(\phi)$ can be extended to a periodic basis in E^n. Thus we have a contradiction, and this completes the proof of Theorem 4.

The statement of Theorem 4 can be used to construct exponentially dichotomous systems (11.1) that are not $C'(T_m)$-block decomposable with an arbitrary size of diagonal blocks.

The following result on $C'(T_m)$-diagonalization of system (11.1) is a trivial consequence of Theorem 2.

Corollary. *In order that the system of equations (11.1) be reduced to the diagonal form*

$$\frac{dy_j}{dt} = -\lambda_j(\phi)y_j, \quad j = 1, \ldots, n$$

by the transformation

$$x = U(\phi)y, \quad U \in C'(T_m), \quad \det U(\phi) \neq 0, \quad \phi \in T_m,$$

it is necessary and sufficient that the operator L of the system have n eigenvalues $\lambda_1(\phi), \ldots, \lambda_n(\phi)$ the eigenfunctions of which form a basis in E^n.

We note that necessary conditions for $C'(T_m)$-diagonalization of system (11.1) are given in [118].

3.12. On smoothness of an exponentially stable invariant torus

In our discussion of the invariant torus

$$x = u(\phi), \quad \phi \in \mathcal{T}_m \tag{12.1}$$

of the system of equations

$$\frac{d\phi}{dt} = a(\phi), \quad \frac{dx}{dt} = P(\phi)x + f(\phi) \tag{12.2}$$

there immediately arises the problem of the dependence of the smoothness of the invariant torus on that of the right hand side of system (12.2). As was noted before, this dependence is non-trivial. Therefore it needs to be studied independently.

In our study of the smoothness of the torus (12.1) we start with the following statement.

Theorem 1. *Let* $a, P \in C^l(\mathcal{T}_m)$, $l \geq 1$. *Then if for an arbitrary function* $f \in C^l(\mathcal{T}_m)$ *a Green's function of system* (11.1) *satisfies the inequality*

$$|G_0(\tau, \phi)f(\phi_\tau(\phi))|_l \leq Ke^{-\gamma|\tau|}|f|_l, \quad \tau \in \mathbf{R} \tag{12.3}$$

where $\phi_t(\phi)$ *is a solution of the first equation of system* (12.2) *for an arbitrary* $\phi \in \mathcal{T}_m$, *and* K *and* γ *are positive constants independent of* ϕ, *then the invariant torus* (12.1) *of system* (12.2) *belongs to the space* $C^l(\mathcal{T}_m)$ *and satisfies the inequality*

$$|u|_l \leq 2K\gamma^{-1}|f|_l. \tag{12.4}$$

The proof of the theorem is obvious if we use estimate (12.3) and the integral representation for the function $u(\phi)$ defining torus (12.1):

$$u(\phi) = \int_{-\infty}^{+\infty} G_0(\tau, \phi)f(\phi_\tau(\phi))d\tau. \tag{12.5}$$

By Theorem 1 the smoothness of invariant torus (12.1) of system (12.2) depends essentially only on the properties of the Green's function $G_0(\tau, \phi)$ and the solutions $\phi_t(\phi)$ of the system defining the trajectory flow (12.2) on torus (12.1). We assume that the functions a, P and f belong to the space $C^l(\mathcal{T}_m)$ with finite $l \geq 1$ and look at conditions ensuring that inequality (12.3) holds.

We shall consider here the case when the torus (12.1) is exponentially stable. According to results of §3.5, this will be the case if the following condition holds:

$$\inf_{\phi \in T_m} \beta(\phi) = \beta_0 > 0, \tag{12.6}$$

where $\beta(\phi)$ is defined by the inequality

$$\inf_{S \in \mathfrak{R}_0} \max_{\|x\|=1} \frac{\langle [S(\phi)P(\phi) + (1/2)\dot{S}(\phi)]x, x \rangle}{\langle S(\phi)x, x \rangle} \leq -\beta(\phi), \tag{12.7}$$

$\mathfrak{R}_0$ is the set of $n \times n$-dimensional positive-definite symmetric matrices belonging to the space $C'(T_m)$.

We define the value of $\alpha(\phi)$ by setting

$$\sup_{S_1 \in \mathfrak{R}_1} \min_{\|\psi\|=1} \frac{\langle [S_1(\phi)\frac{\partial a(\phi)}{\partial \phi} + \frac{1}{2}\dot{S}_1(\phi)]\psi, \psi \rangle}{\langle S_1(\phi)\psi, \psi \rangle} \geq \alpha(\phi), \tag{12.8}$$

where $\mathfrak{R}_1$ is the set of $m \times m$-dimensional positive-definite symmetric matrices belonging to the space $C^1(T_m)$.

Theorem 2. *Suppose that inequality* (12.6) *holds and*

$$\inf_{\phi \in T_m} [\beta(\phi) + l\alpha(\phi)] > 0. \tag{12.9}$$

Then the invariant torus (12.1) *of system* (12.2) *belongs to the space* $C^l(T_m)$.

To prove the theorem we first exploit the coarseness of its conditions, which ensure that there exist matrices $S(\phi) \in \mathfrak{R}_0$, $S_1(\phi) \in \mathfrak{R}_1$ and functions $\beta(\phi)$, $\alpha(\phi)$ satisfying inequalities of the form (12.7), (12.8):

$$\max_{\|x\|=1} \frac{\langle [S(\phi)P(\phi) + (1/2)\dot{S}(\phi)]x, x \rangle}{\langle S(\phi)x, x \rangle} \leq -\beta(\phi),$$

$$\min_{\|\psi\|=1} \frac{\langle [S_1(\phi)\partial a(\phi)/\partial \phi + (1/2)\dot{S}_1(\phi)]\psi, \psi \rangle}{\langle S_1(\phi)\psi, \psi \rangle} \geq \alpha(\phi) \tag{12.10}$$

and conditions of the form (12.6), (12.9):

$$\beta(\phi) \geq \beta_0, \quad \beta(\phi) + l\alpha(\phi) \geq \gamma, \tag{12.11}$$

where β_0 and γ are positive constants, $\beta_0 = \inf_{\phi \in T_m} \beta(\phi)$, $\gamma = \inf_{\phi \in T_m} [\beta(\phi)+l\alpha(\phi)]$. If the first equality of (12.11) is satisfied, then the second is of independent

significance only for those values of ϕ for which $\alpha(\phi) < 0$. Therefore we may
assume that the function $\alpha(\phi)$ satisfies the condition

$$\alpha(\phi) \leq 0, \quad \phi \in \mathcal{T}_m. \tag{12.12}$$

Assuming that inequality (12.12) holds, we prove the estimate

$$\|D_\phi^S \phi_t(\phi)\| \leq K \exp\{-\int_t^0 s\alpha(\phi_\tau(\phi))d\tau - \epsilon t\}, \quad t \leq 0, \tag{12.13}$$

for any $1 \leq s \leq l$ and arbitrarily small $\epsilon > 0$, and some K that depends on s, ϵ
but does not depend on t, ϕ. Here $D_\phi^S \phi_t(\phi)$ is any derivative of order s of the
function $\phi_t(\phi)$ with respect to the variables $\phi = (\phi_1, \ldots, \phi_m)$.

We carry out the proof by induction. To do this we write the system of
equations

$$\frac{d\psi}{dt} = \frac{\partial a(\phi_t(\phi))}{\partial \phi_t} \psi, \tag{12.14}$$

which is satisfied by the matrix of partial derivatives $\partial \phi_t(\phi)/\partial \phi$ of the function
$\phi_t(\phi)$. Setting $\psi_t^i = \partial \phi_t(\phi)/\partial \phi_i$ and considering the function

$$V = \langle S_1(\phi_t(\phi))\psi_t^i, \ \psi_t^i \rangle$$

we find by using inequality (12.10) for $S_1(\phi)$ that

$$\dot{V} \geq 2\alpha(\phi_t(\phi))V. \tag{12.15}$$

Upon integrating this inequality and making the standard estimates we obtain
from (12.15) the inequality:

$$\|\psi_t^i\| \leq K \exp\{-\int_t^0 \alpha(\phi_\tau(\phi))d\tau\}, \quad t \leq 0,$$

which proves estimate (12.13) for $s = 1$.

Suppose that inequality (12.13) holds for all $s \leq l_1$. We prove that it then
holds for $s = l_1 + 1$. To prove this we differentiate l_1 times the identity obtained
by replacing the variable ψ by its value ψ_t^i in equation (12.14), and write the
result in the form of the equality:

$$\frac{d}{dt}(D_\phi^{l_1} \psi_t^i) = \frac{\partial a(\phi_t(\phi))}{\partial \phi_t}(D_\phi^{l_1} \psi_t^i) + R(\phi_t(\phi)),$$

where

$$R(\phi_t(\phi)) = D_\phi^{l_1} \left[\frac{\partial a(\phi_t(\phi))}{\partial \phi_t} \psi_t^i \right] - \frac{\partial a(\phi_t(\phi))}{\partial \phi_t} D_\phi^{l_1} \psi_t^i.$$

Since $R(\phi_t(\phi))$ is a differential expression which contains the terms

$$D_\phi^{l_1-j} \left(\frac{\partial a(\phi_t(\phi))}{\partial \phi_t} \right) D_\phi^j \left(\frac{\partial \phi_t(\phi)}{\partial \phi_\nu} \right), \quad j = 0, 1, \ldots, l_1 - 1, \ \nu = 1, \ldots, m$$

with constant coefficients, inequality (12.13) taken for $s \leq l_1$ enables us to estimate $R(\phi_t(\phi))$ as follows:

$$D_\phi^{l_1-j} \left(\frac{\partial a(\phi_t(\phi))}{\partial \phi_t} \right) = \sum_{\sigma=1}^{l_1-j} D_{\phi_t}^\sigma \left(\frac{\partial a(\phi_t(\phi))}{\partial \phi_t} \right) \times$$

$$\times \sum_\alpha c_{\sigma\alpha} (D_\phi \phi_t(\phi))^{\alpha_1} (D_\phi^2 \phi_t(\phi))^{\alpha_2} \ldots (D_\phi^{l_1-j} \phi_t(\phi))^{\alpha_{l_1-j}},$$

where

$$\alpha_1 + \alpha_2 + \ldots + \alpha_{l_1-j} = \sigma, \quad \alpha_1 + 2\alpha_2 + \ldots + (l_1-j)\alpha_{l_1-j} = l_1 - j.$$

Therefore

$$\left\| D_\phi^{l_1-j} \frac{\partial a(\phi_t(\phi))}{\partial \phi_t} \right\| \leq K_1 \exp\left\{ -\int_t^0 (l_1-j)\alpha(\phi_\tau(\phi))d\tau - (l_1-j)\epsilon t \right\}, \ t \leq 0,$$

which leads to the estimate

$$\left\| D_\phi^{l_1-j} \frac{\partial a(\phi_t(\phi))}{\partial \phi_t} \cdot D_\phi^j \frac{\partial \phi_t(\phi)}{\partial \phi} \right\| \leq$$

$$\leq K_1 K \exp\left\{ -\int_t^0 [(l_1-j)+j+1]\alpha(\phi_\tau(\phi))d\tau \right\} \exp\{-(l_1-j+1)\epsilon t\} \leq$$

$$\leq K_1 K \exp\left\{ -\int_t^0 (l_1+1)\alpha(\phi_\tau(\phi))d\tau - (l_1+1)\epsilon t \right\}, \quad t \leq 0.$$

This proves for $R(\phi_t(\phi))$ the following estimate:

$$\|R(\phi_t(\phi))\| \leq K_2 \exp\left\{ -\int_t^0 (l_1+1)\alpha(\phi_t(\phi))d\tau - (l_1+1)\epsilon t \right\}, \ t \leq 0, \quad (12.16)$$

where K_2 is a positive constant independent of ϕ.

Since

$$\psi_0^i = (\underbrace{0, \ldots, 0}_{i-1}, 1, \underbrace{0, \ldots, 0}_{m-i}),$$

it follows that $[D_\phi^{l_1}\psi_t^i]_{t=0} = 0$ for $l_1 \geq 1$. Consequently the derivative $D_\phi^{l_1}\psi_t^i$ satisfies the relation

$$D_\phi^{l_1}\psi_t^i = -\int_t^0 \Omega_\tau^t \left(\frac{\partial a}{\partial \phi}\right) R(\phi_\tau(\phi)) d\tau, \tag{12.17}$$

where $\Omega_\tau^t(\partial a/\partial \phi)$ denotes the fundamental matrix of solutions of system (12.14), which becomes the identity matrix when $t = \tau$.

We now estimate the integral (12.17). Clearly, the columns of the matrix $\Omega_\tau^t(\partial a/\partial \phi)$ are the solutions $\psi_t^i(\tau, \phi)$ of system (12.14), which at $t = \tau$ take the values of unit vectors e_i in the space E^m. A standard argument using inequalities (12.10) for $S_1(\phi)$ enables us to obtain the estimate for $\psi_t^i(\tau, \phi)$:

$$\|\psi_t^i(\tau, \phi)\| \leq K \exp\left\{-\int_t^\tau \alpha(\phi_\tau(\phi)) d\tau\right\}, \quad t \leq \tau,$$

which shows that $\Omega_\tau^t(\partial a/\partial \phi)$ satisfies an inequality of the form

$$\left\|\Omega_\tau^t \left(\frac{\partial a}{\partial \phi}\right)\right\| \leq K \exp\left\{-\int_t^\tau \alpha(\phi_\tau(\phi)) d\tau\right\}, \quad t \leq \tau.$$

This inequality allows one to obtain an estimate for integral (12.17) and to obtain for $D_\phi^{l_1}\psi_t^i$ an estimate of the form:

$$\|D_\phi^{l_1}\psi_t^i\| \leq K K_2 \int_t^0 \exp\left\{-\int_t^\tau \alpha(\phi_\tau(\phi)d\tau - \int_\tau^0 (l_1+1)\alpha(\phi_\tau(\phi)) d\tau - (l_1+1)\epsilon\tau\right\} d\tau \leq$$

$$\leq K_3 \exp\left\{-\int_t^0 \alpha(\phi_\tau(\phi)) d\tau - (l_1+1)\epsilon t\right\} \times$$

$$\times \int_t^0 \exp\left\{-\int_\tau^0 l_1\alpha(\phi_\tau(\phi)) d\tau\right\} d\tau, \quad t \leq 0. \tag{12.18}$$

On estimating the integral

$$I_1 = \int_t^0 \exp\left\{-\int_\tau^0 l_1\alpha(\phi_\tau(\phi)) d\tau\right\} d\tau, \quad t \leq 0,$$

we find that

$$I_1 = -t \exp\left\{-\int_t^0 l_1\alpha(\phi_\tau(\phi)) d\tau\right\} -$$

$$- \int_t^0 l_1 \tau\alpha(\phi_\tau(\phi)) \exp\left\{-\int_\tau^0 l_1\alpha(\phi_t(\phi)) d\tau\right\} d\tau \leq$$

$$\leq -t \exp\left\{-\int_t^0 l_1\alpha(\phi_\tau(\phi)) d\tau\right\} \leq K(\epsilon) \exp\left\{-\int_t^0 l_1\alpha(\phi_\tau(\phi)) d\tau - \epsilon t\right\}, \tag{12.19}$$

where $K(\epsilon)$ is a constant dependent on ϵ.

Using (12.18), (12.19) we can write the estimate for the function $D_\phi^{l_1} \psi_t^i$ as

$$\|D_\phi^{l_1} \psi_t^i\| \le K_3 K(\epsilon) \exp\left\{-\int_t^0 (l_1 + 1)\alpha(\phi_\tau(\phi))d\tau - (l_1 + 2)\epsilon t\right\}, \quad t \le 0,$$

whence it follows that

$$\|D_\phi^{l_1+1} \phi_t(\phi)\| \le K \exp\left\{-\int_t^0 (l_1 + 1)\alpha(\phi_\tau(\phi))d\tau - \epsilon_1 t\right\}, \quad t \le 0, \quad (12.20)$$

for some positive $K = K(\epsilon_1)$ and arbitrary fixed $\epsilon_1 > 0$. Inequality (12.20) proves (12.13) for $s = l_1 + 1$; hence it proves inequality (12.13).

Now consider the function

$$x_t(\tau, \phi, x_0) = \Omega_\tau^t(\phi)x_0,$$

where $\Omega_\tau^t(\phi)$ is the fundamental matrix of solutions of the system

$$\frac{dx}{dt} = P(\phi_t(\phi))x, \quad (12.21)$$

which becomes the identity matrix for $t = \tau$.

Let us prove the estimate

$$\|D_\phi^s x_t(\tau, \phi, x_0)\| \le K \exp\left\{-\int_\tau^t \beta(\phi_\tau(\phi))d\tau - \int_\tau^0 s\alpha(\phi_\tau(\phi))d\tau - \epsilon\tau\right\}\|x_0\|$$
$$(12.22)$$

for arbitrary $\tau \le t \le 0$, $0 \le s \le l$, arbitrarily small $\epsilon > 0$ and some K which depends on s, ϵ but does not depend on t, τ, ϕ.

We proceed by induction. For $s = 0$ inequality (12.22) follows from condition (12.10) for the matrix $S(\phi)$ by applying standard estimates relating to the function $V = \langle S(\phi_t(\phi))x_t(\tau, \phi, x_0), x_t(\tau, \phi, x_0)\rangle$ and its derivative with respect to t.

Assuming that estimate (12.22) holds for all $0 \le s \le l_1 - 1$, we prove it for $s = l_1$. To do this, we differentiate l_1 times the identity obtained by substituting $x_t(\tau, \phi, x_0)$ for x in equation (12.21) and write the result as:

$$\frac{d}{dt}[D_\phi^{l_1} x_t(\tau, \phi, x_0)] = P(\phi_t(\phi))[D_\phi^{l_1} x_t(\tau, \phi, x_0)] + R_2(\phi_t(\phi), x_t(\tau, \phi, x_0)),$$

where

$$R_2(\phi_t(\phi), x_t(\tau, \phi, x_0)) = D_\phi^{l_1}[P(\phi_t(\phi))x_t(\tau, \phi, x_0)] - P(\phi_t(\phi))D_\phi^{l_1} x_t(\tau, \phi, x_0).$$

The function $R_2(\phi_t(\phi), x_t(\tau, \phi, x_0))$ is a differential expression which contains the terms $D_\phi^{l_1-j} P(\phi_t(\phi)) D_\phi^j x_t(\tau, \phi, x)$ $(j = 0, \ldots, l_1 - 1)$ with constant coefficients. Inequalities (12.13) and (12.22) enable us to estimate $R_2(\phi_t(\phi), x_t(\tau, \phi, x_0))$ as follows.

We have

$$D_\phi^{l_1-j} P(\phi_t(\phi)) = \sum_{\sigma=1}^{l_1-j} D_{\phi_t}^\sigma P(\phi_t(\phi)) \sum_\alpha c_{\sigma\alpha} (D_\phi \phi_t(\phi))^{\alpha_1} \times$$

$$\times (D_\phi^2(\phi_t(\phi))^{\alpha^2} \ldots (D_\phi^{l_1-j} \phi_t(\phi))^{\alpha_{l_1-j}},$$

$$\alpha_1 + \alpha_2 + \ldots + \alpha_{l_1-j} = \sigma, \quad \alpha_1 + 2\alpha_2 + \ldots + (l_1 - j)\alpha_{l_1-j} = l_1 - j.$$

This leads to the estimate

$$\|D_\phi^{l_1-j} P(\phi_t(\phi))\| \le K \exp\left\{-(l_1 - j) \int_t^0 \alpha(\phi_\tau(\phi)) d\tau - \epsilon t\right\}, \quad t \le 0.$$

But then for any $0 \le j \le l_1 - 1$ and $\tau \le t \le 0$

$$\|D_\phi^{l_1-j} P(\phi_t(\phi)) D_\phi^j x_t(\tau, \phi, x_0)\| \le K_4 K \exp\left\{-(l_1 - j) \int_t^0 \alpha(\phi_\tau(\phi)) d\tau\right\} \times$$

$$\times \exp\left\{-\int_\tau^t \beta(\phi_\tau(\phi)) d\tau - j \int_\tau^0 \alpha(\phi_\tau(\phi)) d\tau - \epsilon(t + \tau)\right\} \|x_0\| \le$$

$$\le K_4 K \exp\left\{-\int_\tau^t \beta(\phi_\tau(\phi)) d\tau\right\} \exp\left\{-l_1 \int_\tau^0 \alpha(\phi_\tau(\phi)) d\tau - 2\epsilon\tau\right\} \|x_0\|,$$

from which it follows that

$$\|R_2(\phi_t(\phi), x_t(\tau, \phi, x_0))\| \le$$

$$\le K_5 \exp\left\{-\int_\tau^t \beta(\phi_\tau(\phi)) d\tau - l_1 \int_\tau^0 \alpha(\phi_\tau(\phi)) d\tau - 2\epsilon\tau\right\} \|x_0\| \qquad (12.23)$$

for all $\tau \le t \le 0$ and some positive K_5 independent of t, ϕ, τ.

Since $[D_\phi^s x_t(\tau, \phi, x_0)]_{t=\tau} = 0$ for any $s \ge 1$, it follows from the equation for $D_\phi^{l_1} x_\bullet(\tau, \phi, x_0)$ that

$$D_\phi^{l_1} x_t(\tau, \phi, x_0) = \int_\tau^t \Omega_{\tau_1}^t(\phi) R_2(\phi_{\tau_1}(\phi), x_{\tau_1}(\tau, \phi, x_0)) d\tau_1. \qquad (12.24)$$

Using inequality (12.23) we can obtain the estimate for integral (12.24):

$$\|D_\phi^{l_1} x_t(\tau, \phi, x_0)\| \le K_6 \int_\tau^t \exp\Big\{ - \int_{\tau_1}^t \beta(\phi_\tau(\phi)) d\tau - \int_\tau^{\tau_1} \beta(\phi_\tau(\phi)) d\tau \Big\} \times$$

$$\times \exp\Big\{ -l_1 \int_\tau^0 \alpha(\phi_\tau(\phi)) d\tau - 2\epsilon\tau \Big\} d\tau_1 \|x_0\| =$$

$$= K_6(t - \tau) \exp\Big\{ - \int_\tau^t \beta(\phi_\tau(\phi)) d\tau \Big\} \times$$

$$\times \exp\Big\{ -l_1 \int_\tau^0 \alpha(\phi_\tau(\phi)) d\tau - 2\epsilon\tau \Big\} \|x_0\|, \quad \tau \le t \le 0 \qquad (12.25)$$

It follows from inequality (12.25) that

$$\|D_\phi^{l_1} x_t(\tau, \phi, x_0)\| \le$$

$$\le K(\epsilon_1) \exp\Big\{ - \int_\tau^t \beta(\phi_\tau(\phi)) d\tau - \int_\tau^0 l_1 \alpha(\phi_\tau(\phi)) d\tau - \epsilon_1 \tau \Big\} \|x_0\|$$

for all $\tau \le t \le 0$, arbitrarily small $\epsilon_1 > 0$ and some $K(\epsilon_1) > 0$. This inequality proves estimate (12.22) for $s = l_1$ and consequently, proves it for any $0 \le s \le l$.

If we set $t = 0$ in inequality (12.22) we obtain

$$\|D_\phi^s x_0(\tau, \phi, x_0)\| \le$$

$$\le K \exp\Big\{ - \int_\tau^0 [\beta(\phi_\tau(\phi)) + s\alpha(\phi_\tau(\phi))] d\tau - \epsilon\tau \Big\} \|x_0\| \qquad (12.26)$$

for all $\tau \le 0$. Since we are considering an exponentially stable invariant torus of system (12.2), the homogeneous system of equations corresponding to (12.2) has a Green's function $G_0(\tau, \phi)$ defined by the matrix $\Omega_\tau^0(\phi)$ for $\tau \le 0$ and by the zero matrix for $\tau > 0$. Thus for $\tau > 0$ inequality (12.3) holds, while for $\tau \le 0$ it becomes:

$$|x_0(\tau, \phi, f(\phi_\tau(\phi)))|_l \le Ke^{\gamma\tau} |f|_l. \qquad (12.27)$$

We now prove (12.27).

We have:

$$D_\phi^s x_0(\tau, \phi, f(\phi_\tau(\phi))) = \sum_{p+\sigma \le s} D_\phi^\sigma \frac{\partial^p x_0(\tau, \phi, f)}{\partial f^p} \times$$

$$\times \sum_\alpha c_{\sigma p \alpha} (D_\phi f(\phi_\tau(\phi)))^{\alpha_1} \dots (D_\phi^l f(\phi_\tau(\phi)))^{\alpha_l},$$

$$\alpha_1 + \alpha_2 + \dots + \alpha_l = p, \quad \alpha_1 + 2\alpha_2 + \dots + l\alpha_l + \sigma = s \qquad (12.28)$$

Now since

$$D_\phi^j f(\phi_\tau(\phi)) = \sum_{\sigma_1 \le j} D_{\phi_\tau}^{\sigma_1} f(\phi_\tau(\phi)) \sum_\alpha c_{\sigma_1 \alpha} (D_\phi \phi_\tau(\phi))^{\alpha_1} \dots (D_\phi^j \phi_\tau(\phi))^{\alpha_j},$$

$$\alpha_1 + \alpha_2 + \dots + \alpha_j = \sigma_1, \quad \alpha_1 + 2\alpha_2 + \dots + j\alpha_j = j,$$

we have

$$\|D_\phi^j f(\phi_\tau(\phi))\| \le K \exp\left\{ -j \int_\tau^0 \alpha(\phi_\tau(\phi)) d\tau - \epsilon\tau \right\} |f|_j \qquad (12.29)$$

for all $\tau \le 0$.

The function $x_0(\tau, \phi, f)$ is linear with respect to f, therefore

$$\left\| D_\phi^\sigma \frac{\partial^p x_0(\tau, \phi, f)}{\partial f^p} \right\| \le \text{const} \left[\|D_\phi^\sigma x_0(\tau, \phi, f)\| \delta_{0p} + \sum_{j=1}^n \|D_\phi^\sigma x_0(\tau, \phi, e_j)\| \delta_{1p} \right],$$

where e_j is a unit vector, and $\delta_{\alpha p}$ is equal to 0 for $\alpha \ne p$ and 1 for $\alpha = p$.

In view of inequality (12.26), we have the estimate

$$\left\| D_\phi^\sigma \frac{\partial^p x_0(\tau, \phi, f)}{\partial f^p} \right\| \le$$

$$\le K \left[\|f\| \delta_{0p} + \delta_{1p} \right] \exp\left\{ -\int_\tau^0 \beta(\phi_\tau(\phi)) d\tau \right\} \times$$

$$\times \exp\left\{ -\sigma \int_\tau^0 \alpha(\phi_\tau(\phi)) d\tau - \epsilon\tau \right\}, \quad \tau \le 0. \qquad (12.30)$$

Inequalities (12.29), (12.30) lead to the estimate

$$\|D_\phi^s x_0(\tau, \phi, f(\phi_\tau(\phi)))\| \le$$

$$\le K \exp\left\{ -\int_\tau^0 [\beta(\phi_\tau(\phi)) + s\alpha(\phi_\tau(\phi))] d\tau - \epsilon\tau \right\} |f|_s$$

for any $s \le l$ and $\tau \le 0$. It follows from the second inequality of (12.11) that

$$|x_0(\tau, \phi, f(\phi_\tau(\phi)))|_l \le K e^{(\gamma - \epsilon)\tau} |f|_l, \quad \tau \le 0,$$

which is sufficient for inequality (12.3) to hold for $\epsilon < \gamma$. This completes the proof of Theorem 2.

By looking at the proof of Theorem 2 we see that it holds if the fundamental matrices of the solutions $\Omega^t_\tau(\phi)$ and $\Omega^t_\tau(\partial a(\phi)/\partial\phi)$ of systems (12.21) and (12.14) satisfy the inequalities

$$\|\Omega^t_\tau(\phi)\| \leq K \exp\left\{-\int_\tau^t \beta(\phi_\tau(\phi))d\tau\right\}, \quad \tau \leq t,$$

$$\left\|\Omega^t_\tau\left(\frac{\partial a(\phi)}{\partial\phi}\right)\right\| \leq K \exp\left\{\int_t^\tau \alpha(\phi_\tau(\phi))d\tau\right\}, \quad \tau \geq t \qquad (12.31)$$

respectively, where $\beta(\phi)$ and $\alpha(\phi)$ are subject to conditions (12.6), (12.9).

Since the matrices $\Omega^t_0(\phi)$, $\Omega^t_0(\partial a(\phi)/\partial\phi)$ satisfy the identities of form (4.2):

$$\Omega^t_0(\phi_\theta(\phi)) = \Omega^{t+\theta}_\theta(\phi), \quad \Omega^t_0\left(\frac{\partial a(\phi_\theta(\phi))}{\partial\phi}\right) = \Omega^{t+\theta}_\theta\left(\frac{\partial a(\phi)}{\partial\phi}\right),$$

it follows from the inequalities

$$\|\Omega^t_0(\phi)\| \leq K \exp\left\{-\int_0^t \beta(\phi_\tau(\phi))d\tau\right\}, \quad 0 \leq t,$$

$$\left\|\Omega^t_0\left(\frac{\partial a(\phi)}{\partial\phi}\right)\right\| \leq K \exp\left\{\int_t^0 \alpha(\phi_\tau(\phi))d\tau\right\}, \quad t \leq 0 \qquad (12.32)$$

that inequalities (12.31) hold. Indeed, according to the given identities and inequalities (12.32) we have, for example, the estimate for $\Omega^t_\tau(\phi)$:

$$\|\Omega^t_\tau(\phi)\| = \|\Omega^{t-\tau+\tau}_\tau(\phi)\| = \|\Omega^{t-\tau}_0(\phi_\tau(\phi))\| \leq$$

$$\leq K \exp\left\{-\int_0^{t-\tau} \beta(\phi_{\tau_1}(\phi_\tau(\phi)))d\tau_1\right\} = K \exp\left\{-\int_0^{t-\tau} \beta(\phi_{\tau+\tau_1}(\phi))d\tau_1\right\} =$$

$$= K \exp\left\{-\int_\tau^t \beta(\phi_\tau(\phi))d\tau\right\}, \quad t - \tau \geq 0,$$

which is the same as estimate (12.31) for $\Omega^t_\tau(\phi)$.

This gives rise to the following statement.

Corollary 1. *The statements of Theorem 2 remain true if the matrices $\Omega^t_0(\phi)$ and $\Omega^t_0(\partial a/\partial\phi)$ satisfy inequalities (12.32), where the values $\beta(\phi)$ and $\alpha(\phi)$ are subject to conditions (12.6) and (12.9).*

In particular, if the inequalities

$$\|\Omega^t_0(\phi)\| \leq K e^{-\beta_0 t}, \quad t \geq 0,$$

$$\|\Omega^t_0(\partial a/\partial\phi)\| \leq K e^{-\alpha_0 t}, \quad t \leq 0$$

hold, where $\beta_0 = \text{const} > 0$, $\alpha_0 = \text{const} \geq 0$, then the conditions (12.6), (12.9) are fulfilled for l satisfying $l < \beta_0/\alpha_0$. Then the invariant torus (12.1) of system (12.2) belongs to the space $C^l(\mathcal{T}_m)$ with $l < \beta_0/\alpha_0$.

It should be noted that the number l defined by the conditions of Theorem 2 as the minimal possible smoothness may be attained. This can easily be seen from the example of the system

$$\frac{d\phi}{dt} = -\sin \phi, \quad \frac{dx}{dt} = -bx + \sin^b \phi, \tag{12.33}$$

which consists of two scalar equations. As was shown in [89], the invariant torus (12.1) of this system is defined by the function

$$u(\phi) = -\tan \frac{\phi}{2} \int_{\pi/2}^{\phi/2} \frac{\cos^{2b-1} t}{\sin t} \, dt, \quad 0 < \phi < 2\pi,$$

which behaves as $-c\phi^b \log \phi$ as $\phi \to 0$. Taking b to be a positive integer and taking the bth derivative of $u(\phi)$ we see that this derivative is not bounded. Inequality (12.9) for system (12.33) has the form $\inf_{\phi \in T_1} [b - l \cos \phi] > 0$ and holds for $l = b - 1$. Consequently, the smoothness of torus (12.1) of system (12.33), which is defined by inequality (12.9), is the same as its real smoothness.

3.13. Smoothness properties of Green's functions, the invariant torus and the decomposing transformation of an exponentially dichotomous system

We shall be considering the system of equations (11.1) under the assumption that it is exponentially dichotomous. Let $G_0(\tau, \phi)$ be a Green's function for this system. Suppose that $G_0(\tau, \phi)$ satisfies the inequality

$$\|G_0(\tau, \phi)\| \leq Ke^{-\gamma|\tau|}, \quad \tau \in \mathbf{R}, \tag{13.1}$$

where K, γ are positive constants independent of ϕ.

In order to elucidate the smoothness properties of the function $G_0(\tau, \phi)$ with respect to ϕ, we introduce the constant $\alpha > 0$ defined by the inequality

$$\|\Omega_0^t(\partial a/\partial \phi)\| \leq Ke^{\alpha|t|}, \quad t \in \mathbf{R}, \tag{13.2}$$

where $\Omega_0^t(\partial a/\partial \phi)$ is the fundamental matrix of the solutions of system (12.14).

Theorem 1. *Let $l \geq 0$, $P \in C^l(\mathcal{T}_m) \cap C_{\text{Lip}}(\mathcal{T}_m)$ and*

$$\gamma > l\alpha. \tag{13.3}$$

Then $G_0(\tau, \phi) \in C^l(\mathcal{T}_m)$ for any $\tau \in \mathbf{R}$ and

$$\|D_\phi^s G_0(\tau, \phi)\| \leq K_1 e^{-(\widehat{\gamma} - s\alpha)|\tau|}, \quad \tau \in \mathbf{R}, \tag{13.4}$$

where $\widehat{\gamma} = \gamma - \epsilon$, $0 \leq s \leq l$, ϵ is an arbitrarily small positive number and $K_1 = K_1(\epsilon)$ is a positive constant independent of ϕ, $\phi \in \mathcal{T}_m$.

Because estimate (13.1) holds and the matrix $C(\phi) = G_0(0, \phi)$ belongs to the space $C'(\mathcal{T}_m)$, the theorem holds for $l = 0$. We therefore suppose that $l \geq 1$. Consider the differences

$$Z_t(\tau, \bar{\phi}, \phi) = G_t(\tau, \phi) - G_t(\tau, \bar{\phi}),$$

where $G_t(\tau, \phi) = G_0(\tau - t, \phi_t(\phi))$, $\bar{\phi} = \phi + \Delta e_i$, $e_i = (0, \dots, 1, \dots, 0)$ is the unit vector, and Δ is a scalar constant. Since $G_t(\tau, \phi)$ is a Green's function for system (12.21), $Z_t(\tau, \bar{\phi}, \phi)$ satisfies the matrix equation

$$\frac{dZ}{dt} = P(\phi_t(\bar{\phi}))Z + [P(\phi_t(\phi)) - P(\phi_t(\bar{\phi}))]G_t(\tau, \phi).$$

This has a unique bounded solution on $\mathbf{R}$ given by the expression

$$\int_{-\infty}^{+\infty} G_t(s, \bar{\phi})[P(\phi_s(\phi)) - P(\phi_s(\bar{\phi}))]G_s(\tau, \phi)ds. \tag{13.5}$$

Inequality (13.1) ensures that the function $Z_t(\tau, \bar{\phi}, \phi)$ is bounded on $\mathbf{R}$, so that the function $Z_t(\tau, \bar{\phi}, \phi)$ is equal to the function (13.5). But then

$$G_0(\tau, \phi) - G_0(\tau, \bar{\phi}) = \int_{-\infty}^{+\infty} G_0(s, \bar{\phi})[P(\phi_s(\phi)) - P(\phi_s(\bar{\phi}))]G_s(\tau, \phi)ds.$$

Since

$$\lim_{\Delta \to 0} G_0(s, \bar{\phi}) \frac{P(\phi_s(\phi)) - P(\phi_s(\bar{\phi}))}{\Delta} G_s(\tau, \phi) = G_0(s, \phi) \times$$

$$\times \sum_{\nu=1}^{m} \frac{\partial P(\phi_s(\phi))}{\partial (\phi_s)_\nu} \frac{\partial (\phi_s(\phi))_\nu}{\partial \phi_i} G_s(\tau, \phi) \tag{13.6}$$

uniformly with respect to $\phi \in T_m$ and $(\tau, s) \in D_2$, it follows that

$$\lim_{\Delta \to 0} \frac{G_0(\tau, \phi) - G_0(\tau, \bar{\phi})}{\Delta} = \int_{-\infty}^{+\infty} J(s, \tau, \phi)ds \qquad (13.7)$$

when the integral on the right hand side is uniformly convergent. Here D_2 is any bounded domain of the τ, s plane and $J(s, \tau, \phi)$ is the right hand side of equality (13.6). For $J(s, \tau, \phi)$ we have the estimate

$$\|J(s, \tau, \phi)\| \le K^2 |P|_1 e^{-\gamma|s|-\gamma|s-\tau|+\alpha|s|} = K_1 e^{-(\gamma-\alpha)|s|} e^{-\gamma|s-\tau|}, \qquad (13.8)$$

from which we see that integral (13.7) converges uniformly with respect to $(\tau, \phi) \in \mathbf{R} \times T_m$ and therefore defines the derivative $\partial G_0(\tau, \phi)/\partial \phi_i$.

By formulae (13.7), (13.8) we have the estimate

$$\left\| \frac{\partial G_0(\tau, \phi)}{\partial \phi_i} \right\| \le K_1 \left(\frac{2}{2\gamma - \alpha} + |\tau| \right) e^{-(\gamma-\alpha)|\tau|}, \quad \tau \in \mathbf{R}, \ \phi \in T_m, \qquad (13.9)$$

for $\partial G_0(\tau, \phi)/\partial \phi_i$, and this proves the theorem for $l = 1$.

It follows from equality (13.7) that

$$\frac{\partial G_0(\tau, \phi)}{\partial \phi_i} = \int_{-\infty}^{+\infty} G_0(s, \phi) \sum_{\nu=1}^{m} \frac{\partial P(\phi_s(\phi))}{\partial(\phi_s)_\nu} \frac{\partial(\phi_s(\phi))_\nu}{\partial \phi_i} G_s(\tau, \phi)ds \qquad (13.10)$$

for all $(\tau, \phi) \in \mathbf{R} \times T_m$. Using identity (13.10) and taking derivatives we prove Theorem 1 for any $l \ge 1$.

In particular, for $l = 2$ by using estimates (13.1)–(13.3), (13.9) and the inequality

$$\|D_\phi^2 \phi_t(\phi)\| \le K e^{2\alpha|t|+\epsilon|t|}, \quad t \in \mathbf{R}$$

which follows from (13.2) by reasoning similar to that in §3.11, we obtain the following estimate for the derivative of the integrand expression $J(s, \tau, \phi)$ in formula (13.10):

$$\|D_\phi J(s, \tau, \phi)\| \le$$

$$\le K_2 [e^{-(\gamma-2\alpha-\epsilon)|s|} e^{-\gamma|\tau-s|} + e^{-(\gamma-2\alpha)|s|} e^{-(\gamma-\alpha)|\tau-s|}] \le$$

$$\le K_2 [e^{-(\gamma-\widehat{\alpha}_1)|s|} e^{-\gamma|\tau-s|} + e^{-(\gamma-\widehat{\alpha}_1)|s|} e^{-\gamma_1|s-\tau|}], \qquad (13.11)$$

where we have set $\widehat{\alpha}_1 = 2\alpha + \epsilon$, $\gamma_1 = \gamma - \alpha$.

Estimate (13.11) is of the form (13.8). Consequently the integral

$$\int_{-\infty}^{+\infty} D_\phi J(s, \tau, \phi) ds$$

converges to the function $D_\phi \partial G_0(\tau, \phi)/\partial \phi_i$ uniformly with respect to $(\tau, \phi) \in \mathbf{R} \times \mathcal{T}_m$. This implies the estimate

$$\left\| D_\phi \frac{\partial G_0(\tau, \phi)}{\partial \phi_i} \right\| \leq 2K_2 \left(\frac{1}{2\gamma - \widehat{\alpha}_1} + \frac{1}{2\gamma_1 - \widehat{\alpha}_1} + |\tau| \right) e^{-(\gamma - 2\widehat{\alpha})|\tau|}$$

for all $t \in \mathbf{R}$, $\phi \in \mathcal{T}_m$. Because the number $\epsilon > 0$ is arbitrary we have that the function $D_\phi \partial G_0(\tau, \phi)/\partial \phi_i$ can be estimated by inequality (13.4) for $l = 2$. This completes the proof of the theorem for $l = 2$.

By applying standard estimates for higher derivatives of the integrand function $J(s, \tau, \phi)$ in formula (13.10) and using induction we see that the theorem holds for any $l > 2$.

Theorem 2. *Let the conditions of the previous theorem be satisfied and $f \in C^l(\mathcal{T}_m)$. Then the invariant torus (12.1) of system (12.2) belongs to the space $C^l(\mathcal{T}_m)$ and admits the estimate*

$$|u|_l \leq K|f|_l.$$

To prove the theorem we need to use estimate (13.1) for the Green's function $G_0(\tau, \phi)$ and the estimate for the function $\phi_t(\phi)$ in the form

$$\|D_\phi^s \phi_t(\phi)\| \leq K e^{(s\alpha + \epsilon)|t|}, \quad t \in \mathbf{R},$$

which follows from condition (13.2). These estimates clearly lead to the inequality

$$|G_0(\tau, \phi) f(\phi_\tau(\phi))|_l \leq K e^{-(\widehat{\gamma} - l\alpha)|\tau|} |f|_l \qquad (13.12)$$

for all $(\tau, \phi) \in \mathbf{R} \times \mathcal{T}_m$. If we choose $\epsilon > 0$ sufficiently small, it will follow from inequality (13.3) that the difference $\widehat{\gamma} - l\alpha$ is positive, and this is sufficient for Theorem 2 to hold.

We now find a relation between the index γ of exponential decay of the Green's function $G_0(\tau, \phi)$ for the exponentially dichotomous system (11.1) and the parameters of the matrices $S(\phi)$ and $\widehat{S}(\phi)$ which make the system exponentially dichotomous.

Lemma 1. *Let $a \in C_{\text{Lip}}(\mathcal{T}_m)$, $P \in C(\mathcal{T}_m)$. If the non-singular matrix $S(\phi) \in C'(\mathcal{T}_m)$ satisfies inequalities*

$$\max_{\|x\|=1} |\langle S(\phi)x, x \rangle| \leq \mu,$$

$$\max_{\|x\|=1} \left\langle \left[\tfrac{1}{2}\dot{S}(\phi) + S(\phi)P(\phi)\right]x, x \right\rangle \leq -\beta \tag{13.13}$$

with positive constants μ and β then the Green's function $G_0(\tau, \phi)$ for system (11.1) satisfies inequalities (13.1) with index

$$\gamma = \beta/\mu. \tag{13.14}$$

To prove the lemma, we first of all note that its conditions imply that system (11.1) is exponentially dichotomous. Denote the solution of this system belonging to the separatrix manifold M^+ by $\phi_t(\phi)$, $x_t^+(\phi, x_0)$. Since, according to the Remark in §3.8, this manifold lies in the cone (8.17) for fixed ϕ, we have

$$\langle S(\phi_t(\phi))x_t^+(\phi, x_0),\ x_t^+(\phi, x_0) \rangle \geq \epsilon \|x_t^+(\phi, x_0)\|^2 \tag{13.15}$$

for some positive $\epsilon > 0$. The function $V = \langle S(\phi_t(\phi))x_t^+(\phi, x_0), x_t^+(\phi, x_0) \rangle$ satisfies the inequality

$$\frac{dV}{dt} = 2\left\langle \left[\tfrac{1}{2}\dot{S}(\phi_t(\phi)) + S(\phi_t(\phi))P(\phi_t(\phi))\right] x_t^+(\phi, x_0), x_t^+(\phi, x_0) \right\rangle \leq$$

$$\leq -2\beta \|x_t^+(\phi, x_0)\|^2,$$

and therefore

$$\frac{dV}{dt} \leq -\frac{2\beta \|x_t^+(\phi, x_0)\|^2 V}{|\langle S(\phi_t(\phi))x_t^+(\phi, x_0), x_t^+(\phi, x_0) \rangle|} \leq -2\frac{\beta}{\mu} V.$$

Integrating this inequality, we obtain

$$\langle S(\phi_t(\phi))x_t^+(\phi, x_0), x_t^+(\phi, x_0) \rangle \leq \mu \exp\left\{-2\frac{\beta}{\mu}(t - \tau)\right\} \|x_t^+(\phi, x_0)\| \tag{13.16}$$

for all $t \geq \tau$ and any $(\tau, \phi) \in \mathbf{R} \times \mathcal{T}_m$. Using condition (13.15) and inequality (13.16) we have:

$$\|x_t^+(\phi, x_0)\| \leq \frac{\mu}{\epsilon} \exp\left\{-\frac{\beta}{\mu}(t - \tau)\right\} \|x_\tau(\phi, x_0)\|, \quad t \geq \tau. \tag{13.17}$$

In similar fashion we can prove the following estimate for the motion $\phi_t(\phi)$, $x_t^-(\phi, x_0)$ of system (11.1) belonging to the separatrix manifold M^-:

$$\|x_t^-(\phi, x_0)\| \leq \frac{\mu}{\epsilon} \exp\left\{\frac{\beta}{\mu}(t - \tau)\right\} \|x_\tau^-(\phi, x_0)\|, \quad t \leq \tau. \tag{13.18}$$

But then the Green's function $G_t(\tau, \phi)$ for system (12.21) defined by relation (9.2) for solutions satisfying inequalities (13.17) and (13.18), satisfies the inequality

$$\|G_t(\tau, \phi)\| \leq K \exp\left\{-\frac{\beta}{\mu}|t - \tau|\right\}, \quad (t, \tau, \phi) \in \mathbf{R} \times \mathbf{R} \times \mathcal{T}_m \tag{13.19}$$

with the constant K independent of ϕ. Since $G_0(\tau, \phi) = G_t(\tau, \phi)|_{t=0}$, it follows from (13.19) that estimate (13.1) holds for the function $G_0(\tau, \phi)$ with γ equal to the value given in (13.14).

The index of exponential growth of the matrix $\Omega_0^t(\partial a/\partial \phi)$ can easily be related to the parameters of the matrix $S_1(\phi)$ in the space $\mathfrak{R}_1$. Indeed, if the positive constant α is chosen such that the inequality

$$\sup_{\phi \in \mathcal{T}_m} \inf_{S_1 \in \mathfrak{R}_1} \max_{\|\psi\|=1} \frac{|\langle [\frac{1}{2}\dot{S}_1(\phi) + S_1(\phi)\frac{\partial a(\phi)}{\partial \phi}]\psi, \psi\rangle|}{\langle S_1(\phi)\psi, \psi\rangle} < \alpha \tag{13.20}$$

holds, then estimate (13.2) holds for the matrix $\Omega_0^t(\partial a/\partial \phi)$ with α defined by condition (13.20). This follows at once by making standard estimates of the derivative of the function $V_1 = \langle S_1(\phi_t(\phi))\psi_t(\phi, \psi_0), \psi_t(\phi, \psi_0)\rangle$, where $\psi_t(\phi, \psi_0)$ is a solution of system (12.14).

The next statement follows from what has been said as well as Theorem 1 and Theorem 2.

Corollary 1. *Let $a, f, P \in C^l(\mathcal{T}_m)$, $l \geq 1$. Suppose that the non-singular matrix $S(\phi) \in C^l(\mathcal{T}_m)$ satisfies inequalities (13.13) with positive constants μ and β. If*

$$\beta > l\mu\alpha, \tag{13.21}$$

where α is chosen from condition (13.20), then the Green's function $G_0(\tau, \phi)$ for system (11.1) and the invariant torus (12.1) of system (12.2) belong to the space $C^l(\mathcal{T}_m)$.

It is now easy to answer the question on smoothness of a decomposing transformation of an exponentially dichotomous system when this transformation is decomposing.

Theorem 3. *Let* $a, P \in C^l(\mathcal{T}_m)$, $l \geq 1$. *Suppose that the non-singular symmetric matrix* $S(\phi) \in C'(\mathcal{T}_m)$ *satisfies inequalities* (13.13) *with positive constants* μ, β *and admits the representation*

$$S(\phi) = Q(\phi)\mathrm{diag}\{D_1(\phi), -D_2(\phi)\}Q^*(\phi) \qquad (13.22)$$

with a non-singular matrix $Q(\phi) \in C^l(\mathcal{T}_m)$ *and positive definite blocks* $D_1(\phi)$ *and* $D_2(\phi)$ *for all* $\phi \in \mathcal{T}_m$. *If inequality* (13.3) *holds then system of equations* (11.1) *is* $C^l(\mathcal{T}_m)$*-block decomposable.*

The proof of Theorem 3 follows from the theorem on the $C'(\mathcal{T}_m)$-block decomposability of system (11.1) and the properties of the matrices $C(\phi) = G_0(0, \phi)$ and $Q(\phi)$, namely: $C(\phi) \in C^l(\mathcal{T}_m)$ and $Q(\phi) \in C^l(\mathcal{T}_m)$.

In conclusion, we consider two particular forms of system (11.1) when the smoothness of the Green's function, the invariant torus and the decomposing matrix are the same as that of the functions a, P and f.

The first relates to the case when $a = \omega = $ const. Then we can take an arbitrarily small number α and inequality (13.3) will be satisfied by any finite number l. In this case the smoothness of the Green's function and the invariant torus of the system in question are the same as that of the coefficients a, P and f of the system, while the smoothness of the decomposing transformation is the same as that of the functions a, P, f and the matrix $Q(\phi)$ in representation (13.22).

The second case relates to an exponentially dichotomous system such that its function $C(\phi) = G_0(0, \phi)$ does not depend on ϕ. The smoothness of an invariant torus of such a system is equal to that of the functions a and f, and the decomposing transformation does not depend on ϕ. It is easy to see that we have the second case if the matrix $P(\phi)$ has the following form

$$P(\phi) = S\mathrm{diag}\{D_1(\phi), D_2(\phi)\}S^{-1},$$

where S is a constant matrix.

3.14. Galerkin's method for the construction of an invariant torus

The construction of an invariant torus of system (12.2) requires approximation methods. One such method which is convenient for practical realization is *Galerkin's iteration method*. According to this method, the function $u(\phi)$ that defines the invariant torus (12.1) of system (12.2) is constructed as a limit of successive Galerkin approximations of a periodic solution of the equation

$$Lu(\phi) = f(\phi), \tag{14.1}$$

where L is the differential operator

$$Lu = \sum_{\nu=1}^{m} \frac{\partial u}{\partial \phi_\nu} a_\nu(\phi) - P(\phi)u \tag{14.2}$$

defined on the function space $H^r(\mathcal{T}_m)$. Galerkin's approximations are given by the trigonometric polynomials

$$W_N(\phi) = \sum_{\|k\| \leq N} W_k^{(N)} e^{i(k,\phi)} \tag{14.3}$$

the coefficients $W_k^{(N)}$ of which are solutions of the system of algebraic equations

$$(LW_N(\phi), e^{-i(k,\phi)})_0 = (f(\phi), e^{-i(k,\phi)})_0, \quad \|k\| \leq N. \tag{14.4}$$

The conditions under which the Galerkin's approximations exist for any $N \geq 0$ and converge to a solution of equation (14.1) as $N \to +\infty$ can be obtained from the following statement.

Lemma 1. *Suppose that the following conditions hold:*
1) *the functions $a(\phi)$ and $P(\phi)$ belong to the space $C^r(\mathcal{T}_m)$;*
2) *the matrix*

$$b_0(\phi) = -\frac{P(\phi) + P^*(\phi)}{2} = \frac{1}{2} \sum_{\nu=1}^{m} \frac{\partial a_\nu(\phi)}{\partial \phi_\nu} E, \tag{14.5}$$

where E is the identity matrix, is positive-definite for all $\phi \in \mathcal{T}_m$;
3) *for each $s, r \geq s > 1$ and arbitrary $u \in H^{s+1}(\mathcal{T}_m)$ the inequality*

$$(Lu, u)_s \geq \gamma \|u\|_s^2 - \delta \|u\|_0^2 \tag{14.6}$$

holds, where γ and δ are positive constants independent of u.

Then Galerkin's approximation sequence $\{W_N(\phi)\}$ $(N = 0, 1, 2, \ldots)$ is defined for any function $f \in H^r(\mathcal{T}_m)$ and converges in $H^s(\mathcal{T}_m)$ for $1 \leq s < r$ to a function $u = u_0(\phi)$ which satisfies the inequality

$$\|u_0 - W_N\|_s \leq C N^{-(r-s-1))} \|f\|_r, \tag{14.7}$$

where C is a positive constant independent of N and f.

To prove the lemma, we denote by $S_N f(\phi)$ the partial sum of the Fourier series of the function $f(\phi) \simeq \sum_k f_k e^{i(k,\phi)}$:

$$S_N f(\phi) = \sum_{\|k\| \leq N} f_k e^{i(k,\phi)}.$$

In this notation system (14.4) is equivalent to the single equation

$$S_N L W_N(\phi) = S_N f(\phi)$$

and has a solution if the equation

$$S_N L W_N(\phi) = 0$$

has the only solution $W_N(\phi) \equiv 0$.

Since for any function (14.3) we have

$$(S_N L W_N(\phi),\ W_N(\phi))_0 = (L W_N(\phi), S_N W_N(\phi))_0 =$$

$$= (L W_N(\phi), W_N(\phi))_0 = \left(\frac{L + L^*}{2} W_N(\phi), W_N(\phi) \right)_0 =$$

$$= (b_0(\phi) W_N(\phi), W_N(\phi))_0$$

and the matrix $b_0(\phi)$ is positive-definite, $(b_0(0)x, x) \geq \gamma_0 \|x\|^2$ (we can assume that $\gamma_0 = \gamma$), we have the inequality

$$(S_N L W_N(\phi), W_N(\phi))_0 \geq \gamma(W_N(\phi), W_N(\phi))_0, \qquad (14.8)$$

from which it follows that system (14.4) can be solved for any integer $N \geq 0$ and an arbitrary function $f \in H^0(T_m)$.

Moreover, inequality (14.8) leads to the following estimate for the Galerkin approximations W_N:

$$(f, W_N)_0 \geq \gamma \|W_N\|_0^2,$$

which by Schwarz's inequality yields

$$\|W_N\|_0 \leq \gamma^{-1} \|f\|_0.$$

We use inequalities (14.6) to estimate the Galerkin approximations $W_N(\phi)$ with respect to the $H^r(T_m)$ norm. Since

$$(S_N L W_N, W_N)_r = (L W_N, W_N)_r = (f, W_N)_r,$$

it follows from (14.6) and Schwarz's inequality that

$$\gamma\|W_N\|_r^2 - \delta\|W_N\|_0^2 \le \|f\|_r\|W_N\|_r. \tag{14.9}$$

Solving inequality (14.9) we obtain the estimate

$$\|W_N\|_r \le \gamma_1^{-1}\|f\|_r, \tag{14.10}$$

where

$$\gamma_1 = 2\gamma/(1 + \sqrt{1 + 4\delta/\gamma}).$$

By Sobolev's compactness theorem, inequality (14.10) ensures that the sequence of Galerkin approximations is compact in the space $H^s(\mathcal{T}_m)$ for $s < r$.

Let $\{W_{N_j}\}$ $(j = 1, 2, \ldots)$ be a subsequence of $\{W_N\}$ that converges in $H^s(\mathcal{T}_m)$ to the limit W^0:

$$\lim_{j \to \infty} \|W_{N_j} - W^0\|_s = 0.$$

Consider the product

$$(S_{N_j} L W_{N_j}, g)_0 = (L W_{N_j}, S_{N_j} g)_0 = f(S_{N_j} g)_0, \tag{14.11}$$

where g is a trigonometric polynomial in $\mathcal{P}(\mathcal{T}_m)$. If j is sufficiently large, then $S_{N_j} g = g$ and equality (14.11) takes the form

$$(L W_{N_j}, g)_0 = (f, g)_0.$$

But then

$$(W_{N_j}, L^* g)_0 = (f, g)_0,$$

and by taking the limit we find that

$$(W^0, L^* g)_0 = (f, g)_0. \tag{14.12}$$

For $r > s \ge 1$ $H^s(\mathcal{T}_m) \subset H^1(\mathcal{T}_m)$, therefore $W^0 \in H^1(\mathcal{T}_m)$. It then follows from (14.2) that $(LW^0, g)_0 = (f, g)_0$. Since the function $g \in \mathcal{P}(\mathcal{T}_m)$ is arbitrary, by virtue of the theorem on the completeness of the space $H^0(\mathcal{T}_m)$, it follows that this relation can hold only if

$$LW^0 = f \quad \text{a.e.,} \tag{14.13}$$

that is, when LW^0 is equal to f as elements of the space $H^0(\mathcal{T}_m)$.

We shall be using equality (14.13) to prove that the Galerkin approximations converge. To this end we consider an arbitrary subsequence of $\{W_N\}$ that is convergent in $H^s(\mathcal{T}_m)$. Let $\overline{W}_0 \in H^s(\mathcal{T}_m)$ be its limit. Now equality (14.13) holds for $\overline{W}_0$, therefore

$$L(W^0 - \overline{W}^0) = 0$$

almost everywhere. But then

$$0 = (L(W^0 - \overline{W}^0), W^0 - \overline{W}^0)_0 = (b_0(W^0 - \overline{W}^0), W^0 - \overline{W}^0)_0 \geq \gamma \|W^0 - \overline{W}^0\|_0^2,$$

which is possible only if

$$W^0 = \overline{W}^0 \qquad\qquad (14.14)$$

almost everywhere. Relation (14.14) shows that the sequence of Galerkin approximations has only one accumulation point in $H^0(\mathcal{T}_m)$. Denote it by $u_0(\phi)$. Then $\lim_{N \to \infty} \|W_N - u_0\|_0 = 0$ and, since the sequence $\{W_N\}$ ($N = 0, 1, 2, \ldots$) is compact in $H^s(\mathcal{T}_m)$, we have the limit relation

$$\lim_{N \to \infty} \|W_N - u_0\|_s = 0.$$

To complete the proof of the lemma, it remains to find an estimate for $W_N - u_0$. To do this, using the equality of the form (14.13) for $u = u_0(\phi)$, we write down the value of the operator L at $W_N - u_0$. We have:

$$L(W_N - u_0) = (E - S_N)LW_N - (E - S_N)f,$$

where E is the identity operator.

Then, for $0 \leq s \leq r - 1$,

$$(L(W_N - u_0), W_N - u_0)_s = ((E - S_N)LW_N - (E - S_N)f, W_N - u_0)_s \leq$$
$$\leq (\|(E - S_N)LW_N\|_s + \|(E - S_N)f\|_s)\|W_N - u_0\|_s. \qquad (14.15)$$

For any function $f \in H^r(\mathcal{T}_m)$, $s \leq r$, we clearly have the estimate

$$\|(E - S_N)f\|_s = \left\| \sum_{\|k\| > N} f_k e^{i(k,\phi)} \right\|_s \leq N^{s-r}\|(E - S_N)f\|_r \leq N^{s-r}\|f\|_r,$$

from which it follows that inequality (14.15) can be written in the form

$$(L(W_N - u_0), W_N - u_0)_s \leq N^{s-r+1}(\|LW_N\|_{r-1} + \|f\|_{r-1})\|W_N - u_0\|_s. \quad (14.16)$$

Taking into account the form of the operator L we obtain the following estimate for $\|LW_N\|_{r-1}$:

$$\|LW_N\|_{r-1} \leq \sum_{\nu=1}^{m} \left\| a_\nu(\phi) \frac{\partial W_N(\phi)}{\partial \phi_\nu} \right\|_{r-1} + \|P(\phi)W_N(\phi)\|_{r-1} \leq$$

$$\leq C \left(\sum_{\nu=1}^{m} |a_\nu|_{r-1} \|W_N\|_r + |P|_{r-1}\|W_N\|_{r-1} \right) \leq C_1\|W_N\|_r,$$

where $C_1 = C(\sum_{\nu=1}^{m} |a_\nu|_{r-1} + |P|_{r-1})$, C being a positive constant independent of W_N, a_ν and P.

Now by using estimate (14.10), we find that

$$\|LW_N\|_{r-1} \leq C_1\gamma_1^{-1}\|f\|_r,$$

and from this it follows that we can rewrite inequality (14.16) as:

$$(L(W_N - u_0), W_N - u_0)_s \leq N^{s-r+1}\|f\|_r(1 + C_1\gamma_1^{-1})\|u_0 - W_N\|_s. \quad (14.17)$$

Setting $s = 0$ in (14.17) and using the fact that

$$(L(W_N - u_0), W_N - u_0)_0 \geq \gamma\|W_N - u_0\|_0^2,$$

we have that $\|W_N - u_0\|_0 \leq C'N^{1-r}\|f\|_r$. If we take $1 \leq s \leq r - 1$ in (14.17) and apply estimate (14.6) to the left hand side of (14.17), we find after simple calculations that

$$\|W_N - u_0\|_s \leq C_2 N^{-(r-s-1)}\|f\|_r,$$

where C_2 is a positive constant which depends only on $C_1, \gamma_1, \gamma, \delta$. Inequality (14.7) is proved.

By Sobolev's embedding theorem the limit function u_0 belongs to $C^l(\mathcal{T}_m)$ if

$$s > m/2 + l. \quad (14.18)$$

If we take $l \geq 1$ in inequality (14.18), we obtain the inclusion $u_0 \in C^1(\mathcal{T}_m)$. It follows from the fact that

$$Lu_0(\phi) = f(\phi) \quad (14.19)$$

holds almost everywhere, that the latter equality holds for all $\phi \in \mathcal{T}_m$. This means that $u_0(\phi)$ is a solution of equation (14.1) and belongs to $C^l(\mathcal{T}_m)$. Thus, if the conditions of Lemma 1 are satisfied for

$$r > m/2 + l, \quad l \geq 1, \quad (14.20)$$

then the sequence of Galerkin approximations $\{W_N(\phi)\}$ $(N = 0,1,\ldots)$ is defined for any $f \in H^r(\mathcal{T}_m)$ and is convergent with respect to the norm of the space $C^l(\mathcal{T}_m) \cap H^s(\mathcal{T}_m)$, $s < r$, to the solution of equation (14.1) $u_0(\phi)$ with convergence rate defined by inequality (14.7).

Sufficient conditions for inequalities (14.6) to hold are known from the papers of Friedrichs [149] and Moser [89]. If the assumptions of Lemma 1 on the smoothness of the functions a and P hold, then these conditions can be replaced by the requirement that the inequalities

$$\beta_0(\phi) + s\alpha_0(\phi) - \tfrac{1}{2}\mu(\phi) \geq \gamma, \quad s = 0,1,\ldots,r, \tag{14.21}$$

hold, where γ is an arbitrarily small number,

$$\min_{\|\psi\|=1} \left\langle \frac{\partial a(\phi)}{\partial \phi}\,\psi, \psi \right\rangle \geq \alpha_0(\phi),$$

$$\max_{\|x\|=1} \langle P(\phi)x, x \rangle \leq -\beta_0(\phi), \tag{14.22}$$

$$\mu(\phi) = \sum_{\nu=1}^{m} \frac{\partial a_\nu(\phi)}{\partial \phi_\nu}, \quad \phi \in \mathcal{T}_m.$$

We shall prove this statement in the next section, but here we use it to substantiate Galerkin's procedure for constructing an invariant torus (12.1) of system (12.2).

Theorem 1. *Let* $a, P \in C^r(\mathcal{T}_m)$ *and suppose that the inequalities* (14.21) *hold for* $s = 0$ *and* $s = r$. *If* r *satisfies inequality* (14.20) *then the sequence of Galerkin approximations* $\{W_N(\phi)\}$ $(N = 0,1,\ldots)$ *is defined for any function* $f \in H^r(\mathcal{T}_m)$ *and converges in* $C^l(\mathcal{T}_m) \cap H^{r-1}(\mathcal{T}_m)$ *to an invariant torus* (12.1) *of system* (12.2) *such that the inequality*

$$\|u - W_N\|_s \leq C N^{-(r-s-1)}\|f\|_r \tag{14.23}$$

holds for any $0 \leq s \leq r - 1$ *and some positive constant* C *independent of* N *and* f.

Proof. Taking into account the form of the matrix $b_0(\phi)$ and using inequality (14.22) we can write

$$\langle b_0(\phi)x, x \rangle = -\left\langle \frac{P(\phi) + P^*(\phi)}{2}\,x, x \right\rangle - \tfrac{1}{2}\mu(\phi)\|x\|^2 \geq [\beta_0(\phi) - \tfrac{1}{2}\mu(\phi)]\|x\|^2.$$

But then it follows from inequality (14.21) considered for $s = 0$ that

$$\langle b_0(\phi)x, x \rangle \geq \gamma \|x\|^2.$$

This proves that the matrix $b_0(\phi)$ is positive-definite for all $\phi \in T_m$ and this is sufficient for all the conditions of Lemma 1 to be satisfied.

According to Lemma 1, the sequence of Galerkin approximations $\{W_N(\phi)\}$ $(N = 0, 1, \ldots)$ is defined for any function $f \in H^r(T_m)$ and converges with respect to the norm of the space $H^s(T_m)$, $s < r$, to a function $u_0(\phi)$ satisfying inequality (14.7). The limit function $u_0(\phi)$ satisfies equation (14.1) almost everywhere. If inequalities (14.20) are satisfied, then $u_0(\phi) \in C^l(T_m) \subset C^1(T_m)$, so that the function $u_0(\phi)$ is a solution of equation (14.1). Consequently, the function $u = u_0(\phi)$ defines an invariant torus (12.1) of system (12.2). The limit relation

$$\lim_{N \to \infty} \|u_0 - W_N\|_s = 0, \quad 0 \leq s \leq r - 1$$

implies the convergence of the sequence $\{W_N(\phi)\}$ $(N = 0, 1, \ldots)$ in the space $C^l(T_m) \cap H^s(T_m)$ to a function $u_0(\phi)$. This completes the proof of Theorem 1.

Inequality (14.23) gives an estimate of the deviation of the Nth Galerkin approximation from the exact solution of equation (14.1) with respect to the norm of the space $H^s(T_m)$.

By using Sobolev's embedding theorem we can estimate the difference between W_N and u_0 in the uniform metric as follows:

$$|W_N - u_0|_s \leq C N^{-(l-s-1)} \|f\|_r, \quad 0 \leq s \leq l.$$

It should be noted that condition (14.20) is engendered by the method used to substantiate Galerkin's method. It is typical for functional methods used in mathematical physics [14], [62]. It can be weakened by making it independent of the dimension m of the invariant torus by combining the convergence conditions for the sequence of Galerkin approximations with the conditions for solvability in $C(T_m)$ of equation (14.1). This leads us to the following statement.

Theorem 2. *Let* $a, P \in C^2(T_m)$ *and suppose that the inequalities*

$$\inf_{\phi \in T_m} \beta(\phi) > 0, \quad \inf_{\phi \in T_m} [\beta(\phi) + \alpha(\phi)] > 0,$$

$$\inf_{\phi \in T_m} \left[\beta_0(\phi) - \tfrac{1}{2}\mu(\phi) \right] > 0, \tag{14.24}$$

$$\inf_{\phi \in T_m} \left[\beta_0(\phi) + 2\alpha_0(\phi) - \tfrac{1}{2}\mu(\phi) \right] > 0$$

hold, where $\beta(\phi), \alpha(\phi), \beta_0(\phi), \alpha_0(\phi)$ and $\mu(\phi)$ are the quantities defined by inequalities (12.7), (12.8), (14.22).

Then the sequence of Galerkin approximations $\{W_N(\phi)\}$ $(N = 0, 1, \ldots)$ is defined for any function $f \in H^2(T_m)$ and converges in $H^1(T_m)$ to a function $u(\phi)$ which defines an invariant torus (12.1) of system (12.2).

Proof. According to Theorem 12.2, the first two inequalities of (14.24) imply that there exists an invariant torus (12.1) of system (12.2); furthermore, it belongs to the space $C^1(T_m)$. The last two inequalities of (14.24) imply that the sequence of Galerkin approximations $\{W_N(\phi)\}$ $(N = 0, 1, \ldots)$ exists and converges in $H^1(T_m)$ for any function $f \in H^2(T_m)$. The limit function $u_0(\phi)$ of the sequence $\{W_N(\phi)\}$ $(N = 0, 1, \ldots)$ satisfies equation (14.1) almost everywhere, as follows from the statements of Lemma 1 for $r = 2$ and $s = 1$.

Since the torus (12.1) is smooth, it follows that

$$Lu(\phi) = f(\phi), \quad \phi \in T_m.$$

Then the difference $u_0(\phi) - u(\phi)$ satisfies the equality

$$L(u_0(\phi) - u(\phi)) = 0$$

almost everywhere. By the third inequality of (14.24),

$$(L(u(\phi) - u_0(\phi)), \ u(\phi) - u_0(\phi))_0 =$$
$$= (b_0(\phi)(u(\phi) - u_0(\phi)), u(\phi) - u_0(\phi))_0 \geq \gamma \|u(\phi) - u_0(\phi)\|_0^2,$$

where $\gamma = \text{const} > 0$. But then

$$\|u - u_0\|_0 = 0,$$

which implies that the functions u and u_0 are equal for all $\phi \in T_m$. This proves Theorem 2.

It should be noted that the inequality

$$\|W_N(\phi) - u(\phi)\|_0 \leq CN^{-1}\|f\|_2,$$

which determines the rate of convergence of the Galerkin approximations $\{W_N(\phi)\}$ $(N = 0, 1, 2, \ldots)$ as $N \to \infty$ under the conditions of Theorem 2, follows from estimate (14.7) taken for $s = 0$ and $r = 2$ together with the equality $u_0(\phi) = u(\phi)$, $\phi \in T_m$, which holds if the conditions of Theorem 2 are satisfied.

We also note that the first two inequalities of (14.24) can be replaced by the requirement that an invariant torus (12.1) of system (12.2) should exist. This does not affect the validity of Theorem 2.

3.15. Proof of the main inequalities for the substantiation of Galerkin's method

To render the statements of the preceding section mathematically rigorous, it remains to show that condition 3 of Lemma 1 of §13 holds when inequality (14.21) is true for $s = 0$ and $s = r$. To prove this fact we shall assume that the functions a and P belong to the space $C^r(T_m)$.

For an arbitrary function $u \in H^{s+1}(T_m)$ consider the product $(Lu, u)_s$. For $s = 0$ we have

$$(Lu, u)_0 = \left(\frac{L + L^*}{2} u, u\right)_0 = \left(-\left(\frac{P + P^*}{2} + \frac{1}{2}\sum_{\nu=1}^{m} \frac{\partial a_\nu}{\partial \phi_\nu} E\right)u, u\right)_0 = (b_0 u, u)_0.$$

As was shown in the proof of Theorem 1 of §3.13, it follows from inequality (14.21) considered for $s = 0$ that the matrix $b_0(\phi)$ is positive-definite for all $\phi \in T_m$.

Using the relation given at the end of §1.3 we have for $s = 1$:

$$(Lu, u)_1 = (Lu, Ku)_0 = (Lu, u)_0 + (Lu, -\Delta u)_0 =$$

$$= (b_0 u, u)_0 + \sum_j \left(\frac{\partial}{\partial \phi_j}, Lu, \frac{\partial u}{\partial \phi_j}\right)_0 = (b_0 u, u)_0 +$$

$$+ \sum_j \left(L\frac{\partial u}{\partial \phi_j}, \frac{\partial u}{\partial \phi_j}\right)_0 + \sum_j \left(\sum_\nu \frac{\partial a_\nu}{\partial \phi_j}\frac{\partial u}{\partial \phi_\nu} - \frac{\partial P}{\partial \phi_j}u, \frac{\partial u}{\partial \phi_j}\right)_0 =$$

$$= \sum_j \left(b_0 \frac{\partial u}{\partial \phi_j} + \sum_\nu \frac{\partial a_\nu}{\partial \phi_j}\frac{\partial u}{\partial \phi_\nu}, \frac{\partial u}{\partial \phi_j}\right)_0 + \sum_j \left(-\frac{\partial P}{\partial \phi_j}u, \frac{\partial u}{\partial \phi_j}\right)_0 + (b_0 u, u)_0 =$$

$$= \sum_{\nu,j} \left(\left[b_0 E_{\nu j} + \frac{\partial a_\nu}{\partial \phi_j} E\right]\frac{\partial u}{\partial \phi_\nu}, \frac{\partial u}{\partial \phi_j}\right)_0 + \Phi_1 = I_1 + \Phi_1,$$

where $E_{\nu j}$ denotes the $n \times n$ matrix equal to E for $\nu = j$ and 0 for $\nu \neq j$.

We estimate the functional Φ_1. Taking its form into account, we have:

$$|\Phi_1\| \leq \sum_j \left\|\frac{\partial P}{\partial \phi_j} u\right\|_0 \left\|\frac{\partial u}{\partial \phi_j}\right\|_0 + \|b_0 u\|_0 \|u\|_0 \leq$$

$$\leq 2C[|P|_1 + \sum_\nu |a_\nu|_1] \|u\|_1 \|u\|_0 \leq$$

$$\leq C\Big(\sum_\nu |a_\nu|_1 + |P|_1\Big)\Big(\mu^2 \|u\|_1^2 + \frac{1}{\mu^2}\|u\|_0^2\Big) =$$

$$= C'\Big(\mu^2 \|u\|_1^2 + \frac{1}{\mu^2}\|u\|_0^2\Big), \qquad (15.1)$$

where μ is a sufficiently small positive constant and C is a constant independent of u and the coefficients of L.

We impose the condition that the inequality

$$\sum_{\nu,j} \Big\langle \Big[b_0 E_{\nu j} + \frac{\partial a_\nu}{\partial \phi_j} E\Big] \eta_\nu, \eta_j \Big\rangle \geq \gamma_1 \sum_\nu \|\eta_\nu\|^2 \qquad (15.2)$$

holds for any collection $\eta_1, \ldots, \eta_m$ of n-dimensional vectors, $\eta_\nu = (\eta_\nu^1, \ldots, \eta_\nu^n)$ $(\nu = 1, \ldots, m)$ and some $\gamma_1 > 0$.

We defer until the end of the section the proof that by taking γ_1 sufficiently small we can make the above inequality hold .

If inequality (15.2) holds, then we have the following lower estimate for I_1:

$$I_1 = \frac{1}{(2\pi)^m} \int_0^{2\pi} \cdots \int_0^{2\pi} \sum_{\nu,j} \Big\langle \Big[b_0 E_{\nu j} + \frac{\partial a_\nu}{\partial \phi_j} E\Big] \frac{\partial u}{\partial \phi_\nu}, \frac{\partial u}{\partial \phi_j} \Big\rangle d\phi_1 \ldots d\phi_m \geq$$

$$\geq \gamma_1 \sum_\nu \frac{1}{(2\pi)^m} \int_0^{2\pi} \cdots \int_0^{2\pi} \left\|\frac{\partial u}{\partial \phi_\nu}\right\|^2 d\phi_1 \ldots d\phi_m =$$

$$= \gamma_1 \sum_\nu \left\|\frac{\partial u}{\partial \phi_\nu}\right\|_0^2 \geq \gamma_1 C_1 \|u\|_1^2 - \gamma_1 C_2 \|u\|_0^2, \qquad (15.3)$$

where C_1 and C_2 are positive constants independent of u and the coefficients of L.

Using both inequalities (15.1) and (15.3) we obtain estimate (14.6) for $s = 1$:

$$(Lu, u)_1 \geq (\gamma_1 C_1 - C'\mu^2)\|u\|_1^2 - (\gamma_1 C_2 + C'/\mu^2)\|u\|_0^2 = \gamma\|u\|_1^2 - \delta\|u\|_0^2,$$

where $\gamma = \gamma_1 C_1 - C'/\mu^2 > 0$, $\delta = \gamma_1 C_2 + C'/\mu^2 > 0$.

For $s = 2$ we use the relation given at the end of §1.3 to obtain

$$(Lu, u)_2 = (Lu, K^2 u)_0 = (KLu, Ku)_0 = (LKu, Ku)_0 +$$

$$+((KL - LK)u, Ku)_0 = (b_0 Ku, Ku)_0 + ((KL - LK)u, Ku)_0.$$

But

$$(KL - LK)u = -(\Delta L - L\Delta)u = -\Big[\Delta\Big(\sum_\nu a_\nu \frac{\partial u}{\partial \phi_\nu}\Big) - \sum_\nu a_\nu \frac{\partial \Delta u}{\partial \phi_\nu}\Big] +$$

$$+[\Delta(Pu) - P\Delta u] = -\Big[\sum_{\nu,j} 2\frac{\partial a_\nu}{\partial \phi_j} \frac{\partial^2 u}{\partial \phi_j \partial \phi_\nu} + \sum_\nu (\Delta a_\nu)\frac{\partial u}{\partial \phi_\nu}\Big] +$$

$$+\Big[2\sum_j \frac{\partial P}{\partial \phi_j} \frac{\partial u}{\partial \phi_j} + (\Delta P)u\Big] = -\sum_{\nu,j} 2\frac{\partial a_\nu}{\partial \phi_j} \frac{\partial^2 u}{\partial \phi_j \partial \phi_\nu} + L_1 u,$$

so that we have the following representation for $(Lu, u)_2$:

$$(Lu, u)_2 = -\Big(b_0 \Delta u + 2\sum_{\nu,j} \frac{\partial a_\nu}{\partial \phi_j} \frac{\partial^2 u}{\partial \phi_\nu \partial \phi_j}, Ku\Big)_0 +$$

$$+(b_0 u + L_1 u, Ku)_0 = -\sum_{\nu,j}\Big(\Big[b_0 E_{\nu j} + 2\frac{\partial a_\nu}{\partial \phi_j}E\Big]\frac{\partial^2 u}{\partial \phi_\nu \partial \phi_j}, u\Big)_1 +$$

$$+(L_1' u, u)_1 = (L_2 u, u)_1 + (L_1' u, u)_1 = (L_2 u, u)_1 + \Phi_2(u), \qquad (15.4)$$

where L_2 is the second order differential operator:

$$L_2 = -\sum_{\nu,j}\Big(b_0 E_{\nu j} + 2\frac{\partial a_\nu}{\partial \phi_j}E\Big)\frac{\partial^2}{\partial \phi_\nu \partial \phi_j},$$

and L_1' is the first order differential operator, the coefficients of which are the functions a_ν, P and their derivatives up to second order.

Taking into account the form of the operator L_1' and the inequality for $2\|u\|_2\|u\|_1$ given in §1.10, we estimate $(L_1' u, u)_1 = \Phi_2(u)$ as follows:

$$\|\Phi_2(u)\| = |(L_1' u, u)_1| \le \|L_1' u\|_0 \|u\|_2 \le 2C\Big(\sum_\nu |a_\nu|_2 + |P|_2\Big)\|u\|_2\|u\|_1 =$$

$$= 2C'\|u\|_2\|u\|_1 \le C'\Big(\frac{3}{2}\mu^2\|u\|_2^2 + \frac{1}{2\mu^6}\|u\|_0^2\Big), \qquad (15.5)$$

where μ is a sufficiently small positive constant and C is a positive constant independent of u and the coefficients of L.

We require that the operator L_2 be elliptic. For this it is sufficient that the inequality

$$\sum_{\nu,j}\left\langle\left[b_0 E_{\nu j}+2\frac{\partial a_\nu}{\partial\phi_j}E\right]\eta_\nu,\eta_j\right\rangle\geq\gamma_2\sum_\nu\|\eta_\nu\|^2$$

holds for any collection $\eta_1,\ldots,\eta_m$ of n-dimensional vectors $\eta_\nu=(\eta_\nu^1,\ldots,\eta_\nu^n)$ $(\nu=1,\ldots,m)$ and some $\gamma_2>0$.

We defer to the end of this section the proof that this inequality is satisfied by taking γ_2 sufficiently small.

Then by using Gårding's inequality we can make an estimate for $(L_2 u,u)_1$ from below as follows:

$$(L_2 u,u)_1\geq\gamma_1\|u\|_2^2-\delta_1\|u\|_0^2,\tag{15.6}$$

where γ_1 and δ_1 are positive constants that depend only on $\sum_{\nu=1}^m|a_\nu|_2+|P|_2$.

Combining inequalities (15.5) and (15.6), we obtain an estimate for $(Lu,u)_2$ in the required form:

$$(Lu,u)_2\geq\gamma\|u\|_2^2-\delta\|u\|_0^2,$$

where γ and δ are positive constants independent of u.

Passing to the general case, we denote by L_2^s the operator

$$L_2^s=-\left(b_0\Delta+s\sum_{\nu,j}\frac{\partial a_\nu}{\partial\phi_j}\frac{\partial^2}{\partial\phi_\nu\partial\phi_j}\right)=-\sum_{\nu,j}\left(b_0 E_{\nu j}+s\frac{\partial a_\nu}{\partial\phi_j}E\right)\frac{\partial^2}{\partial\phi_\nu\partial\phi_j}$$

and prove that for $s\geq 2$

$$(Lu,u)_s=(L_2^s u,u)_{s-1}+\Phi_s(u),\tag{15.7}$$

where $\Phi_s(u)$ is a functional admitting the estimate

$$|\Phi_s(u)|\leq C'\|u\|_s\|u\|_{s-1}.\tag{15.8}$$

Here $C'=C(\sum_\nu|a_\nu|_s+|P|_s)$, C being a positive constant independent of u and of the coefficients of the operator L.

Since representation (15.7) and estimate (15.8) have been proved for $s=2$, we can assume that formula (15.7) and estimate (15.8) are proved for all $s = 2, \ldots, p$ because (15.7) is equal to (15.4) and (15.8) to (15.5). Consider the product $(Lu, u)_{p+1}$ under this assumption. Using the relation given at the end of §1.3, we have:

$$(Lu, u)_{p+1} = (Lu, Ku)_p = (Lu, u)_p + (Lu, -\Delta u)_p =$$

$$= \sum_j \left(\frac{\partial}{\partial \phi_j} Lu, \frac{\partial u}{\partial \phi_j} \right)_p + \Phi'_{p+1}(u),$$

where

$$|\Phi'_{p+1}(u)| = |(Lu, u)_p| \le \|Lu\|_{p-1} \|u\|_{p+1} \le$$

$$\le \left(\sum_\nu \left\| a_\nu \frac{\partial u}{\partial \phi_\nu} \right\|_{p-1} + \|Pu\|_{p-1} \right) \|u\|_{p+1} \le$$

$$\le C_1 \left(\sum_\nu |a_\nu|_{p-1} + |P|_{p-1} \right) \|u\|_{p+1} \|u\|_p = C' \|u\|_{p+1} \|u\|_p.$$

Since the following chain of equalities holds:

$$\sum_j \left(\frac{\partial}{\partial \phi_j} Lu, \frac{\partial u}{\partial \phi_j} \right)_p = \sum_j \left(\sum_\nu a_\nu \frac{\partial^2}{\partial \phi_j \partial \phi_\nu} - P \frac{\partial u}{\partial \phi_j} + \sum_{\nu,j} \frac{\partial a_\nu}{\partial \phi_j} \frac{\partial u}{\partial \phi_\nu} - \right.$$

$$- \frac{\partial P}{\partial \phi_j} u, \frac{\partial u}{\partial \phi_j} \right)_p = \sum_j \left(L \frac{\partial u}{\partial \phi_j}, \frac{\partial u}{\partial \phi_j} \right)_p + \sum_{\nu,j} \left(\frac{\partial a_\nu}{\partial \phi_j} \frac{\partial u}{\partial \phi_\nu}, \frac{\partial u}{\partial \phi_j} \right)_p -$$

$$- \sum_j \left(\frac{\partial P}{\partial \phi_j} u, \frac{\partial u}{\partial \phi_j} \right)_p = \sum_j \left(L_2^p \frac{\partial u}{\partial \phi_j}, \frac{\partial u}{\partial \phi_j} \right)_{p-1} - \sum_{\nu,j} \left(\frac{\partial a_\nu}{\partial \phi_j} \frac{\partial^2 u}{\partial \phi_\nu \partial \phi_j}, u \right)_p +$$

$$+ \sum_j \Phi_p \left(\frac{\partial u}{\partial \phi_j} \right) - \sum_{\nu,j} \left(\frac{\partial^2 a_\nu}{\partial \phi_j^2} \frac{\partial u}{\partial \phi_\nu}, u \right)_p - \sum_j \left(\frac{\partial P}{\partial \phi_j} u, \frac{\partial u}{\partial \phi_j} \right)_p =$$

$$= \sum_j \left(\frac{\partial}{\partial \phi_j} (L_2^p u), \frac{\partial u}{\partial \phi_j} \right)_{p-1} - \sum_{\nu,j} \left(\frac{\partial a_\nu}{\partial \phi_j} \frac{\partial^2 u}{\partial \phi_\nu \partial \phi_j}, u \right)_p + \Phi''_{p+1}(u) =$$

$$= - \sum_j \left(L_2^p u, \frac{\partial^2 u}{\partial \phi_j^2} \right)_{p-1} - \sum_{\nu,j} \left(\frac{\partial a_\nu}{\partial \phi_j} \frac{\partial^2 u}{\partial \phi_\nu \partial \phi_j}, u \right)_p + \Phi''_{p+1}(u),$$

where

$$|\Phi''_{p+1}(u)| \le \sum_j \left| \Phi_p \left(\frac{\partial u}{\partial \phi_j} \right) \right| + \sum_{\nu,j} \left\| \frac{\partial^2 a_\nu}{\partial \phi_j^2} \frac{\partial u}{\partial \phi_\nu} \right\|_{p-1} \|u\|_{p+1} +$$

$$+ \sum_j \Big\| \frac{\partial P}{\partial \phi_j} u \Big\|_p \Big\| \frac{\partial u}{\partial \phi_j} \Big\|_p + \Big| \sum_{j_1} \Big(\frac{\partial b_0}{\partial \phi_{j_1}} \Delta u + P \sum_{\nu,j} \frac{\partial^2 a_\nu}{\partial \phi_{j_1} \partial \phi_j} \frac{\partial^2 u}{\partial \phi_\nu \partial \phi_j}, \frac{\partial u}{\partial \phi_{j_1}} \Big)_{p-1} \Big| \le$$

$$\le C \sum_j \Big\| \frac{\partial u}{\partial \phi_j} \Big\|_p \Big\| \frac{\partial u}{\partial \phi_j} \Big\|_{p-1} + C_2 \sum_\nu |a_\nu|_{p+1} \|u\|_p \|u\|_{p+1} +$$

$$+ C_2 |P|_{p+1} \|u\|_p \|u\|_{p+1} + C_2 |P|_p \|u\|_{p+1} \|u\|_p +$$

$$+ C_2 \sum_\nu |a_\nu|_{p+1} \|u\|_{p+1} \|u\|_p \le$$

$$\le \overline{C}_2 \Big(\sum_\nu |a_\nu|_{p+1} + |P|_{p+1} \Big) \|u\|_{p+1} \|u\|_p \le C'' \|u\|_{p+1} \|u\|_p,$$

we finally have for $(Lu, u)_{p+1}$ the following expression:

$$(Lu, u)_{p+1} = -(L_2^p u, \Delta u)_{p-1} - \sum_{\nu,j} \Big(\frac{\partial a_\nu}{\partial \phi_j} \frac{\partial^2 u}{\partial \phi_\nu \partial \phi_j}, u \Big)_p + \Phi'_{p+1}(u) +$$

$$+ \Phi''_{p+1}(u) = (L_2^p u, Ku)_{p-1} - \sum_{\nu,j} \Big(\frac{\partial a_\nu}{\partial \phi_j} \frac{\partial^2 u}{\partial \phi_\nu \partial \phi_j}, u \Big)_p -$$

$$- (L_2^p u, u)_{p-1} + \Phi'_{p+1}(u) + \Phi''_{p+1}(u) =$$

$$= \Big(\Big(L_2^p - \sum_{\nu,j} \frac{\partial a_\nu}{\partial \phi_j} \frac{\partial^2}{\partial \phi_\nu \partial \phi_j} \Big) u, u \Big)_p + \Phi'_{p+1}(u) + \Phi''_{p+1}(u) +$$

$$+ \Phi'''_{p+1}(u) = (L_2^{p+1} u, u)_p + \Phi_{p+1}(u),$$

where $\Phi_{p+1}(u)$ is a functional which can be represented as the sum of three terms:

$$\Phi'_{p+1}(u) + \Phi''_{p+1}(u) + \Phi'''_{p+1}(u).$$

To estimate the functional $\Phi_{p+1}(u)$ it only remains to estimate the term $\Phi'''_{p+1}(u) = -(L_2^p u, u)_{p-1}$. Since

$$|\Phi'''_{p+1}(u)| \le \|L_2^p u\|_{p-2} \|u\|_{p+1} \le$$

$$\le C_3 \Big(\sum_\nu |a_\nu|_{p-1} + |P|_{p-2} \Big) \|u\|_{p+1} \|u\|_p,$$

it follows from the estimates for the functionals $\Phi'_{p+1}(u)$, $\Phi''_{p+1}(u)$ that

$$|\Phi_{p+1}(u)| \le C \Big(\sum_\nu |a_\nu|_{p+1} + |P|_{p+1} \Big) \|u\|_{p+1} \|u\|_p,$$

where C is a positive constant independent of u and the coefficients of the operator L.

By induction we see that representation (15.7) and estimate (15.8) hold for any $2 \le s \le r$.

We require that the operator L_2^s be elliptic. For this it is sufficient that the inequality

$$\sum_{\nu,j}\left\langle \left[b_0 E_{\nu j} + s\frac{\partial a_\nu}{\partial \phi_j}E\right]\eta_\nu, \eta_j\right\rangle \ge \gamma_s \sum_\nu \|\eta_\nu\|^2 \qquad (15.9)$$

holds for any collection $\eta_1,\ldots,\eta_m$ of n-dimensional vectors $\eta_\nu - (\eta_\nu^1,\ldots,\eta_\nu^n)$ ($\nu = 1,\ldots,m$) and some $\gamma_s > 0$.

The proof that this inequality can be satisfied by taking γ_s sufficiently small will be given at the end of the section.

Under the assumption that inequality (15.9) holds we apply Gårding's inequality to the product $(L_2^s u, u)_{s-1}$, which leads to the estimate

$$(L_2^s u, u)_{s-1} \ge \gamma_s\|u\|_s^2 - \delta_s\|u\|_0^2,$$

where γ_s and δ_s are positive constants that depend on $\sum_\nu |a_\nu|_s + |P|_s$.

But then by applying the inequality for $2\|u\|_s\|u\|_{s-1}$ given in §1.10, we obtain:

$$(Lu, u)_s = (L_2^s u, u)_{s-1} + \Phi_s(u) \ge \gamma_s\|u\|_s^2 - \delta_s\|u\|_0^2-$$

$$-2C'\|u\|_s\|u\|_{s-1} \ge \gamma_s\|u\|_s^2 - \delta_s\|u\|_0^2 - C'\left(\frac{5}{2}\mu^2\|u\|_s^2+\right.$$

$$\left.+\frac{1}{2\mu^{2(4s-1)}}\|u\|_0^2\right) \ge \gamma\|u\|_s^2 - \delta\|u\|_0^2,$$

as required. Thus, if inequality (15.9) holds for $s = 0, 1,\ldots,r$ then condition 3 of Lemma 13.1 holds.

It remains to show that inequalities (15.9) hold for $s = 0, 1,\ldots,r$ whenever inequalities (14.21) hold for $s = 0$ and $s = r$.

For any $0 \le s \le r$ we have:

$$\sum_{\nu,j}\left\langle \left[b_0 E_{\nu j} + s\frac{\partial a_\nu}{\partial \phi_j}E\right]\eta_\nu, \eta_j\right\rangle = \sum_\nu\langle b_0\eta_\nu, \eta_\nu\rangle+$$

$$+s\sum_{\nu,j}\frac{\partial a_\nu}{\partial \phi_j}\langle\eta_\nu, \eta_j\rangle = \sum_\nu\langle b_0\eta_\nu, \eta_\nu\rangle + s\sum_{\nu,j}\frac{\partial a_\nu}{\partial \phi_j}\sum_i\eta_\nu^i\eta_j^i =$$

$$= \sum_{\nu} \langle b_0 \eta_\nu, \eta_\nu \rangle + s \sum_i \Big(\sum_{\nu,j} \frac{\partial a_\nu}{\partial \phi_j} \eta_j^i \eta_\nu^i \Big) =$$

$$= \sum_{\nu} \langle b_0 \eta_\nu, \eta_\nu \rangle + s \sum_i \langle \frac{\partial a}{\partial \phi} \Psi_i, \Psi_i \rangle,$$

where $\Psi_i = (\eta_1^i, \ldots, \eta_m^i)$ is an m-dimensional vector. But then

$$\sum_{\nu,j} \Big\langle \Big\{ b_0 E_{\nu j} + s \frac{\partial a_\nu}{\partial \phi_j} E \Big] \eta_\nu, \eta_j \Big\rangle \geq$$

$$\geq \Big(\beta_0(\phi) - \tfrac{1}{2}\mu(\phi) \Big) \sum_{\nu} \|\eta_\nu\|^2 + s\alpha_0(\phi) \sum_i \|\Psi_i\|^2 =$$

$$= \Big(\beta_0(\phi) + s\alpha_0(\phi) - \tfrac{1}{2}\mu(\phi) \Big) \sum_{\nu} \|\eta_\nu\|^2,$$

and this leads to inequalities (15.9) whenever inequalities (14.21) hold. It is clear that if inequalities (14.21) hold for $s = 0$ and $s = r$, then they also hold for any $0 < s < r$. This completes the proof of the required statement.

Chapter 4. Perturbation theory of an invariant torus of a non-linear system

4.1. Introductory remarks. The linearization process

The problems of the linear theory considered in the previous chapter are also important for the general theory of invariant tori of non-linear systems. This deals with the invariant surface

$$h = u(\phi), \quad \phi \in T_m, \ u \in C(T_m) \tag{1.1}$$

of the system of equations

$$\frac{d\phi}{dt} = a(\phi, h), \quad \frac{dh}{dt} = F(\phi, h), \tag{1.2}$$

the right hand side of which is defined, continuous in ϕ, h in the domain

$$\|h\| \leq d, \quad \phi \in T_m \tag{1.3}$$

and periodic in ϕ_ν ($\nu = 1, \ldots, m$) with period 2π. The study of system (1.2) is complicated by the fact that the functions a and F are non-linear with respect to the variable h and are accessible by modern methods of investigation only within the framework of perturbation theory, where it is assumed that the quantities $\|a(\phi, h) - a(\phi, 0)\|$ and $\|F(\phi, h) - F(\phi, 0)\|$ are small in domain (1.3). Here progress is achieved by using iteration procedures that linearize the problem at each stage of the iteration.

We shall consider a linearization procedure that enables one to find an invariant surface (1.1) of system (1.2). To do this we separate the "linear terms" in the second equation of system (1.2) by writing this system in the form

$$\frac{d\phi}{dt} = a(\phi, h), \quad \frac{dh}{dt} = P(\phi, h)h + f(\phi), \tag{1.4}$$

211

where

$$f(\phi) = F(\phi, 0), \quad P(\phi, h) = \int_0^1 \frac{\partial F(\phi, th)}{\partial(th)} dt.$$

We define the zeroth iteration of the procedure by a function $u^0 \in C(\mathcal{T}_m)$ which takes values in the domain (1.3) such that

$$\|u^0(\phi)\| \leq d, \quad \phi \in \mathcal{T}_m$$

and define a sequence of tori

$$h = u^{i+1}(\phi), \quad \phi \in \mathcal{T}_m, \ i = 0, 1, \ldots, \tag{1.5}$$

each of which is an invariant torus of the system

$$\frac{d\phi}{dt} = a(\phi, u^i(\phi)), \quad \frac{dh}{dt} = P(\phi, u^i(\phi))h + f(\phi). \tag{1.6}$$

Lemma 1. *Let the functions* a, P *and* f *be defined and continuous in* ϕ, h *in the domain* (1.3), *and periodic with respect to* ϕ_ν ($\nu = 1, \ldots, m$) *with period* 2π. *Suppose that for any* $i = 0, 1, \ldots$ *system* (1.6) *has an invariant torus* (1.5) *lying in the domain* (1.3).

If

$$\lim_{i \to \infty} u^i(\phi) = u(\phi)$$

uniformly with respect to $\phi \in \mathcal{T}_m$, *then the limit function* $u(\phi)$ *defines an invariant surface* (1.1) *of system* (1.4).

Proof. We denote by

$$\phi = \phi_t^i, \quad h = h_t^i$$

the solution of system (1.6) that takes the value

$$h_0^i = u^{i+1}(\phi), \quad \phi_0^i = \phi$$

at $t = 0$. Then

$$\phi_t^i = \phi + \int_0^t a(\phi_\tau^i, u^i(\phi_\tau^i)) d\tau,$$

$$u^{i+1}(\phi_t^i) = u^{i+1}(\phi) + \int_0^t (P(\phi_\tau^i, u^i(\phi_\tau^i))u^{i+1}(\phi_\tau^i) + f(\phi_\tau^i)) d\tau \tag{1.7}$$

for all $t \in \mathbf{R}$.

Taking into account the fact that the functions $u^i(\phi)$ are continuous and periodic and that their values lie in the domain (1.3) for all $i = 1, 2, \ldots$, we see that the first equation of system (1.7) implies that the sequence of functions

$$\phi_t^1, \ \phi_t^2, \ldots, \ \phi_t^i, \ldots \tag{1.8}$$

is uniformly bounded and equicontinuous for t from any bounded interval I on the $\mathbf{R}$ axis. This is sufficient for the sequence (1.8) to be compact in the space of continuous functions on I. Hence we can choose a subsequence that is uniformly convergent on I. Let $\phi_t^{i_\nu}$ ($\nu = 1, 2, \ldots$) be such a subsequence and ϕ_t its limit:

$$\lim_{\nu \to \infty} \phi_t^{i_\nu} = \phi_t, \quad t \in I.$$

Replacing i by i_ν in (1.7) and then passing to the limit, we find that

$$\phi_t = \phi + \int_0^t a(\phi_\tau, u(\phi_\tau)) d\tau,$$

$$u(\phi_t) = u(\phi) + \int_0^t (P(\phi_\tau, u(\phi_\tau)) u(\phi_\tau) + f(\phi_\tau)) d\tau \tag{1.9}$$

for all $t \in I$. Because the interval I was chosen arbitrarily, equality (1.9) holds for any $t \in \mathbf{R}$. Hence the surface (1.1) is an invariant set of system (1.4), as required.

To realize the given linearization procedure, we need to find conditions under which the system of equations

$$\frac{d\phi}{dt} = a(\phi) + a_1(\phi) \quad \frac{dh}{dt} = (P(\phi) + P_1(\phi))h + f(\phi) \tag{1.10}$$

has an invariant torus

$$h = u(\phi), \quad \phi \in \mathcal{T}_m \tag{1.11}$$

for arbitrary sufficiently small functions $a_1(\phi), P_1(\phi)$ (with respect to the $C^r(\mathcal{T}_m)$ norm). These conditions naturally presuppose a certain coarseness of the "generating" system of equations

$$\frac{d\phi}{dt} = a(\phi), \quad \frac{dh}{dt} = P(\phi)h, \tag{1.12}$$

which is expressed in terms of the "coarseness" of a Green's function $G_0(\tau, \phi)$ for system (1.12).

We shall call a Green's function $G_0(\tau, \phi)$ for system (1.12) *coarse* if there exist a constant $\delta > 0$ and an integer $r \geq 0$ such that whenever $a_1 \in C^{r_0}(T_m)$ (where $r_0 = \max(1, r)$) and

$$\|a_1\|_{r_0} \leq \delta, \tag{1.13}$$

the system of equations

$$\frac{d\phi}{dt} = a(\phi) + a_1(\phi), \quad \frac{dh}{dt} = P(\phi)h, \tag{1.14}$$

has a Green's function $\overline{G}_0(\tau, \phi)$, for which

$$|\overline{G}_0(\tau, \phi)f(\phi_\tau(\phi))|_r \leq K e^{-\gamma|\tau|}|f|_r, \tag{1.15}$$

where f is an arbitrary function in $C^r(T_m)$, $\phi_t(\phi)$ a solution of the first equation of (1.14), and K and γ are positive constants independent of ϕ, δ and f.

Lemma 2. *Suppose that the system* (1.12) *has a coarse Green's function* $G_0(\tau, \phi)$. *Then we can find* $\rho = \rho(\delta) > 0$, $\rho(\delta) \to 0$ *as* $\delta \to 0$ *such that for any* $a_1 \in C^{r_0}(T_m)$, $P_1 \in C^r(T_m)$ *satisfying the condition*

$$|a_1|_{r_0} + |P_1|_r \leq \rho, \tag{1.16}$$

and an arbitrary function $f \in C^r(T_m)$, *the system of equations* (1.10) *has invariant torus* (1.11) *satisfying the condition*

$$|u|_r \leq K_1|f|_r, \tag{1.17}$$

where K_1 *is a positive constant independent of* ρ *and* f.

Proof. We choose ρ so small that inequality (1.13) follows from inequality (1.16) and system (1.14) has a Green's function $\overline{G}_0(\tau, \phi)$ that satisfies condition (1.15). Using the operator G defined on functions $f \in C^r(T_m)$ by the inequality

$$Gf(\phi) = \int_{-\infty}^{+\infty} \overline{G}_0(\tau, \phi)f(\phi_\tau(\phi))d\tau,$$

we define the equation

$$u = GP_1 u + Gf. \tag{1.18}$$

For sufficiently small ρ, the norm of the operator GP_1 in the space $C^r(T_m)$ satisfies the inequality

$$|GP_1|_r \leq \frac{2cK}{\gamma}|P_1|_r \leq \frac{2cK}{\gamma}\rho \leq d < 1,$$

therefore equation (1.18) has the unique solution

$$u = u_f = \sum_{k=0}^{\infty} (GP_1)^k Gf$$

in $C^r(\mathcal{T}_m)$, and this solution satisfies the inequality

$$|u_f|_r \leq \frac{2K}{\gamma(1-d)} |f|_r. \tag{1.19}$$

It follows from the definition of the operator G that the torus (1.11) with function $u = u_f$ is an invariant set of system (1.10). Inequality (1.19) implies that the estimate (1.17) holds for it.

As in the linear case, an invariant surface (1.1) of system (1.2) will be called an m-dimensional invariant torus of this system. By the smoothness of torus (1.1) we mean the smoothness of the function u on the right hand side of its equation (1.1). The number r in the definition of a coarse Green's function will be called the *smoothness index* of this function.

4.2. Main theorem

We shall be considering the following system of differential equations:

$$\frac{d\phi}{dt} = a(\phi, h, \epsilon),$$

$$\frac{dh}{dt} = P(\phi, h, \epsilon)h + f(\phi, \epsilon), \tag{2.1}$$

where a, P and f are periodic functions in ϕ_ν ($\nu = 1, \ldots, m$) with period 2π which are defined and continuous with respect to all the variables ϕ, h, ϵ in the domain

$$\|h\| \leq d, \quad \phi \in \mathcal{T}_m, \quad \epsilon \in [0, \epsilon_0] \tag{2.2}$$

and are such that

$$f(\phi, 0) \equiv 0, \quad \phi \in \mathcal{T}_m. \tag{2.3}$$

Condition (2.3) ensures that there exists a trivial invariant torus

$$h = 0, \quad \phi \in \mathcal{T}_m \tag{2.4}$$

of system (2.1) for $\epsilon = 0$.

We write down the variational equation corresponding to torus (2.4). It has the form

$$\frac{d\phi}{dt} = a_0(\phi), \quad \frac{dh}{dt} = P_0(\phi)h, \tag{2.5}$$

where

$$a_0(\phi) = a(\phi, 0, 0), \quad P_0(\phi) = P_0(\phi, 0, 0).$$

For a function $u \in C_{\mathrm{Lip}}^s(\mathcal{T}_m)$, the quantity $|u|_s + K$ will be denoted by $|u|_{s,\mathrm{Lip}}$ where K is the Lipschitz constant for derivatives of order s of the function $u(\phi)$.

Theorem 1. *Suppose that the functions a, P and f have partial derivatives with respect to ϕ, h up to order r which are continuous with respect to ϕ, h, ϵ in the domain (2.2). Suppose further that system (2.5) has a coarse Green's function with smoothness index r.*

If $r \geq 1$, then there exists a sufficiently small $\epsilon_0 > 0$ such that for any $\epsilon \in [0, \epsilon_0]$ system (2.1) has an invariant torus

$$h = u(\phi, \epsilon), \quad \phi \in \mathcal{T}_m, \tag{2.6}$$

where the function u belongs to the space $C_{\mathrm{Lip}}^{r-1}(\mathcal{T}_m)$ and satisfies the inequality

$$|u|_{r-1,\mathrm{Lip}} \leq K|f|_r, \tag{2.7}$$

K being a positive constant independent of ϵ.

Proof. We write the system of equations (2.1) in the form

$$\frac{d\phi}{dt} = a_0(\phi) + a_1(\phi, h, \epsilon), \quad \frac{dh}{dt} = [P_0(\phi) + P_1(\phi, h, \epsilon)]h + f(\phi, \epsilon),$$

where

$$a_1(\phi, h, \epsilon) = a(\phi, h, \epsilon) - a(\phi, 0, 0),$$
$$P_1(\phi, h, \epsilon) = P(\phi, h, \epsilon) - P(\phi, 0, 0),$$

and apply the iteration procedure discussed in the previous section to find an invariant torus (2.6) of this system. We define the zeroth iteration by the function $u^0(\phi) \equiv 0$, $\phi \in \mathcal{T}_m$. We define the first approximation for the system of equations

$$\begin{aligned}
\frac{d\phi}{dt} &= a_0(\phi) + a_1(\phi, 0, \epsilon), \\
\frac{dh}{dt} &= [P_0(\phi) + P_1(\phi, 0, \epsilon)]h + f(\phi, \epsilon)
\end{aligned} \tag{2.8}$$

as the invariant torus

$$h = u_1(\phi, \epsilon), \quad \phi \in T_m. \tag{2.9}$$

We choose $\epsilon_0 > 0$ so small that the inequality

$$|a_1|_r + |P_1|_r \le \rho \tag{2.10}$$

holds for any functions $a_1 = a_1(\phi, u(\phi, \epsilon), \epsilon)$, $P_1 = P_1(\phi, u(\phi, \epsilon), \epsilon)$, where $u(\phi, \epsilon)$ is an arbitrary function in $C^r(T_m)$ satisfying the inequality

$$|u|_r \le K_1 |f|_r. \tag{2.11}$$

Here ρ and K_1 are the positive constants defined by Lemma 2 of §1, and $f = f(\phi, \epsilon)$. Such a choice of the value ϵ_0 is possible since the right hand side of system (2.1) is smooth and condition (2.3) holds.

Then the right hand side of the system of equations (2.8) satisfies the conditions of Lemma 2 of §1 for all $\epsilon \in [0, \epsilon_0]$. Consequently an invariant torus (2.9) of this system exists and satisfies the inequality

$$|u_1|_r \le K_1 |f|_r, \tag{2.12}$$

where K_1 is the constant in estimate (2.11). This is sufficient to find the second iteration satisfying estimate (2.12) by using the first one (2.9).

Suppose that for the chosen $\epsilon_0 > 0$ all the iterations have been found for $i = 1, \ldots, p - 1$, all of them satisfying inequality (2.12). The next pth approximation will then be defined from the system of equations

$$\frac{d\phi}{dt} = a_0(\phi) + a_1(\phi, u_{p-1}(\phi, \epsilon), \epsilon),$$

$$\frac{dh}{dt} = [P_0(\phi) + P_1(\phi, u_{p-1}(\phi, \epsilon), \epsilon)]h + f(\phi, \epsilon) \tag{2.13}$$

as the invariant torus

$$h = u_p(\phi, \epsilon), \quad \phi \in T_m. \tag{2.14}$$

Since the functions $a_1 = a_1(\phi, u_{p-1}(\phi, \epsilon), \epsilon)$, $P_1 = P_1(\phi, u_{p-1}(\phi, \epsilon), \epsilon)$ satisfy inequality (2.10) for all $\epsilon \in [0, \epsilon_0]$, the torus (2.14) of system (2.13) exists and satisfies the condition

$$|u_p|_r \le K_1 |f|_r, \tag{2.15}$$

where K_1 is the constant from estimates (2.11), (2.12).

The mathematical induction principle now guarantees that for all $\epsilon \in [0, \epsilon_0]$, any iteration of the procedure is defined and satisfies inequality (2.15).

We now prove the convergence of the iteration procedure. To this end, we consider the difference

$$w_{i+1}(\phi, \epsilon) = u_{i+1}(\phi, \epsilon) - u_i(\phi, \epsilon).$$

Since the functions $u_i(\phi, \epsilon)$ belong to the space $C^r(T_m)$ with $r \geq 1$, the invariant torus of the corresponding system of equations determined by these functions is smooth. Therefore

$$\frac{\partial u_i(\phi, \epsilon)}{\partial \phi}(a_0(\phi) + a_1(\phi, u_{i-1}(\phi, \epsilon), \epsilon)) =$$

$$= [P_0(\phi) + P_1(\phi, u_{i-1}(\phi, \epsilon), \epsilon)]u_i(\phi, \epsilon) + f(\phi, \epsilon)$$

for every $\phi \in T_m$ and $i = 1, 2, \ldots$. Hence it follows that the function $w_{i+1}(\phi, \epsilon)$ satisfies the equation

$$\frac{\partial w}{\partial \phi} = (a_0(\phi) + a_1(\phi, u_i(\phi, \epsilon), \epsilon)) =$$

$$= [P_0(\phi) + P_1(\phi, u_1(\phi, \epsilon), \epsilon)]w + f_i(\phi, \epsilon),$$

where

$$f_i(\phi, \epsilon) = [P_1(\phi, u_i(\phi, \epsilon), \epsilon) - P_1(\phi, u_{i-1}(\phi, \epsilon), \epsilon)]u_i(\phi, \epsilon) -$$

$$- \frac{\partial u_i(\phi, \epsilon)}{\partial \phi}[a_1(\phi, u_i(\phi, \epsilon), \epsilon) - a_1(\phi, u_{i-1}(\phi, \epsilon), \epsilon)].$$

Therefore the torus given by the equation

$$h = w_{i+1}(\phi, \epsilon), \quad \phi \in T_m,$$

is an invariant torus of the system of equations

$$\frac{d\phi}{dt} = a_0(\phi) + a_1(\phi, u_i(\phi, \epsilon), \epsilon),$$

$$\frac{dh}{dt} = [P_0(\phi) + P_1(\phi, u_i(\phi, \epsilon), \epsilon)] + f_i(\phi, \epsilon) \tag{2.16}$$

for any $\epsilon \in [0, \epsilon_0]$. Since the functions

$$a_1 = a_1(\phi, u_i(\phi, \epsilon), \epsilon), \quad P_1 = P_1(\phi, u_i(\phi, \epsilon), \epsilon)$$

satisfy inequality (2.10) for $\epsilon \in [0, \epsilon_0]$ and $f_i \in C^{r-1}(T_m)$, it follows that the system of equations (2.16) satisfies all the conditions of Lemma 1.2 with r equal to the value of $r - 1$. Estimate (1.17) for $w_{i+1}(\phi, \epsilon)$ takes the form

$$|w_{i+1}|_{r-1} \leq K_1 |f_i|_{r-1}.$$

For $r = 1$, this estimate leads to the inequality

$$|w_{i+1}|_0 \leq K_1 |f_i|_0 \leq K(|u_i|_0 + |u_i|_1)|w_i|_0, \tag{2.17}$$

where K is some constant independent of i and ϵ. Inequality (2.17) and estimates (2.15) for the iterations show that

$$|w_{i+1}|_0 \leq K K_1 |f|_1 |w_i|_0 \leq \tfrac{1}{2} |w_i|_0$$

for all $\epsilon \in [0, \epsilon_0]$ and sufficiently small $\epsilon_0 > 0$. But then

$$|w_i|_0 \leq (1/2)^{i-1} |w_1|_0, \quad i = 1, 2, \ldots,$$

which ensures that that the sequence $u_i(\phi, \epsilon)$, $i = 1, 2, \ldots$, converges in $C(T_m)$. Consequently for each $\epsilon \in [0, \epsilon_0]$ there exists a function $u(\phi, \epsilon)$ in $C(T_m)$ such that

$$\lim_{i \to \infty} u_i(\phi, \epsilon) = u(\phi, \epsilon) \tag{2.18}$$

uniformly with respect to $\phi \in T_m$, $\epsilon \in [0, \epsilon_0]$.

Since the sequence $u_i(\phi, \epsilon)$ $(i = 1, 2, \ldots)$ is bounded in $C^r(T_m)$, it is compact in the space $C^{r-1}(T_m)$ and this leads to the relation

$$\lim_{i \to \infty} |u_i(\phi, \epsilon) - u(\phi, \epsilon)|_{r-1} = 0, \quad \epsilon \in [0, \epsilon_0].$$

On passing to the limit in inequality (2.15) with r equal to the value of $r - 1$ we obtain

$$|u(\phi, \epsilon)|_{r-1} \leq K_1 |f|_{r-1}. \tag{2.19}$$

Finally, since for any $(r - 1)$th order derivative $D^{r-1}f(\phi, \epsilon)$ of the function $f \in C^r(T_m)$ the estimate

$$\|D^{r-1}f(\phi, \epsilon) - D^{r-1}f(\phi', \epsilon)\| \leq c|f|_r \|\phi - \phi'\|$$

holds with a constant c independent of f and $\phi, \phi' \in \mathcal{T}_m$, it follows that

$$\|D^{r-1}u_i(\phi, \epsilon) - D^{r-1}u_i(\phi', \epsilon)\| \le c|u_i|_r\|\phi - \phi'\| \le cK_1|f|_r\|\phi - \phi'\| \quad (2.20)$$

for all $\phi, \phi' \in \mathcal{T}_m$, $\epsilon \in [0, \epsilon_0]$. Passing to the limit in inequality (2.20), we find that

$$\|D^{r-1}u(\phi, \epsilon) - D^{r-1}u(\phi', \epsilon)\| \le cK_1|f|_r\|\phi - \phi'\|.$$

Inequalities (2.19), (2.20) are equivalent to estimate (2.7). Theorem 1 is true by virtue of Lemma 1.1.

It should be noted that inequality (2.7) implies the limit relation

$$\lim_{\epsilon \to 0} |u|_{r-1,\mathrm{Lip}} = 0,$$

which proves that the functions $u(\phi, \epsilon)$ and its derivatives with respect to ϕ are continuous with respect to ϵ at the point $\epsilon = 0$ uniformly in $\phi \in \mathcal{T}_m$.

4.3. Exponential stability of an invariant torus and conditions for its preservation under small perturbations of the system

We consider the system of equations (1.2). Suppose that it has an invariant torus

$$h = u(\phi), \quad \phi \in \mathcal{T}_m, \quad (3.1)$$

a neighbourhood of which lies in the domain (1.3). We denote by ϕ_t, h_t the solution of system of equations (1.2), which at $t = 0$ takes the value ϕ_0, h_0.

We set

$$d_t = \min_{\phi \in \mathcal{T}_m} [\|\phi_t - \phi\|_{\mathcal{T}_m}^2 + \|h_t - u(\phi)\|^2]^{1/2}, \quad (3.2)$$

where we have denoted by $\|\psi\|_{\mathcal{T}_m}$ the Euclidian length of the vector ψ on the torus $\mathcal{T}_m : \|\psi\|_{\mathcal{T}_m} = \min \|\psi \bmod 2\pi\|$.

In the direct product of the m-dimensional torus $\mathcal{T}_m$ and n-dimensional Euclidian space E^n, where n is the number of components of the vector $h = (h_1, \ldots, h_n)$, the function d_t gives the distance between the point (ϕ_t, h_t) and the torus (3.1). Using this distance one can define in the usual way *stability, asymptotic stability, instability* of the torus (3.1) considered as an invariant set of system (1.2).

An invariant torus (3.1) of system of equations (1.2) will be called *exponentially stable* if it is stable and there exists $\delta_0 > 0$ such that for any solution ϕ_t, h_t satisfying the condition $d_0 < \delta_0$ we have the inequality

$$d_t \leq K e^{-\gamma(t-\tau)} d_\tau, \quad t \geq \tau, \tag{3.3}$$

where K and γ are positive constants independent of τ and d_0, and τ is an arbitrary number of the half-axis $\mathbf{R}^+ = [0, +\infty)$. It follows from the definition that exponential stability of torus (3.1) implies its asymptotic stability.

We denote the expression

$$\rho_t = \|h_t - u(\phi_t)\|,$$

by $\rho_t = \rho_t(\phi_0, h_0)$. It is clear that $d_t \leq \rho_t$ and that $d_t = \rho_t$ whenever $u(\phi) =$ const.

Lemma 1. *If $u \in C_{\text{Lip}}(\mathcal{T}_m)$, then*

$$\rho_t \leq \sqrt{2}(1+K)d_t, \tag{3.4}$$

where K is the Lipschitz constant for the function $u(\phi)$.

Indeed, since $\mathcal{T}_m$ is compact, it follows that

$$d_t = [\|\phi_t + \phi_t^*\|_{\mathcal{T}_m}^2 + \|h_t - u(\phi_t^*)\|^2]^{1/2}$$

for some value $\phi_t^* \in \mathcal{T}_m$. But then

$$\rho_t \leq \|h_t - u(\phi_t^*)\| + \|u(\phi_t^*) - u(\phi_t)\| \leq (1+K)\times$$

$$\times [\|h_t - u(\phi_t^*)\| + \|\phi_t^* - \phi_t\|_{\mathcal{T}_m}] \leq \sqrt{2}(1+K)\times$$

$$\times [\|h_t - u(\phi_t^*)\|^2 + \|\phi_t^* - \phi_t\|_{\mathcal{T}_m}^2]^{1/2} = \sqrt{2}(1+K)d_t,$$

which proves estimate (3.4).

The number γ in inequality (3.3) is called the *index of exponential attraction* of the trajectories to the torus (3.1).

Lemma 2. *Let $u \in C_{\text{Lip}}(\mathcal{T}_m)$ and suppose that for some $\delta_0 > 0$ any solution ϕ_t, h_t of system (1.2) for which $\rho_0 < \delta_0$ satisfies the inequality*

$$\rho_t \leq K_1 e^{-\gamma t} \rho_0, \quad t \geq 0, \tag{3.5}$$

with constants $K_1 > 0$ and $\gamma > 0$ independent of ρ_0. Then the invariant torus
(3.1) of system (1.2) is exponentially stable with index of exponential attraction
of the trajectories to the torus (3.1) equal to γ.

To prove the lemma we consider the solutions of system (1.2) for which
$\rho_0 < \delta_0/K_1$. For these solutions $\rho_t \le K_1 e^{-\gamma \tau}\rho_0 < \delta_0$ for any $\tau \in \mathbf{R}^+$. Therefore
the solution $\phi_{t+\tau}$, $h_{t+\tau}$ of system (1.2) that takes the value ϕ_τ, h_τ at $t = 0$
satisfies the conditions of Lemma 2 for any $t \in \mathbf{R}^+$, and consequently, estimate
(3.5):

$$\rho_t(\phi_\tau, h_\tau) \le K_1 e^{-\gamma t}\rho_t, \quad t \ge 0, \ \tau \in \mathbf{R}^+ \tag{3.6}$$

holds for it.

Since $\rho_t(\phi_\tau, h_\tau) = \|h_{t+\tau} - u(\phi_{t+\tau})\| = \rho_{t+\tau}$, inequality (3.6) can be written
in the form

$$\rho_{t+\tau} \le K_1 e^{-\gamma[t+\tau-\tau]}\rho_\tau, \quad t + \tau \ge \tau, \ \tau \in \mathbf{R}^+,$$

and, on replacing $t + \tau$ by t, we obtain

$$\rho_t \le K_1 e^{-\gamma(t-\tau)}\rho_\tau, \quad t \ge \tau, \ \tau \in \mathbf{R}^+. \tag{3.7}$$

Inequalities (3.4), (3.7) lead to the estimate

$$d_t \le \rho_t \le K_1 e^{-\gamma(t-\tau)}\rho_\tau \le \sqrt{2}K_1(1 + K)e^{-\gamma(t-\tau)}d_\tau, \tag{3.8}$$

which holds for all $t \ge \tau$, any $\tau \in \mathbf{R}^+$ and those solutions ϕ_t, h_t of system (1.2)
for which $\rho_0 < \delta_0/K_1$.

It follows from Lemma 1 considered for $t = 0$ that all the values ϕ_0, h_0 for
which $d_0 < \delta_0/(\sqrt{2}K_1(1 + K))$ satisfy the condition $\rho_0 < \delta_0/K_1$. Therefore
inequality (3.8) holds for all solutions ϕ_t, h_t of system of equations (1.2) such
that

$$d_0 < \frac{\delta_0}{\sqrt{2}K_1(1 + K)} \tag{3.9}$$

and this proves Lemma 2.

Lemma 2 simplifies the verification that the invariant torus (3.1) of system
(1.2) is exponentially stable by reducing it to the verification of inequality (3.5).

We now return to our consideration of system of equations (2.1). We
denote by $\beta(\phi)$ and $\alpha(\phi)$ the quantities defined by inequalities (12.7) and (12.8)
of Chapter 3, where we now set $a(\phi) = a_0(\phi)$, $P(\phi) = P_0(\phi)$ and the inf and

sup are taken over the subsets $\mathfrak{R}_0$ and $\mathfrak{R}_1$ consisting of matrices in the space $C^1(T_m)$.

Theorem 1. *Suppose that functions a, P and f have partial derivatives with respect to ϕ, h up to order r, which are continuous relatively to ϕ, h, ϵ in the domain (2.2). Suppose that the inequalities*

$$\beta_0 = \inf_{\phi \in T_m} \beta(\phi) > 0,$$

$$\inf_{\phi \in T_m} [\beta(\phi) + r\alpha(\phi)] > 0 \tag{3.10}$$

hold.

If $r \geq 1$, then we can find a sufficiently small $\epsilon_0 > 0$ such that for any $\epsilon \in [0, \epsilon_0]$ system (2.1) has an exponentially stable invariant torus (2.6) with the function belonging to the space $C^{r-1}_{\mathrm{Lip}}(T_m)$ and satisfying inequality (2.7).

Proof. Since the quantities $\beta(\phi)$ and $\alpha(\phi)$ are defined in terms of the matrices P_0, $\partial a_0/\partial \phi$, S, S_1 in the space $C^1(T_m)$, we see that the similar quantities $\beta_\delta(\phi)$ and $\alpha_\delta(\phi)$, defined in terms of the matrices P_0 and $\partial a_0/\partial \phi + \partial a_1/\partial \phi$ for an arbitrary $a_1 \in C^r(T_m)$ satisfying inequality (1.14), are related to $\beta(\phi)$ and $\alpha(\phi)$ via the inequality

$$|\beta(\phi) - \beta_\delta(\phi)| + |\alpha(\phi) - \alpha_\delta(\phi)| < K\delta, \tag{3.11}$$

where K is a constant independent of δ. For sufficiently small $\delta > 0$, inequalities (3.10) and (3.11) lead to inequalities for $\beta_\delta(\phi)$ and $\alpha_\delta(\phi)$ similar to (3.10). This suffices for any of the systems (1.13) with $a = a_0$, $P = P_0$ to be exponentially stable and for its Green's function $\overline{G}_0(\tau, \phi)$ to satisfy inequality (12.3) of Chapter 3. The latter is equivalent to the condition that the Green's function for system (2.5) be coarse with smoothness index r.

This means that the conditions of our theorem satisfy all the assumptions of the main theorem, so that there exists an invariant torus (2.6) of system (2.1) for all $\epsilon \in [0, \epsilon_0]$, the function u defining this torus belongs to the space $C^{r-1}_{\mathrm{Lip}}(T_m)$ and equality (2.7) holds.

We now prove that the torus under consideration is exponentially stable. Since the torus (2.6) is, in general, merely Lipschitzian, it is impossible to write down the variational equation and to conclude on the basis of Theorem 7.1 of

Chapter 2 and the estimates obtained in its proof that it is exponentially stable. This forces us to use in our proof properties of the invariant torus (2.6) related to the iteration procedure for its construction. According to this procedure:

$$u(\phi, \epsilon) = \lim_{i \to \infty} u_i(\phi, \epsilon)$$

uniformly with respect to $\phi \in T_m$, $\epsilon \in [0, \epsilon_0]$, where

$$u_i \in C^1(T_m), \quad |u_i|_1 \leq K|f|_1, \quad i = 1, 2, \ldots,$$

$$\frac{\partial u_i(\phi, \epsilon)}{\partial \phi} a(\phi, u_{i-1}(\phi, \epsilon), \epsilon) = P(\phi, u_{i-1}(\phi, \epsilon), \epsilon) u_i(\phi, \epsilon) + f(\phi, \epsilon). \qquad (3.12)$$

We use the properties of the functions $u_i(\phi, \epsilon)$ to estimate the quantity $\rho_t = \|h_t - u(\phi_t, \epsilon)\|$, where ϕ_t, h_t is the solution of system of equations (2.1), that at $t = 0$ takes the value (ϕ_0, h_0) in the domain

$$\rho_0 < \delta_0.$$

We choose $\epsilon_0 > 0$ and $\delta_0 > 0$ so small that for all $\epsilon \in [0, \epsilon_0]$ we have the inequality

$$\delta_0 + K|f|_0 \leq d/2,$$

where d defines the domain (2.2) and K is the constant in inequality (3.12). With such a choice, a neighbourhood of the point (ϕ_0, h_0) lies in the domain (2.2), and the solution ϕ_t, h_t is defined for $t > 0$ until the moment when (ϕ_t, h_t) leaves the domain (2.2).

We fix an arbitrary δ, $\delta_0 < \delta < d$ and define $T > 0$ to be the greatest value such that

$$\rho_t < \delta \quad \forall t \in [0, T).$$

Consider the function $z_t = h_t - u(\phi_t, \epsilon)$ for $t \in [0, T)$. Since $u \in C_{\text{Lip}}(T_m)$, z_t satisfies a Lipschitz condition with respect to the variable t and therefore it has a derivative with respect to t for almost all $t \in [0, T)$. But then for almost all $t \in [0, T)$:

$$\frac{du(\phi_t, \epsilon)}{dt} = \lim_{\Delta t \to 0} \lim_{i \to \infty} \frac{u_i(\phi_t + \Delta t, \epsilon) - u_i(\phi_t, \epsilon)}{\Delta t} =$$

$$= \lim_{\Delta t \to 0} \lim_{i \to \infty} \left[\frac{\partial u_i(\phi_\theta, \epsilon)}{\partial \phi} a(\phi_\theta, u(\phi_\theta, \epsilon) + z_\theta, \epsilon) \right],$$

where $\theta = t + \tau \Delta t$, $\tau \in [0,1]$. We set

$$\Phi_i(\phi_\theta, z_\theta, \epsilon) = \frac{\partial u_i(\phi_\theta, \epsilon)}{\partial \phi}[a(\phi_\theta, u(\phi_\theta, \epsilon) + z_\theta, \epsilon) - a(\phi_\theta, u_{i-1}(\phi_\theta, \epsilon), \epsilon)].$$

It follows from equality (3.12) that

$$\frac{du(\phi_t, \epsilon)}{dt} = \lim_{\Delta t \to 0} \lim_{i \to \infty} \left[\frac{\partial u_i(\phi_\theta, \epsilon)}{\partial \phi} a(\phi_\theta, u_{i-1}(\phi_\theta, \epsilon), \epsilon) + \Phi_i(\phi_\theta, z_\theta, \epsilon) \right] =$$

$$= P(\phi_t, u(\phi_t, \epsilon), \epsilon) u(\phi_t, \epsilon) + f(\phi_t, \epsilon) + \lim_{\Delta t \to 0} \lim_{i \to \infty} \Phi_i(\phi_\theta, z_\theta, \epsilon) \quad (3.13)$$

for almost all $t \in [0, T]$.

Now we have

$$\|\Phi_i(\phi_\theta, z_\theta, \epsilon)\| \leq K_1 |f|_1 (|u - u_{i-1}|_0 + \|z_\theta\|)$$

with the obvious value of the constant K_1. From this it follows that

$$\| \lim_{\Delta t \to 0} \lim_{i \to \infty} \Phi_i(\phi_\theta, z_\theta, \epsilon)\| \leq K_1 |f|_1 \|z_t\| \quad (3.14)$$

for almost all $t \in [0, T]$. Using relation (3.13), we can write for dz_t/dt the expression:

$$\frac{dz}{dt} = P_0(\phi_t) z_t + R(\phi_t, u(\phi_t, \epsilon), z_t, \epsilon), \quad (3.15)$$

where

$$R(\phi_t, u(\phi_t, \epsilon), z_t, \epsilon) = [P(\phi_t, u(\phi_t, \epsilon) + z_t, \epsilon) - P(\phi_t, 0, 0)] z_t +$$

$$+ [P(\phi_t, u(\phi_t, \epsilon) + z_t, \epsilon) - P(\phi_t, u(\phi_t, \epsilon), \epsilon)] u(\phi_t, \epsilon) - \lim_{\Delta t \to 0} \lim_{i \to \infty} \Phi_i(\phi_\theta, z_\theta, \epsilon).$$

The inequalities (3.12), (3.14) enable us to obtain the following estimate for the function $R = R(\phi_t, u(\phi_t, \epsilon), z_t, \epsilon)$:

$$\|R\| \leq K_2(|f|_1 + \|z_t\| + \epsilon_0)\|z_t\| \leq \mu(\delta, \epsilon_0)\|z_t\|, \quad (3.16)$$

where $\mu(\delta, \epsilon_0)$ is monotonically decreasing to zero as $\delta \to 0$, $\epsilon_0 \to 0$.

Inequality (3.16) enables us to estimate $\|z_t\| = \rho_t$ in standard fashion by using the condition $\beta_0 > 0$ and, as a result, obtain an estimate in the form of (5.8) of Chapter 3:

$$\rho_t \leq K e^{-(\hat{\beta}_0 - 3\mu(\delta, \epsilon_0))t} \rho_0 \quad (3.17)$$

where $\hat{\beta}_0$ is an arbitrary fixed number in the interval $(0, \beta_0)$. When δ and ϵ_0 are sufficiently small, inequality (3.17) leads to the estimate (3.5) with

$$\gamma = \hat{\beta}_0 - 3\mu(\delta, \epsilon_0) > 0. \tag{3.18}$$

But then the conditions of Lemma 2 are satisfied and the invariant torus (2.6) of system (2.1) is exponentially stable.

It should be noted that the index of exponential attraction of the trajectories to torus (2.6), as can be seen from formula (3.18), depends on the distance from the point (ϕ_0, h_0) to the torus (2.6) and increases as this distance decreases.

We further note that inequalities (2.12), (3.17) together with inequality (3.9) guarantee that for a sufficiently small $\delta_0 > 0$ it is possible to choose $\epsilon_0 = \epsilon_0(\delta_0) > 0$ such that for all $\epsilon \in [0, \epsilon_0]$ the inequality

$$\|u(\phi, \epsilon)\| < \delta_0/2, \quad \phi \in \mathcal{T}_m$$

holds and that any solution ϕ_t, h_t of system (2.1) for which

$$\|h_0\| < \delta_0$$

is attracted to the torus (2.6) as $t \to +\infty$ according to the exponential law (3.3).

4.4. Theorem on exponential attraction of motions in a neighbourhood of an invariant torus of a system to its motions on the torus

The notions of asymptotic and exponential stability of an invariant torus of a dynamical system presuppose that the attraction of semi-trajectories of the system from a small neighbourhood of the torus to the torus itself is "coarse", namely, the attraction to the torus is considered as an attraction to a set of points in the phase space of the system. Being invariant, the torus is, apart from anything else, a set of trajectories of the dynamical system. In this connection there arises the question of when the attraction to the torus has the characteristics of an attraction between the motions of the system that start on the torus and in a small neighbourhood of it.

We shall answer this question for system (1.2), assuming that the torus (1.1) is an invariant set of it.

We set

$$a_0(\phi) = a(\phi, u(\phi)), \quad P_0(\phi) = \frac{\partial F(\phi, u(\phi))}{\partial h} - \frac{\partial u(\phi)}{\partial \phi} \frac{\partial a(\phi, u(\phi))}{\partial h} \qquad (4.1)$$

and define the quantities $\beta(\phi)$ and $\alpha(\phi)$ in terms of the functions a_0 and P_0, as in the previous section.

Let

$$d(\phi_0, h_0) = \min_{\phi \in T_m} [\|\phi_0 - \phi\|_{T_m}^2 + \|h_0 - u(\phi)\|^2]^{1/2}.$$

We adjoin to the system of equations (1.2) the equation

$$\frac{d\psi}{dt} = a_0(\psi) \qquad (4.2)$$

which defines the trajectory flow of this system on the torus (1.1).

Theorem 1. *Let the functions a, F, u and their first order derivatives with respect to ϕ, h be defined and satisfy a Lipschitz condition in the domain (1.3). Suppose that the torus (1.1) along with a d_0-neighbourhood of it lies in the domain (1.3) and is an invariant set of system (1.2). If the inequalities*

$$\inf_{\phi \in T_m} \beta(\phi) = \beta_0 > 0, \quad \inf_{\phi \in T_m} [\beta(\phi) + \alpha(\phi)] > 0 \qquad (4.3)$$

hold, then for any sufficiently small $\epsilon > 0$, $\mu > 0$ and $\rho > 0$ the solution ϕ_t, h_t of system (1.2), for which

$$d(\phi_0, h_0) \le \mu\rho,$$

satisfies the inequality

$$\|\phi_t - \psi_t\| + \|h_t - u(\psi_t)\| \le K \exp\left\{-\int_0^t \hat{\beta}(\psi_\tau) d\tau\right\}, \quad t \ge 0, \qquad (4.4)$$

where $\psi_t = \psi_t(\psi_0)$ is a solution of system (4.2) for which $\psi_0 = \psi_0(\phi_0, h_0)$, $\|\phi_0 - \psi_0\| \le K\mu\rho$, $\hat{\beta}(\psi) = \beta(\psi) - \epsilon - K\mu$, K being a positive constant independent of μ and ρ.

Proof. We make a change of variables in system (1.2), (4.2) by setting

$$\phi = \psi + \mu\theta, \quad h = u(\phi) + \mu\rho z. \qquad (4.5)$$

Equations (1.2) of system (1.2), (4.2) are written in terms of the new variables as:

$$\mu \frac{d\theta}{dt} = a(\psi + \mu\theta, u(\psi + \mu\theta) + \mu\rho z) - a(\psi, u(\psi)),$$

$$\mu\rho \frac{dz}{dt} = F(\psi + \mu\theta, u(\psi + \mu\theta) + \mu\rho z) -$$

$$- \frac{\partial u(\psi + \mu\theta)}{\partial \phi} a(\psi + \mu\theta, u(\psi + \mu\theta) + \mu\rho z). \qquad (4.6)$$

We set

$$A(\phi, \theta) = \int_0^1 \frac{\partial a_0(\phi + \tau\theta)}{\partial \phi} d\tau,$$

$$A_1(\phi, z) = \int_0^1 \frac{\partial a(\phi, u(\phi) + \tau z)}{\partial h} d\tau, \qquad (4.7)$$

$$P(\phi, z) = \int_0^1 \frac{\partial F(\phi, u(\phi) + \tau z)}{\partial h} d\tau - \frac{\partial u(\phi)}{\partial \phi} A_1(\phi, z)$$

and write system of equations (4.6) in the form:

$$\frac{d\theta}{dt} = A(\psi, \mu\theta)\theta + A_1(\psi + \mu\theta, \mu\rho z)\rho z,$$

$$\frac{dz}{dt} = P(\psi + \mu\theta, \mu\rho z)z. \qquad (4.8)$$

It follows from the assumptions of the theorem and formulae (4.7) that the functions A, A_1 and P are defined and continuous and satisfy a Lipschitz condition with respect to the variables $\psi, \theta, z, \mu, \rho$ in the region

$$\|\theta\| \le 1, \ \|z\| \le 1, \ \psi \in \mathcal{T}_m, \ 0 \le \mu \le \mu_0, \ 0 \le \rho \le \rho_0, \qquad (4.9)$$

where μ_0 and ρ_0 are sufficiently small constants, $\mu_0 \le 1$, $\rho_0 \le 1$. It can also be seen from formulae (4.7) that

$$A(\phi, 0) = \frac{\partial a_0(\phi)}{\partial \phi},$$

$$P(\phi, 0) = P_0(\phi),$$

where a_0 and P_0 are the functions (4.1).

We shall denote by $\Omega_\tau^t(P)$ the fundamental matrix of solutions of the homogeneous linear system of equations in normal form, the coefficient matrix of which is $P = P(t)$, such that $\Omega_\tau^\tau(P) = E$, where E is the identity matrix.

Lemma 1. *Let the matrices $P(t)$ and $P_1(t)$ be defined and continuous for $t \geq \tau$ (or $t \leq \tau$). If for any $t \geq \tau$ (or $t \leq \tau$) and x_0 the following inequality holds:*

$$\|\Omega_\tau^t(P)x_0\| \leq \mathcal{L} \exp\left\{ \int_\tau^t \beta(s)ds \right\} \|x_0\|, \qquad (4.10)$$

where $\mathcal{L}$ is a constant and $\beta(t)$ is an integrable function for $t \geq \tau$ (or $t \leq \tau$), then

$$\|\Omega_\tau^t(P + P_1)\| \leq \mathcal{L} \exp\left\{ \int_\tau^t \beta(s)ds + \mathcal{L} \left| \int_\tau^t \|P_1(s)\| ds \right| \right\} \|x_0\| \qquad (4.11)$$

for any $t \geq \tau$ (or $t \leq \tau$), where $\|P_1\| = \max_{\|x\|=1} \|P_1 x\|$.

We prove this lemma as follows. Using the function

$$x_1(t, \tau, x_0) = \Omega_\tau^t(P)x_0,$$

which satisfies inequality (4.10), we construct a sequence of functions $x_n(t, \tau, x_0)$ $(n = 1, 2, \dots)$ by setting

$$x_{n+1}(t, \tau, x_0) = \Omega_\tau^t(P)x_0 + \int_\tau^t \Omega_s^t(P)P_1(s)x_n(s, \tau, x_0)ds. \qquad (4.12)$$

Suppose that for $n = 1, 2, \dots, k$ the inequality

$$\|x_n(t, \tau, x_0)\| \leq \mathcal{L} \exp\left\{ \int_\tau^t \beta(s)ds \right\} \sum_{\nu=0}^{n-1} \frac{1}{\nu!} \left| \mathcal{L} \int_\tau^t \|P_1(s)\| ds \right|^\nu \|x_0\| \qquad (4.13)$$

holds for all $t \geq \tau$ (or $t \leq \tau$). Then it follows from formula (4.12) that for $t \geq \tau$ (or $t \leq \tau$)

$$\|x_{k+1}(t, \tau, x_0)\| \leq \mathcal{L} \exp\left\{ \int_\tau^t \beta(s)ds \right\} \|x_0\| +$$

$$+ \left| \int_\tau^t \mathcal{L} \exp\left\{ \int_s^t \beta(s)ds \right\} \|P_1(s)\| \mathcal{L} \exp\left\{ \int_\tau^s \beta(s)ds \right\} \times \right.$$

$$\times \left. \sum_{\nu=0}^{k-1} \frac{1}{\nu!} \left| \mathcal{L} \int_\tau^s \|P_1(s)\| ds \right|^\nu ds \right| \|x_0\| \leq \mathcal{L} \exp\left\{ \int_\tau^t \beta(s)ds \right\} \times$$

$$\times \Big(1 + \sum_{\nu=0}^{k-1} \frac{1}{\nu!} \mathcal{L}^{\nu+1} \Big| \int_\tau^t \|P_1(s)\| \Big| \int_\tau^s \|P_1(s)\| ds \Big|^\nu ds \Big| \Big) \|x_0\| =$$

$$= \mathcal{L} \exp\Big\{ \int_\tau^t \beta(s) ds \Big\} \sum_{\nu=0}^{k} \frac{1}{\nu!} \Big| \mathcal{L} \int_\tau^t \|P_1(s)\| ds \Big|^\nu \|x_0\|.$$

The inequality obtained for $\|x_{k+1}(t, \tau, x_0)\|$ has the form of inequality (4.13) for $n = k + 1$. This is sufficient for inequality (4.13) to hold for any $n \geq 1$. For all $t \geq \tau$ (or $t \leq \tau$), $n = 1, 2, \ldots$, we can estimate the differences $x_{n+1}(t, \tau, x_0) - x_n(t, \tau, x_0)$ as follows:

$$\|x_2(t, \tau, x_0) - x_1(t, \tau, x_0)\| \leq \Big| \int_\tau^t \|\Omega_s^t(P) P_1(s) x_1(s, \tau, x_0)\| ds \Big| \leq$$

$$\leq \mathcal{L} \exp\Big\{ \int_\tau^t \beta(s) ds \Big\} \Big| \mathcal{L} \int_\tau^t \|P_1(s)\| ds \Big| \|x_0\|,$$

$$\|x_3(t, \tau, x_0) - x_2(t, \tau, x_0)\| \leq \Big| \int_\tau^t \|\Omega_s^t(P) P_1(s)(x_2(s, \tau, x_0) -$$

$$- x_1(s, \tau, x_0))\| ds \Big| \leq \mathcal{L} \exp\Big\{ \int_\tau^t \beta(s) ds \Big\} \frac{1}{2!} \Big| \mathcal{L} \int_\tau^t \|P_1(s)\| ds \Big|^2 \|x_0\|,$$

$$\cdots \cdots \cdots \cdots \cdots \cdots \cdots \cdots \cdots \cdots \cdots \cdots$$

$$\|x_{n+1}(t, \tau, x_0) - x_n(t, \tau, x_0)\| \leq \mathcal{L} \exp\Big\{ \int_\tau^t \beta(s) ds \Big\} \frac{1}{n!} \Big| \mathcal{L} \int_\tau^t \|P_1(s)\| ds \Big|^n \|x_0\|.$$

It follows from this that the sequence of functions $x_n(t, \tau, x_0)$ $(n = 1, 2, \ldots)$ converges uniformly with respect to t, τ, x_0.

We set

$$x(t, \tau, x_0) = \lim_{n \to \infty} x_n(t, \tau, x_0).$$

On passing to the limit in equality (4.12) we obtain

$$x(t, \tau, x_0) = \Omega_\tau^t(P + P_1) x_0.$$

Similarly we use inequality (4.13) to find that

$$\|x(t, \tau, x_0)\| \leq \mathcal{L} \exp\Big\{ \int_\tau^t \beta(s) ds \Big\} \exp\Big\{ \mathcal{L} \Big| \int_\tau^t \|P_1(s)\| ds \Big| \Big\} \|x_0\|$$

for all $t \geq \tau$ (or $t \leq \tau$). The last two inequalities prove inequality (4.11) of the lemma.

From the conditions of the theorem it follows that

$$\|\Omega_t^s(P_0)z_0\| \leq \mathcal{L} \exp\left\{-\int_t^s \beta(\psi_\tau)d\tau + \epsilon(s-t)\right\}\|z_0\|,$$

$$\left\|\Omega_s^t\left(\frac{\partial a_0}{\partial \phi}\right)\theta_0\right\| \leq \mathcal{L}_1 \exp\left\{\int_s^t \alpha(\psi_\tau)d\tau + \epsilon(s-t)\right\}\|\theta_0\|$$

(4.14)

for all $s \geq t \geq 0$, arbitrary $\epsilon > 0$, some $\mathcal{L} \geq 1$, $\mathcal{L}_1 \geq 1$ and any $z_0 \in E^n$, $\theta_0 \in E^m$ and $P_0 = P_0(\psi_t)$. Here, $\partial a_0/\partial \phi = \partial a_0(\psi_t)/\partial \phi$, where ψ_t is the solution of system (4.2) that takes the value $\psi_0 \in T_m$ at $t = 0$.

For arbitrary ψ, z, μ, ρ in the region (4.9) the functions $A(\psi, \mu\theta)$ and $P(\psi + \mu\theta, \mu\rho z)$ satisfy the estimate

$$\|A(\psi, \mu\theta) - A(\psi, 0)\| \leq K\mu,$$

$$\|P(\psi + \mu\theta, \mu\rho z) - P(\psi, 0)\| \leq K\mu(1 + \rho),$$

where K is a positive constant independent of μ and ρ. Therefore, for any functions $\theta(t)$, $z(t)$ that are continuous for $t \geq 0$ and satisfy the condition

$$\|\theta(t)\| \leq 1, \quad \|z(t)\| \leq 1, \quad t \geq 0,$$

(4.15)

the matrices $\Omega_t^s(P_0 + \overline{P})$ and $\Omega_s^t(\partial a_0/\partial \phi + \overline{A})$, where

$$\overline{P} = P(\psi_t + \mu\theta(t), \mu\rho z(t)) - P_0(\psi_t),$$
$$\overline{A} = A(\psi_t, \mu\theta(t)) - A(\psi_t, 0),$$

admit the estimates

$$\|\Omega_t^s(P_0 + \overline{P})z_0\| \leq \mathcal{L} \exp\left\{-\int_t^s \beta(\psi_t)d\tau + (\epsilon + \mathcal{L}K(1+\rho)\mu)(s-t)\right\}\|z_0\|,$$

$$\left\|\Omega_s^t\left(\frac{\partial a_0}{\partial \phi} + \overline{A}\right)\theta_0\right\| \leq \mathcal{L}_1 \exp\left\{\int_s^t \alpha(\psi_\tau)d\tau + (\epsilon + \mathcal{L}_1 K\mu)(s-t)\right\}\|\theta_0\|$$

(4.16)

for all $s \geq t \geq 0$ and arbitrary z_0, θ_0, μ, ρ in the region (4.9).

We define the sequences of functions $\theta_n(t)$, $z_n(t)$ by setting $\theta_0(t) = 0$, $z_0(t) = z_0$ and

$$\theta_n(t) = -\rho \int_t^{+\infty} \Omega_s^t(A_{n-1})A_1(\psi_s + \mu\theta_{n-1}(s), \mu\rho z_n(s))z_n(s)ds,$$

$$z_n(t) = \Omega_0^t(P_{n-1})z_0, \quad n = 1, 2, \ldots, \quad t \geq 0,$$

(4.17)

where

$$A_{n-1} = A(\psi_t, \mu\theta_{n-1}(t)), \quad P_{n-1} = P(\psi_t + \mu\theta_{n-1}(t), \mu\rho z_{n-1}(t)).$$

We now show that for an appropriate choice of the values μ_0, ρ_0 the functions (4.17) are defined for any $n \geq 1$ and all μ, ρ, z_0 in the region

$$0 \leq \mu \leq \mu_0, \quad 0 \leq \rho \leq \rho_0, \quad \|z_0\| \leq 1/\mathcal{L}. \tag{4.18}$$

For $n = 1$ it follows from formulae (4.17) and inequalities (4.14) that

$$\|z_1(t)\| \leq \mathcal{L}\exp\Big\{-\int_0^t \beta(\psi_\tau)d\tau + \epsilon t\Big\}\|z_0\| \leq 1, \quad t \geq 0,$$

if μ, ρ, z_0 belong to region (4.18) and $\epsilon < \beta_0$.

For $n = 1$ we have the following estimate for the integrand in the first of equalities (4.17):

$$I = \|\Omega_s^t(A_0)A_1(\psi_s, \mu\rho z_1(s))z_1(s)\| \leq \mathcal{L}\mathcal{L}_1 M_1 \exp\Big\{\int_s^t \alpha(\psi_\tau)d\tau\Big\} \times$$

$$\times \exp\Big\{-\int_0^s \beta(\psi_\tau)d\tau + \epsilon(s-t) + \epsilon s\Big\}\|z_0\|, \quad s \geq t \geq 0,$$

where M_1 is a constant chosen from the condition that

$$\|A_1(\psi, \mu\rho z)\| \leq \max_{\phi \in \mathcal{T}_m} \|A_1(\psi, 0)\| + K\mu\rho\|z\| \leq M_1$$

in the domain (4.9).

For $s \geq t \geq 0$ the inequality

$$\int_s^t \alpha(\psi_\tau)d\tau - \int_0^s \beta(\psi_\tau)d\tau = -\int_t^s [\alpha(\psi_\tau) + \beta(\psi_\tau)]d\tau - \int_0^t \beta(\psi_\tau)d\tau \leq$$

$$\leq -\gamma(s-t) - \int_0^t \beta(\psi_\tau)d\tau$$

holds, where $\gamma = \inf_{\phi \in \mathcal{T}_m} [\beta(\phi) + \alpha(\phi)] > 0$. Therefore

$$I \leq \mathcal{L}\mathcal{L}_1 M_1 \exp\Big\{-(\gamma - 2\epsilon)(s-t) - \int_0^t \beta(\psi_\tau)d\tau + \epsilon t\Big\}\|z_0\|$$

for $s \geq t \geq 0$, which leads to the uniform convergence of the integral (4.17) for $\epsilon < \gamma/2$, $n = 1$, and to the following estimate for its value $\theta_1(t)$:

$$\|\theta_1(t)\| \leq \frac{\mathcal{L}\mathcal{L}_1 M_1 \rho}{\gamma - 2\epsilon} \exp\left\{-\int_0^t \beta(\psi_\tau)d\tau + \epsilon t\right\}\|z_0\|, \quad t \geq 0. \tag{4.19}$$

We set

$$\bar{\epsilon} = \epsilon + K\mu \max(2\mathcal{L}, \mathcal{L}_1), \quad \gamma' = \gamma - 2\bar{\epsilon}$$

and require that the quantities ϵ, μ_0, ρ_0 satisfy the conditions

$$\epsilon + K\mu_0 \max(2\mathcal{L}, \mathcal{L}_1) \leq \beta_0/2,$$

$$2(\epsilon + K\mu_0 \max(2\mathcal{L}, \mathcal{L}_1)) \leq \gamma/2,$$

$$\rho_0 \leq \gamma \max(2\mathcal{L}\mathcal{L}_1 M_1). \tag{4.20}$$

Inequality (4.19) leads to the estimate

$$\|\theta_1(t)\| \leq \frac{2\mathcal{L}\mathcal{L}_1 M_1 \rho}{\gamma} \exp\left\{-\int_0^t \beta(\psi_\tau)d\tau + \epsilon t\right\}\|z_0\|$$

for all $t \geq 0$ and all μ, ρ, z_0 in the region (4.18), (4.20).

Suppose that the functions $\theta_n(t)$, $z_n(t)$ are defined and satisfy the inequalities

$$\|\theta_n(t)\| \leq \frac{2\mathcal{L}\mathcal{L}_1 M_1 \rho}{\gamma} \exp\left\{-\int_0^t \beta(\psi_\tau)d\tau + \bar{\epsilon} t\right\}\|z_0\|,$$

$$\|z_n(t)\| \leq \mathcal{L} \exp\left\{-\int_0^t \beta(\psi_\tau)d\tau + \bar{\epsilon} t\right\}\|z_0\| \tag{4.21}$$

for $t \geq 0$, $n = 1, 2, \ldots, k$ and all μ, ρ, z_0 in the region (4.18), (4.20). Then $\|\theta_k(t)\| \leq 1$, $\|z_k(t)\| \leq 1$ for $t \geq 0$ and μ, ρ, z_0 in the region (4.18), (4.20), and the function $z_{k+1}(t)$ is defined and satisfies the inequality

$$\|z_{k+1}(t)\| \leq \mathcal{L} \exp\left\{-\int_0^t \beta(\psi_\tau)d\tau + (\epsilon + 2\mathcal{L}K\mu)t\right\}\|z_0\|,$$

while the function $\theta_{k+1}(t)$ is defined and satisfies the inequality

$$\|\theta_{k+1}(t)\| \leq \mathcal{L}\mathcal{L}_1 M_1 \rho \int_t^{+\infty} \exp\left\{\int_0^t \alpha(\psi_\tau)d\tau + (\epsilon + \mathcal{L}_1 K\mu)(s - t)\right\} \times$$

$$\times \exp\left\{-\int_0^s \beta(\psi_\tau)d\tau + (\epsilon + 2\mathcal{L}K\mu)s\right\}ds\|z_0\| \leq$$

$$\leq \mathcal{L}\mathcal{L}_1 M_1 \rho \exp\left\{-\int_0^t \beta(\psi_\tau)d\tau + \bar{\epsilon}t\right\} \times$$

$$\times \int_t^{+\infty} \exp\left\{-[\gamma - 2\epsilon - (\mathcal{L}_1 + 2\mathcal{L})K\mu](s - t)\right\}ds\|z_0\| \leq$$

$$\leq \frac{\mathcal{L}\mathcal{L}_1 M_1 \rho}{\gamma'} \exp\left\{-\int_0^t \beta(\psi_t)d\tau + \bar{\epsilon}t\right\}\|z_0\| \leq$$

$$\leq \frac{2\mathcal{L}\mathcal{L}_1 M_1 \rho}{\gamma} \exp\left\{-\int_0^t \beta(\psi_\tau)d\tau + \bar{\epsilon}t\right\}\|z_0\|$$

for $t \geq 0$ and all μ, ρ, z_0 in the region (4.18), (4.20).

According to the mathematical induction principle, it is sufficient that the functions $\theta_n(t)$, $z_n(t)$ be defined and satisfy inequalities (4.21) for $t \geq 0$, any $n \geq 1$ and all μ, ρ, z_0 in the region (4.18), (4.20).

We now prove that sequence (4.17) converges. To do this, we consider the difference

$$r_{n+1} = z_{n+1}(t) - z_n(t) = [\Omega_0^t(P_n) - \Omega_0^t(P_{n-1})]z_0.$$

Since $r_{n+1}\big|_{t=0} = 0$ and

$$\frac{dr_{n+1}}{dt} = P_n z_{n+1}(t) - P_{n-1}z_n(t) = P_n r_{n+1} + (P_n - P_{n-1})z_n,$$

we have the following representation for r_{n+1}:

$$r_{n+1} = \int_0^t \Omega_s^t(P_n)(P_n - P_{n-1})\Omega_0^s(P_{n-1})z_0 ds,$$

from which, using estimates for the matrices in the integrand, we obtain the inequality

$$\|r_{n+1}\| \leq \mathcal{L}^2 \int_0^t \exp\left\{-\int_s^t \beta(\psi_\tau)d\tau + (\epsilon + 2\mathcal{L}K\mu)(t - s)\right\}\|P_n - P_{n-1}\| \times$$

$$\times \exp\left\{-\int_0^s \beta(\psi_\tau)d\tau + (\epsilon + 2\mathcal{L}K\mu)s\right\}ds\|z_0\| \leq$$

$$\leq \mathcal{L}^2 t \exp\left\{-\int_0^t \beta(\psi_\tau)d\tau + \bar{\epsilon}t\right\} \max_{s \in [0,t]}\|P_n - P_{n-1}\|\|z_0\|, \quad t \geq 0. \quad (4.22)$$

According to our assumptions and the adopted notation,

$$\|P_n - P_{n-1}\| = \|P(\psi_t + \mu\theta_n(t), \mu\rho z_n(t)) - P(\psi_t + \mu\theta_{n-1}(t), \mu\rho z_{n-1}(t))\| \leq$$

$$\leq K\mu[\|\theta_n(t) - \theta_{n-1}(t)\| + \rho\|r_n\|].$$

This inequality enables us to obtain from (4.22) the estimate

$$\|r_{n+1}\| \le \mathcal{L}^2 K \mu t \exp\left\{-\int_0^t \beta(\psi_t)d\tau + \bar\epsilon t\right\} \times$$
$$[\|\theta_n - \theta_{n-1}\|_0 + \|r_n\|_0]\|z_0\|, \quad t \ge 0, \qquad (4.23)$$

where $\|\cdot\|_0 = \sup\limits_{t\in\mathbf{R}^+} \|\cdot\|$, $\mathbf{R}^+ = [0, +\infty)$.

We set

$$\Psi_n = \Omega_s^t(A_{n-1})\theta_0$$

and consider the difference $\Psi_{n+1} - \Psi_n = [\Omega_s^t(A_n) - \Omega_s^t(A_{n-1})]\theta_0$. Since $[\Psi_{n+1} - \Psi_n]_{t=s} = 0$ and

$$\frac{d(\Psi_{n+1} - \Psi_n)}{dt} = A_n(\Psi_{n+1} - \Psi_n) + (A_n - A_{n-1})\Psi_n,$$

we have the following representation for $\Psi_{n+1} - \Psi_n$:

$$\Psi_{n+1} - \Psi_n = \int_s^t \Omega_\tau^t(A_n)(A_n - A_{n-1})\Omega_s^\tau(A_{n-1})\theta_0 d\tau,$$

from which, using estimates for the matrices in the integrand, we obtain

$$\|\Psi_{n+1} - \Psi_n\| \le \mathcal{L}_1^2 \int_t^s \exp\left\{\int_\tau^t \alpha(\psi_t)d\tau + (\epsilon + \mathcal{L}_1 K\mu)(\tau - t)\right\}\|A_n - A_{n-1}\| \times$$
$$\times \exp\left\{\int_s^\tau \alpha(\psi_\tau)d\tau + (\epsilon + \mathcal{L}_1 K\mu)(s - \tau)\right\}d\tau\|\theta_0\|, \quad s \ge t \ge 0. \qquad (4.24)$$

Since

$$\|A_n - A_{n-1}\| \le \sup_{t\in\mathbf{R}^+} \|A(\psi_t, \mu\theta_n(t)) - A(\psi_t, \mu\theta_{n-1}(t))\| \le K\mu\|\theta_n - \theta_{n-1}\|_0,$$

we obtain from inequality (4.24) the final estimate:

$$\|\Psi_{n+1} - \Psi_n\| \le \mathcal{L}_1^2 K\mu(s - t)\exp\left\{\int_s^t \alpha(\psi_\tau)d\tau + (\epsilon + \right.$$
$$\left. + \mathcal{L}_1 K\mu)(s - t)\right\}\|\theta_n - \theta_{n-1}\|_0\|\theta_0\|$$

for $s \ge t \ge 0$. By using this in combination with (4.23), we estimate the difference $\mathcal{C}_{n+1} = \theta_{n+1}(t) - \theta_n(t)$. For $t \ge 0$ we have:

$$\|\mathcal{C}_{n+1}\| \le \rho \int_t^{+\infty} \|[\Omega_s^t(A_n) - \Omega_s^t(A_{n-1})] \times$$

$$\times A_1(\psi_s + \mu\theta_n(s), \mu\rho z_{n+1}(s))z_{n+1}(s)\|ds+$$

$$\rho\int_t^{+\infty} \|\Omega_s^t(A_{n-1})\|\,\|A_1(\psi_s + \mu\theta_n(s), \mu\rho z_{n+1}(s))(z_{n+1}(s) - z_n(s))\|ds+$$

$$+\rho\int_t^{+\infty} \|\Omega_s^t(A_{n-1})\|\,\|A_1(\psi_s + \mu\theta_n(s), \mu\rho z_{n+1}(s))-$$

$$-A_1(\psi_s + \mu\theta_{n-1}(s), \mu\rho z_n(s))\|\,\|z_n(s)\|ds. \qquad (4.25)$$

The first term I_1 in inequality (4.25) has the following estimate for $t \geq 0$:

$$I_1 \leq \rho\mathcal{L}_1^2\mathcal{L}KM_1\mu\int_t^{+\infty} \exp\Big\{\int_s^t \alpha(\psi_\tau)d\tau + (\epsilon + \mathcal{L}_1K\mu)(s-t)\Big\}\times$$

$$\times\exp\Big\{-\int_0^s \beta(\psi_\tau)d\tau + (\epsilon + 2\mathcal{L}K\mu)s\Big\}ds\|C_{n+1}\|_0\|z_0\| \leq$$

$$\leq \rho\mathcal{L}_1^2\mathcal{L}KM_1\mu\int_t^{+\infty} \exp\{-\gamma'(s-t)\}ds\times$$

$$\times\exp\Big\{-\int_0^t \beta(\psi_\tau)d\tau + \bar{\epsilon}t\Big\}\|C_{n+1}\|_0\|z_0\| \leq$$

$$\leq \frac{2\mathcal{L}_1^2\mathcal{L}KM_1\rho\mu}{\gamma}\exp\Big\{-\int_0^t \beta(\psi_\tau)d\tau + \bar{\epsilon}t\Big\}\|C_{n+1}\|_0\|z_0\|,$$

where

$$\max_{\|z\|\leq 1}\|A_1(\phi, \mu\rho z)\| \leq M_1.$$

For the second term I_2 in inequality (4.25) we have the following estimate for $t \geq 0$:

$$I_2 \leq \rho\mathcal{L}_1\mathcal{L}^2KM_1\mu\int_t^{+\infty} s\exp\Big\{\int_s^t \alpha(\psi_\tau)d\tau + (\epsilon + \mathcal{L}_1K\mu)(s-t)\Big\}\times$$

$$\times\exp\Big\{-\int_0^s \beta(\psi_\tau)d\tau + (\epsilon + 2\mathcal{L}K\mu)s\Big\}ds\times$$

$$\times[\|C_n\|_0 + \|r_n\|_0]\|z_0\| \leq \rho\mathcal{L}_1\mathcal{L}^2KM_1\mu\int_t^{+\infty} s\exp\Big\{-\gamma'(s-t)\Big\}ds\times$$

$$\times\exp\Big\{-\int_0^t \beta(\psi_\tau)d\tau + \bar{\epsilon}t\Big\}[\|C_n\|_0 + \|r_n\|_0]\|z_0\| \leq$$

$$\leq \frac{2\mathcal{L}_1\mathcal{L}^2KM_1}{\gamma}\mu\rho\Big(\frac{2}{\gamma} + t\Big)\exp\Big\{-\int_0^t \beta(\psi_\tau)d\tau + \bar{\epsilon}t\Big\}[\|C_n\|_0 + \|r_n\|_0]\|z_0\|.$$

The third term I_3 in inequality (4.25) has the following estimate for $t \geq 0$:

$$I_3 \leq \rho \mathcal{L} \mathcal{L}_1 K_1 \mu \int_t^{+\infty} \exp\left\{-\gamma'(s-t)\right\} ds \exp\left\{-\int_0^t \beta(\psi_\tau) d\tau + \bar{\epsilon}t\right\} \times$$

$$\times [\|\mathcal{C}_n\|_0 + \rho\|r_{n+1}\|_0]\|z_0\| \leq \frac{2\mathcal{L}\mathcal{L}_1 K_1}{\gamma} \mu\rho \exp\left\{-\int_0^t \beta(\psi_\tau) d\tau + \bar{\epsilon}t\right\} \times$$

$$\times [\|\mathcal{C}_n\|_0 + \rho\|r_{n+1}\|_0]\|z_0\|,$$

where K_1 is the Lipschitz constant for the function $A_1(\phi, z)$.

Inequality (4.25) now leads to the estimate

$$\|\mathcal{C}_{n+1}\| \leq \mu\rho C(1+t) \exp\left\{-\int_0^t \beta(\psi_\tau) d\tau + \bar{\epsilon}t\right\} \|z_0\| \times$$

$$\times (\|\mathcal{C}_n\|_0 + \|r_n\|_0 + \rho\|r_{n+1}\|_0), \quad t \geq 0, \qquad (4.26)$$

where C is a constant independent of μ, ρ, ϵ chosen in obvious fashion.

Let $J = \sup_{t \in \mathbf{R}^+} (1+t)e^{-(\beta/2)t}$. It follows from inequality (4.23) that

$$\|r_{n+1}\|_0 \leq \mathcal{L}^2 K J \mu (\|\mathcal{C}_n\|_0 + \|r_n\|_0)\|z_0\|. \qquad (4.27)$$

On applying (4.26) and (4.27) to $\|\mathcal{C}_{n+1}\|$ we obtain a final estimate:

$$\|\mathcal{C}_{n+1}\|_0 \leq \mu\rho C J \|z_0\|(1 + \mu\rho\mathcal{L}^2 K\|z_0\|)(\|\mathcal{C}_n\|_0 + \|r_n\|_0). \qquad (4.28)$$

Inequalities (4.27), (4.28) prove that

$$\|r_{n+1}\|_0 + \|\mathcal{C}_{n+1}\|_0 \leq (1/2)(\|r_n\|_0 + \|\mathcal{C}_n\|_0) \qquad (4.29)$$

for all μ, ρ, z_0 in the region (4.18) with sufficiently small μ_0 and ρ_0.

Inequality (4.29) holds for any $\psi_t = \psi_t(\psi_0)$. Thus it follows from this that the sequences of functions $\theta_n(t)$ and $z_n(t)$ ($n = 1, 2, \ldots$) converge uniformly with respect to $t, \psi_0, z_0, \mu, \rho$ in the region

$$t \in \mathbf{R}^+, \; \psi_0 \in \mathcal{T}_m, \; \|z_0\| \leq 1/\mathcal{L}, \; 0 \leq \mu \leq \mu_0, \; 0 \leq \rho \leq \rho_0. \qquad (4.30)$$

We set

$$\theta(t) = \lim_{n \to \infty} \theta_n(t), \quad z(t) = \lim_{n \to \infty} z_n(t) \qquad (4.31)$$

and look at certain properties of the functions $\theta(t)$, $z(t)$. First of all by taking the limit in inequalities (4.21) we find that

$$\|\theta(t)\| \leq \frac{2\mathcal{L}\mathcal{L}_1 M_1 \rho}{\gamma} \exp\left\{-\int_0^t \beta(\psi_\tau)d\tau + \bar{\epsilon}t\right\}\|z_0\|,$$

$$\|z(t)\| \leq \mathcal{L} \exp\left\{-\int_0^t \beta(\psi_\tau)d\tau + \bar{\epsilon}t\right\}\|z_0\| \tag{4.32}$$

for all $t, \psi_0, z_0, \mu, \rho$ in the region (4.30).

It follows from formulae (4.17) that

$$\frac{d\theta_n(t)}{dt} = A(\psi_t, \mu\theta_{n-1}(t))\theta_n(t) + \rho A_1(\psi_t + \mu\theta_{n-1}(t), \mu\rho z_n(t))z_n(t),$$

$$\frac{dz_n(t)}{dt} = P(\psi_t + \mu\theta_{n-1}(t), \mu\rho z_{n-1}(t))z_n(t), \quad t \geq 0. \tag{4.33}$$

Since the right hand side of equalities (4.33) has a limit as $n \to \infty$ uniformly with respect to $t, \psi_0, z_0, \mu, \rho$ in the region (4.30), the functions $\theta(t), z(t)$ have derivatives with respect to t given by:

$$\frac{d\theta(t)}{dt} = \lim_{n\to\infty} \frac{d\theta_n(t)}{dt}, \quad \frac{dz(t)}{dt} = \lim_{n\to\infty} \frac{dz_n(t)}{dt}.$$

By taking the limit in equalities (4.33) we see that the functions $\theta(t), z(t)$ are solutions of equations (4.8) for $\psi = \psi_t$, $t \geq 0$

Let us study the dependence of the functions $\theta(t), z(t)$ on the parameters μ, ρ and the values ψ_0, z_0. Since the limit (4.31) is uniform with respect to $t, \psi_0, z_0, \mu, \rho$ in the region (4.30), the functions $\theta(t), z(t)$ are continuous with respect to these same variables $t, \psi_0, z_0, \mu, \rho$ in the region (4.30) whenever the functions $\theta_n(t), z_n(t)$ are continuous with respect to $t, \psi_0, z_0, \mu, \rho$ in the region (4.30) for $n = 1, 2, \ldots$.

Under our assumptions concerning the torus (1.1) and the right hand side of system (1.2), the function $\psi_t = \psi_t(\psi_0)$ is continuous in $(t, \psi_0) \in \mathbf{R}^+ \times \mathcal{T}_m$. But then the matrix $P_0 = P(\psi_t(\psi_0), \mu\rho z_0)$ is continuous with respect to $t, \psi_0, z_0, \mu, \rho$ in the region (4.30). It follows from the theorem on the continuous dependence of solutions of differential equations on the initial conditions and parameters that the function $z_1(t)$ is continuous with respect to $t, \psi_0, z_0, \mu, \rho$ in the region (4.30). Formulae (4.17) taken for $n = 1$ imply that the integrand function defining $\theta_1(t)$ is continuous with respect to $t, \psi_0, z_0, \mu, \rho$ in the region (4.30) and $s \geq t$. Since the convergence of integral (4.17) is uniform for $n = 1$, the function $\theta_1(t)$ is continuous with respect to the variables $t, \psi_0, z_0, \mu, \rho$ in the region (4.30).

Continuing in this manner, we prove by induction that the functions $\theta_n(t), z_n(t)$ are continuous with respect to the variables $t, \psi_0, z_0, \mu, \rho$ in the region (4.30) for $n \geq 1$.

We set

$$\theta(0) = \theta(\psi_0, z_0, \mu, \rho). \tag{4.34}$$

As was proved before, the solution $\phi(t), h(t)$ of system (1.2) for which

$$\phi(0) = \psi_0 + \mu\theta(\psi_0, z_0, \mu, \rho), \quad h(0) = u(\phi(0)) + \mu\rho z_0, \tag{4.35}$$

satisfies the inequalities

$$\|\phi(t) - \psi_t\| \leq \mathcal{L}_2\mu\rho \exp\left\{-\int_0^t \beta(\psi_\tau)d\tau + \bar{\varepsilon}t\right\}\|z_0\|,$$

$$\|h(t) - u(\phi(t))\| \leq \mathcal{L}\mu\rho \exp\left\{-\int_0^t \beta(\psi_\tau)d\tau + \bar{\varepsilon}t\right\}\|z_0\|, \quad t \geq 0 \tag{4.36}$$

where $\mathcal{L}_2 = 2\mathcal{L}\mathcal{L}_1 M_1/\gamma$.

Using the fact that the function $u(\phi)$ in (4.36) is smooth we obtain the new inequality

$$\|h(t) - u(\psi_t)\| \leq \|h(t) - u(\phi(t))\| + K_2\|\phi(t) - \psi_t\| \leq$$

$$\leq (\mathcal{L} + K_2\mathcal{L}_2)\mu\rho \exp\left\{-\int_0^t \beta(\psi_\tau)d\tau + \bar{\varepsilon}t\right\}\|z_0\|, \quad t \geq 0. \tag{4.37}$$

According to the second equality of (4.35) we have

$$\mu\rho\|z_0\| = \|h(0) - u(\phi(0))\| \leq \|h(0) - u(\psi_0)\| + \|u(\phi(0)) - u(\psi_0)\| \leq$$

$$\leq (1 + K_2)[\|h(0) - u(\psi_0)\| + \|\phi(0) - \psi_0\|]. \tag{4.38}$$

Inequalities (4.36)–(4.38) lead to the estimate

$$\|\phi(t) - \psi_t\| + \|h(t) - u(\psi_t)\| \leq$$

$$\leq (1 + K_2)(\mathcal{L} + \mathcal{L}_2 + K_2\mathcal{L}_2) \exp\left\{-\int_0^t \beta(\psi_\tau)d\tau + \bar{\varepsilon}t\right\}\times$$

$$\times[\|\phi(0) - \psi_0\| + \|h(0) - u(\psi_0)\|] \tag{4.39}$$

for all $t, \psi_0, z_0, \mu, \rho$ in the region (4.30).

We fix the values μ, ρ and take any ϕ_0, h_0 such that

$$\|h_0 - u(\phi_0)\| \leq \mu\rho/\mathcal{L}. \tag{4.40}$$

We claim that for such a choice of ϕ_0, h_0 the equation

$$\phi_0 = \psi + \mu\theta\left(\psi, \frac{h_0 - u(\phi_0)}{\mu\rho}, \mu, \rho\right) \tag{4.41}$$

has a solution. To prove this, we replace ψ by the variable $\vartheta = \phi_0 - \psi$ and rewrite equation (4.41) as

$$\vartheta = \mu\theta\left(\phi_0 - \vartheta, \frac{h_0 - u(\phi_0)}{\mu\rho}, \mu, \rho\right). \tag{4.42}$$

It follows from inequalities (4.32) and (4.41) that

$$\left\|\theta\left(\phi_0 - \vartheta, \frac{h_0 - u(\phi_0)}{\mu\rho}, \mu, \rho\right)\right\| \leq \mathcal{L}_2\rho.$$

Therefore for ϑ in the region

$$\|\vartheta\| \leq \mathcal{L}_2\mu\rho \tag{4.43}$$

the right hand side of equation (4.42) defines an operator which maps the ball (4.43) into itself. Since this operator is continuous in ϑ, it follows from Brouwer's theorem [1] that it has a fixed point in (4.43). This is equivalent to the fact that equation (4.42) has a solution in the region (4.43). We denote it by $\vartheta_0 = \vartheta_0(\phi_0, h_0)$. The quantity

$$\psi_0 = \phi_0 - \vartheta_0(\phi_0, h_0)$$

is now a solution of equation (4.41), which satisfies the condition

$$\|\phi_0 - \psi_0\| \leq \mathcal{L}_2\mu\rho.$$

But then each value ϕ_0, h_0 for which inequality (4.40) holds admits the representation

$$\phi_0 = \psi_0 + \mu\theta(\psi_0, z_0, \mu, \rho), \quad h_0 = u(\phi_0) + \mu\rho z_0,$$

where $\psi_0 = \phi_0 - \vartheta_0(\phi_0, h_0)$, $\quad z_0 = [h_0 - u(\phi_0)]/(\mu\rho)$.

Thus the solution ϕ_t, h_t of system (1.2), for which ϕ_0, h_0 satisfies inequality (4.40), satisfies the inequality of the form (4.39):

$$\|\phi_t - \psi_t\| + \|h_t - u(\psi_t)\| \leq (1 + K_2)(\mathcal{L} + \mathcal{L}_2 + K_2\mathcal{L}_2) \times$$

$$\times \exp\left\{-\int_0^t \beta(\psi_\tau(d\tau + \bar{\epsilon}t)\right\}[\|\phi_0 - \psi_0\| + \|h_0 - u(\psi_0)\|], \quad t \geq 0, \qquad (4.44)$$

where ψ_t is the solution of system (4.2) for which $\psi_0 = \phi_0 - \vartheta_0(\phi_0, h_0)$.

It follows from inequality (3.4) that all points (ϕ_0, h_0) in the neighbourhood of torus (1.1) defined by the inequality

$$d(\phi_0, h_0) \leq \frac{\mu\rho}{\sqrt{2}(1 + K_2)\mathcal{L}} \qquad (4.45)$$

satisfy estimate (4.40). But then any solution ϕ_t, h_t of system (1.2), for which (ϕ_0, h_0) belongs to domain (4.45) satisfies inequality (4.44). Obvious transformations of the constants $\mu, \rho, \mathcal{L}, \mathcal{L}_1, \mathcal{L}_2, K_2$ complete the proof of the theorem.

An example of a system of equations satisfying the conditions of the above theorem is system (2.1) when it satisfies the conditions of Theorem 1 in §3 for $r = 2$ and ϵ_0 sufficiently small. There is an attraction described by inequality (4.4) between the motions of such a system that start on the torus (2.6) and those that start in a small neighbourhood of it.

4.5. Exponential dichotomy of invariant torus and conditions for its preservation under small perturbations of the system

An invariant torus (1.1) of system (1.2) will be called *exponentially dichotomous* if for any sufficiently small $\epsilon > 0$ there exists $\delta = \delta(\epsilon) > 0$ such that the ϵ-neighbourhood of torus (1.1) contains one-side invariant sets M^+ and M^- of system (1.2) (where for at least one of them its intersection with the n-dimensional plane $\phi = \phi_0$ is non-empty for each $\phi_0 \in T_m$) with the following property. Let ϕ_t, h_t be a solution of system (1.2), that starts in a

δ-neighbourhood of torus (1.1). Then: if $(\phi_0, h_0) \in M^+$, then $(\phi_t, h_t) \in M^+$ for all $t \geq 0$ and

$$d_t \leq K e^{-\gamma(t-\tau)} d_\tau \quad \forall t \geq \tau \ \forall \tau \in \mathbf{R}^+; \tag{5.1}$$

if $(\phi_0, h_0) \in M^-$, then $(\phi_t, h_t) \in M^-$ for all $t \leq 0$ and

$$d_t \leq K e^{\gamma(t-\tau)} d\tau \quad \forall t \leq \tau \ \forall \tau \in \mathbf{R}^-; \tag{5.2}$$

while if $(\phi_0, h_0) \notin M^+ \cup M^-$, then

$$d_t \geq \epsilon \tag{5.3}$$

for some $t = T_1 < 0$ and some $t = T_2 > 0$, where $\mathbf{R}^- = (-\infty, 0]$, K and γ are positive constants independent of ϕ_0, h_0, ϵ, and

$$d_t = \min_{\phi \in T_m} [\|\phi_t - \phi\|^2_{T_m} + \|h_t - u(\phi)\|^2]^{1/2}$$

is the distance from (ϕ_t, h_t) to the torus (1.1).

We write down the system of variational equations corresponding to the invariant torus (1.1) of system (1.2):

$$\frac{d\psi}{dt} = a(\psi, u(\psi)), \quad \frac{d\delta h}{dt} = P_0(\psi, u(\psi))\delta h, \tag{5.4}$$

where

$$P_0(\psi, u(\psi)) = \frac{\partial F(\psi, u(\psi))}{\partial h} - \frac{\partial u(\psi)}{\partial \psi} \frac{\partial a(\psi, u(\psi))}{\partial h}.$$

Theorem 1. *Suppose that the right hand side of system (1.2) is defined and continuous and has continuous first order partial derivatives with respect to ϕ, h in the region (1.3) which satisfy a Lipschitz condition. Suppose further that the torus (1.1) along with a neighbourhood of it lies in the region (1.3) and is an invariant set of system (1.2), and that $u \in C^1_{\text{Lip}}(T_m)$. If there exists a non-singular symmetric matrix $S(\phi) \in C^1(T_m)$ for which the matrix*

$$\hat{S}(\phi) = \frac{\partial S(\phi)}{\partial \phi} a(\phi, u(\phi)) + S(\phi) P_0(\phi, u(\phi)) + P_0^*(\phi, u(\phi)) S(\phi)$$

constructed in accordance with the variational equation (5.4) is negative-definite for all $\phi \in T_m$, then torus (1.1) is exponentially dichotomous.

Proof. By means of the change of variables

$$h = u(\phi) + z \tag{5.5}$$

we transform system (1.2) to the form

$$\frac{d\phi}{dt} = a(\phi, u(\phi) + z),$$
$$\frac{dz}{dt} = P(\phi, u(\phi), z)z, \tag{5.6}$$

where

$$P(\phi, u(\phi), z) = \int_0^1 \left[\frac{\partial F(\phi, u(\phi) + tz)}{\partial h} - \frac{\partial u(\phi)}{\partial \phi}\frac{\partial a(\phi, u(\phi) + tz)}{\partial h}\right]dt.$$

The right hand side of system (5.6) is defined and continuous and satisfies a Lipschitz condition with respect to ϕ, z in the domain

$$\|z\| \le \delta_0, \quad \phi \in \mathcal{T}_m, \tag{5.7}$$

where δ_0 is a sufficiently small positive number. We choose a function $\chi_1(\theta_j)$ from the conditions that

$$\chi_1(\theta_j) \in C_{\text{Lip}}(\mathcal{T}_1), \quad \|\chi_1(\theta_j)\| \le \delta_0,$$
$$\chi_1(\theta_j) = \theta_j, \quad |\theta_j| \le \epsilon_1 < \delta_0, \tag{5.8}$$

and set $\chi(\theta) = (\chi_1(\theta_1), \dots, \chi_1(\theta_n))$, $\theta = (\theta_1, \dots, \theta_n)$.

Instead of system of equations (5.6) we consider its 'extension":

$$\frac{d\theta}{dt} = P(\phi, u(\phi), \chi(\theta))\dot\chi(\theta),$$
$$\frac{d\phi}{dt} = a(\phi, u(\phi) + \chi(\theta)), \tag{5.9}$$
$$\frac{dz}{dt} = P(\phi, u(\phi), \chi(\theta))z.$$

This extension has the locally invariant set

$$\theta = z, \|z\| < \epsilon_1, \tag{5.10}$$

the motion of which is given by system of equations (5.6).

It follows from the conditions of the theorem that the trivial torus $\delta h = 0$, $\psi \in T_m$ of system (5.4) is exponentially dichotomous.

Since

$$\|P(\phi, u(\phi), \chi(\theta))\chi(\theta)\| + \|a(\phi, u(\phi) + \chi(\theta)) - a(\phi, u(\phi))\| +$$

$$+\|P(\phi, u(\phi), \chi(\theta)) - P_0(\phi, u(\phi))\| < K\delta_0, \quad \phi \in T_m, \ \theta \in T_m,$$

where K does not depend on ϕ, θ, δ_0, the matrix $\hat{S}(\phi, 0)$ constructed from $S(\phi)$ according to equations (5.9) and having the form

$$\hat{S}(\phi, \theta) = \frac{\partial S(\phi)}{\partial \phi}[a(\phi, u(\phi) + \chi(\theta)) - a(\phi, u(\phi))] +$$

$$+S(\phi)[P(\phi, u(\phi), \chi(\theta)) - P_0(\phi, u(\phi))] +$$

$$+[P^*(\phi, u(\phi), \chi(\theta)) - P_0^*(\phi, u(\phi))]S(\phi) + \hat{S}(\phi),$$

is negative-definite for all $\phi \in T_m$, $\theta \in T_n$ and any sufficiently small $\delta_0 > 0$. By Theorem 1 of §3.8, this implies that the trivial torus $z = 0$, $\phi \in T_m$, $\theta \in T_n$ of system (5.9) is exponentially dichotomous.

It then follows from Theorem 2 of §3.9 that there exists a unique Green's function $G_0(\tau, \phi, \theta)$ for system (5.9), which satisfies the estimate

$$\|G_0(\tau, \phi, \theta)\| \leq K_1 e^{-\gamma|\tau|}, \quad \tau \in \mathbf{R},$$

where K_1 and γ are positive constants independent of $(\phi, \theta) \in T_m \times T_n$. The matrix $C(\phi, \theta) = G_0(0, \phi, \theta) \in C(T_m \times T_n)$ is a projection of rank r where r is the number of positive eigenvalues of the matrix $S(\phi)$. The separatrix manifolds M_0^+ and M_0^- of system (5.9) are defined by the equations $C(\phi, \theta)z = z$ and $C(\phi, \theta)z = 0$, $\phi \in T_m$, $\theta \in T_n$ respectively. The solutions ϕ_t, θ_t, z_t of system (5.9) satisfy for some $T = T(\phi_0, \theta_0, z_0, \epsilon_1) > 0$:

$$\|z_t\| \leq K_1 e^{-\gamma(t-\tau)}\|z_\tau\|, \ t \geq \tau, \ \tau \in \mathbf{R}^+ \text{ for } (\phi_0, \theta_0, z_0) \in M_0^+, \quad (5.11)$$

$$\|z_t\| \leq K_1 e^{\gamma(t-\tau)}\|z_\tau\|, \ t \leq \tau, \ \tau \in \mathbf{R}^- \text{ for } (\phi_0, \theta_0, z_0) \in M_0^-, \quad (5.12)$$

$$\|z_t\| \geq \epsilon_1, |t| \geq T \text{ for } (\phi_0, \theta_0, z_0) \notin M_0^+ \cup M_0^-. \quad (5.13)$$

The restriction of these manifolds to the locally invariant set (5.10) gives the sets M_1^+ and M_1^- consisting of the points (ϕ, z) of the region

$$\|z\| < \epsilon_1, \ \phi \in T_m, \tag{5.14}$$

defined by the equation

$$C(\phi, z)z = z \tag{5.15}$$

and the equation

$$C(\phi, z)z = 0 \tag{5.16}$$

respectively. It follows from inequalities (5.11) and (5.12) that the solution ϕ_t, $\theta_t = z_t$, z_t of system (5.9) for which $\|z_0\| < \epsilon_1/K_1$ takes values in the region (5.14) for $t \geq 0$ if $(\phi_0, \theta_0 = z_0, z_0) \in M_0^+$, and for $t \leq 0$ if $(\phi_0, \theta_0 = z_0, z_0) \in M_0^-$. But then $(\phi_t, \theta_t = z_t, z_t) \in M_0^+$ for all $t \geq 0$ if $(\phi_0, \theta_0 = z_0, z_0) \in M_0^+$, and $(\phi_t, \theta_t = z_t, z_t) \in M_0^-$ for all $t \leq 0$ if $(\phi_0, \theta_0 = z_0, z_0) \in M_0^-$.

It follows from this that the following statements hold for the solutions ϕ_t, z_t of system (5.6) for which $\|z_0\| < \epsilon_1/K_1$:

1) if $(\phi_0, z_0) \in M_1^+$, then $(\phi_t, z_t) \in M_1^+$ for all $t \geq 0$ and ϕ_t, z_t satisfies inequality (5.11),

2) if $(\phi_0, z_0) \in M_1^-$, then $(\phi_t, z_t) \in M_1^-$ for all $t \leq 0$ and ϕ_t, z_t satisfies inequality (5.12).

Since $u(\phi) \in C_{\text{Lip}}^1(T_m)$, it follows from Lemma 3.1 that the inequality $\|z_0\| < \epsilon_1$ holds whenever

$$d_0 < \frac{\epsilon_1}{\sqrt{2}(1 + K_2)},$$

where K_2 is the Lipschitz constant for the function $u(\phi)$.

We set $\epsilon = \epsilon_1/(\sqrt{2}(1 + K_2))$. Then the ϵ-neighbourhood of torus (1.1) is contained in the region $\|z_0\| < \epsilon_1$, $\phi \in T_m$. We denote by M^+ and M^- the restrictions of the sets M_1^+ and M_1^- to this ϵ-neighbourhood. The sets M^+ and M^- are locally invariant sets of system (1.2). This follows from the fact that some trajectory arc of the motion ϕ_t, h_t of system (1.2) which starts in an ϵ-neighbourhood of torus (1.1) belongs to this neighbourhood and hence, in view of the local invariance of M_1^+ and M_1^-, it belongs to M^+ if $(\phi_0, h_0) \in M^+$ and to M^- if $(\phi_0, h_0) \in M^-$.

We set $\delta(\epsilon) = \epsilon/(\sqrt{2}(1 + K_2)K_1)$ and consider the solutions ϕ_t, h_t of system (1.2), for which (ϕ_0, h_0) belongs to the domain $d_0 < \delta(\epsilon)$. Since for such

solutions $\|z_0\| = \rho_0$ satisfies the inequality $\|z_0\| < \epsilon/K_1 < \epsilon_1/K_1$, it follows
from the inclusion $(\phi_0, h_0) \in M^+$ that $(\phi_t, h_t) \in M^+$ for all $t \geq 0$ and

$$d_t \leq \rho_t \leq K_1 e^{-\gamma t} \rho_0 < \epsilon, \quad t \geq 0, \tag{5.17}$$

while it follows from the inclusion $(\phi_0, h_0) \in M^-$ that $(\phi_t, h_t) \in M_1^-$ for all
$t \leq 0$ and

$$d_t \leq \rho_t \leq K_1 e^{\gamma t} \rho_0 < \epsilon, \quad t \leq 0. \tag{5.18}$$

Inequalities (5.17) and (5.18) are sufficient for $(\phi_0, h_0) \in M^+$ to imply
the inclusion $(\phi_t, h_t) \in M^+$ for all $t \geq 0$, and for $(\phi_0, h_0) \in M^-$ to imply the
inclusion $(\phi_t, h_t) \in M^-$ for all $t \leq 0$. Then inequalities (5.11) and (5.12) lead
to the estimates (5.1) and (5.2) with $K = \sqrt{2}(1 + K_2)K_1$.

The required properties of the motions ϕ_t, h_t of system (1.2) that start in
a $\delta(\epsilon)$-neighbourhood of torus (1.1) on the manifolds M^+ and M^- are established.

Let ϕ_t, h_t be a solution of system (1.2) for which $(\phi_0, h_0) \notin M^+ \cup M^-$,
$d_0 < \delta$. We consider a solution $\phi_t(\phi_0, z_0, z_0)$, $\theta_t(\phi_0, z_0, z_0)$, $z_t(\phi_0, z_0, z_0)$ of
system of equations (5.9), that takes the value ϕ_0, z_0, z_0 at $t = 0$. Since
$(\phi_0, z_0, z_0) \notin M_0^+ \cup M_0^-$ and $\|z_0\| < \delta$, there exist numbers $T_1 < 0$ and $T_2 > 0$
such that

$$\|z_t(\phi_0, z_0, z_0)\| < \epsilon_1, \quad t \in (T_1, T_2),$$
$$\|z_{T_1}(\phi_0, z_0, z_0)\| = \|z_{T_2}(\phi_0, z_0, z_0)\| = \epsilon_1. \tag{5.19}$$

The functions $\phi_t(\phi_0, z_0, z_0)$, $z_t(\phi_0, z_0, z_0)$ define a solution ϕ_t, z_t of system of
equations (5.6) for $t \in (T_1, T_2)$. The inequality

$$\|z_t\| \leq \sqrt{2}(1 + K_2)d_t$$

together with relations (5.19) proves that

$$d_t \geq \frac{\epsilon_1}{\sqrt{2}(1 + K_2)} = \epsilon$$

for $t = T_1$, and $t = T_2$ and hence establishes relation (5.3).

To complete the proof it remains to show that at least one of the sets
M^+, M^- is non-empty for any $\phi \in T_m$. Consider the set M_1^+ for an arbitrary
fixed value of $\phi = \phi_0 \in T_m$. Let $S(\phi_0, z)$ be the matrix transforming $C(\phi_0, z)$
to the Jordan canonical form $\mathfrak{F} = \text{diag}\{E_1, 0\}$, where E_1 is the $r \times r$ identity

matrix and $0 < r < n$. For a sufficiently small ϵ_1 in the region (5.14) the matrix $S(\phi_0, z)$ is continuous in z and satisfies the inequality

$$\|S(\phi_0, z)\| \leq M,$$

where $M = \max\limits_{\phi \in T_m, \|z\| \leq \epsilon_1} \|S(\phi, z)\|$ does not depend on ϕ_0.

We set

$$x = S^{-1}(\phi_0, z)z \qquad (5.20)$$

and write equation (5.15) in terms of the variables $x = (x_1, x_2)$, where x_1 is an r-dimensional and x_2 an $(n - r)$-dimensional vector. Using the representation $C(\phi_0, z) = S(\phi_0, z)\mathfrak{F}S^{-1}(\phi_0, z)$ we obtain instead of (5.15) the equation

$$\mathfrak{F}x = x$$

or, taking into account the form of the matrix $\mathfrak{F}$,

$$x_2 = 0. \qquad (5.21)$$

In the x coordinates the set M_1^+ for $\phi = \phi_0$ consists of points satisfying equation (5.21) and belonging to the image D of the ball $\|z\| \leq \epsilon_1$ under the map $z \to x$ given by equality (5.20). These points x have the form $x = (c, 0)$ where $c = (c_1, \ldots, c_r)$ is an r-dimensional vector. We show that M_1^+ contains the set of the $\overline{M}_1$ which is the inverse image of the set of points $x = (c, 0)$ of the ball $\|c\| < \delta_1$, under the map $z \to x$ given by equality (5.20), and that this set $\overline{M}_1$ is non-empty for sufficiently small $\delta_1 > 0$.

Accordingly, the set $\overline{M}_1$ consists of points z belonging to the ball $\|z\| \leq \bar{\epsilon}_1 < \epsilon_1$ that are solutions of equation (5.20) for $x = (c, 0)$, where $\|c\| < \delta_1$. We write the equation for the definition of the points of the set $\overline{M}_1$ in the form

$$z = S_1(\phi_0, z)c, \qquad (5.22)$$

where $S_1(\phi_0, z)$ is the block of the matrix $S(\phi_0, z) = (S_1(\phi_0, z), \; S_2(\phi_0, z))$ of appropriate dimension.

We choose $\delta_1 \leq \bar{\epsilon}_1/M$. Then the right hand side of equation (5.22) defines a continuous map $z \to z$ of the ball $\|z\| \leq \bar{\epsilon}_1$ into itself, since

$$\|S_1(\phi_0, z)c\| \leq M\delta_1 \leq \bar{\epsilon}_1.$$

This suffices for equation (5.22) to have a solution $z = z(c)$ in the ball $\|z\| \leq \bar{\epsilon}_1$. Since the matrix $S(\phi_0, z)$ is non-singular, $S_1(\phi_0, z) \neq 0$ and consequently, $z(c) \neq 0$ for $c \neq 0$. The set $\overline{M}_1$ contains the points $z(c)$ for all $\|c\| \leq \delta_1$. These points satisfy the condition $\|z(c)\| \leq \bar{\epsilon}_1$ and the inequality $z(c) \neq 0$ for $c \neq 0$. Thus M_1^+ is non-empty for any $\phi = \phi_0 \in T_m$.

We set $h(c) = u(\phi_0) + z(c)$ and consider the point $(\phi_0, h(c))$. The distance $d_0 = d_0(\phi_0, h(c))$ from this point to the torus (1.1) satisfies the estimate

$$d_0 \leq \|z(c)\| \leq \bar{\epsilon}_1.$$

For $\bar{\epsilon}_1 < \epsilon$ the point $(\phi_0, h(c))$ belongs to the ϵ-neighbourhood of torus (1.1). This proves that the set $\overline{M}_1$ considered for $\bar{\epsilon}_1 < \epsilon$ is contained in the set M^+. Since $h(c) \neq u(\phi_0)$, it follows that $M^+ \neq \varnothing$ for any $\phi = \phi_0 \in T_m$.

By writing equation (5.16) for $\phi = \phi_0$ in the form

$$C_1(\phi_0, z)z = z,$$

where $C_1(\phi, z) = E - C(\phi, z)$, ($E$ is the identity matrix) and repeating the argument used to prove that $M^+ \neq \varnothing$ for $\phi = \phi_0$ we see that M^- is non-empty for any $\phi \in T_m$.

The cases $r = n$ and $r = 0$ are included in the case under consideration with the sole difference that for $r = n$ we have the equality $C(\phi_0, z) = E$, and for $r = 0$ the equality $C_1(\phi_0, z) = E$. In the first case M^+ contains a neighbourhood of the torus (1.1) and $M^- = \varnothing$, while in the second case $M^+ = \varnothing$ and M^- contains a neighbourhood of the torus (1.1).

Next we consider system (2.1). We shall assume that the invariant torus (2.4) of the generating system (2.1) is exponentially dichotomous for $\epsilon = 0$ by virtue of the variational equations (2.5) and that this is established by using a non-singular symmetric matrix $S(\phi) \in C^1(T_m)$ for which the matrix

$$\hat{S}(\phi) = \frac{\partial S(\phi)}{\partial \phi} a_0(\phi) + S(\phi) P_0(\phi) + P_0^*(\phi) S(\phi),$$

formed by using the variational equations (2.5), is negative-definite for all $\phi \in T_m$. We define the positive constants β, μ, α from the conditions

$$\max_{\|h\|=1} \langle \hat{S}(\phi)h, h \rangle \leq -2\beta, \quad \max_{\|h\|=1} |\langle S(\phi)h, h \rangle| \leq \mu,$$

$$\inf_{S_1 \in \mathfrak{R}_1} \max_{\|\psi\|=1} \frac{|\langle [S_1(\phi)\frac{\partial a_0(\phi)}{\partial \phi} + \frac{1}{2}\frac{\partial S_1(\phi)}{\partial \phi} a_0(\phi)]\psi, \psi \rangle|}{\langle S_1(\phi)\psi, \psi \rangle} \leq \alpha, \qquad (5.23)$$

where $\mathfrak{R}_1$ is the set of $m \times m$ positive-definite symmetric matrices belonging to the space $C^1(T_m)$.

Theorem 2. *Suppose that the functions a, P and f have rth order partial derivatives with respect to ϕ, h that are continuous with respect to ϕ, h, ϵ in the region (2.2). Suppose further that the inequality*

$$\beta/\mu > r\alpha \qquad (5.24)$$

holds. If $r > 1$, then there exists a sufficiently small $\epsilon_0 > 0$ such that for any $\epsilon \in [0, \epsilon_0]$ system (2.1) has an invariant torus (2.6) with function u belonging to the space $C^{r-1}_{\text{Lip}}(T_m)$ and satisfying inequality (2.7). For $r \geq 2$ this torus is exponentially dichotomous, while for $r = 1$ it is unstable if the matrix $S(\phi)$ has negative eigenvalues and is asymptotically stable otherwise.

Proof. By Lemma 1, of §3.13, the system of variational equations (2.5) has a Green's function $G_0(\tau, \phi)$ satisfying the inequality

$$\|G_0(\tau, \phi)\| \leq Ke^{-\gamma|\tau|}, \quad \tau \in \mathbf{R}, \qquad (5.25)$$

where $\gamma = \beta/\mu$ and $K = $ const does not depend on $\phi \in T_m$.

Using the inequality (5.23) which defines the constant α, one can derive via a standard argument an estimate for the fundamental matrix of the solution, $\Omega_0^t(\partial a_0/\partial \phi)$ of this system of linear homogeneous equations with the coefficient matrix $\partial a_0/\partial \phi = \partial a_0(\phi_t(\phi))/\partial \phi$, where $\phi_t(\phi)$ is a solution of the first equation of system (2.5):

$$\left\| \Omega_0^t \left(\frac{\partial a_0}{\partial \phi} \right) \right\| \leq K_1 e^{\hat{\alpha}|t|}, \quad t \in \mathbf{R} \qquad (5.26)$$

where $\hat{\alpha} = \alpha + \epsilon_1$ and $K_1 = K_1(\epsilon_1)$ is a positive constant which depends only on ϵ_1 (ϵ_1 is an arbitrarily small positive number).

In view of Theorem 1 of §3.13, inequalities (5.24)–(5.26) lead to the estimate

$$|G_0(\tau, \phi)f(\phi_\tau(\phi))\|_r \leq K_2 e^{-(\hat{\gamma} - r\hat{\alpha})|\tau|}|f|_r, \quad \tau \in \mathbf{R} \qquad (5.27)$$

for any function $f \in C^r(T_m)$, $\hat{\gamma} = \gamma - \epsilon_1$, and some constant $K_2 = K_2(\epsilon_1)$ which depends only on ϵ_1.

We introduce a perturbation into the first equation of system (2.5) by adding to its right hand side the function $a_1(\phi) \in C^r(T_m)$ satisfying the inequality

$$|a_1|_r \leq \delta.$$

Then the matrix $\hat{S}_1(\phi) = \hat{S}(\phi) + (\partial S(\phi)/\partial\phi)a_1(\phi)$ satisfies the inequality

$$\max_{\|h\|=1} \langle \hat{S}_1(\phi)h, h \rangle \leq -2\beta_\delta$$

with constant $\beta_\delta = \beta - \overline{K}\delta$, where $\overline{K} = \max_{\phi \in T_m} \left\| \frac{\partial S(\phi)}{\partial\phi} \right\|$.

Replacing a_0 by $a_0 + a_1$ in the left hand side of inequality (5.23), we obtain a similar inequality with α equal to $\alpha_\delta = \hat{\alpha} + \overline{K}_1\delta$, where $\overline{K}_1$ is a positive constant independent of ϵ_1, δ, $\hat{\alpha} = \alpha + \epsilon_1$. It follows from the form of the constants β_δ, a_δ that δ and ϵ_1 can be fixed so small that the inequality

$$\beta_\delta/\mu > r\alpha_\delta \qquad (5.28)$$

holds. Then system (2.5) perturbed by the term $a_1(\phi)$ has a Green's function $\overline{G}_0(\tau, \phi)$ that satisfies inequality (5.25) with γ equal to $\gamma_\delta = \beta_\delta/\mu$. The matrix $\Omega_0^t \left(\frac{\partial a_0}{\partial\phi} + \frac{\partial a_1}{\partial\phi} \right)$, for which

$$\frac{\partial a_0}{\partial\phi} + \frac{\partial a_1}{\partial\phi} = \frac{\partial a_0(\bar{\phi}_t(\phi))}{\partial\phi} + \frac{\partial a_1(\bar{\phi}_t(\phi))}{\partial\phi},$$

where $\bar{\phi}_t(\phi)$ is a solution of the first equation of system (2.5) perturbed by the term $a_1(\phi)$ on its right hand side, now satisfies inequality (5.26) with $\hat{\alpha}$ equal to $\hat{\alpha}_\delta = \alpha_\delta + \epsilon_1$. This leads to estimate (5.27) for $\overline{G}_0(\tau, \phi)$ with $\hat{\gamma}$ equal to $\hat{\gamma}_\delta$ and $\hat{\alpha}$ equal to $\hat{\alpha}_\delta + \epsilon_1$.

For sufficiently small ϵ_1, inequality (5.28) ensures that the exponent in inequality (5.27) for the Green's function is negative. This is sufficient for the Green's function $\overline{G}_0(\tau, \phi)$ for the system (2.5) to be coarse with smoothness index r. The main theorem ensures that there exists an invariant torus (2.6) of system (2.1) with the function u belonging to the space $C_{\text{Lip}}^{r-1}(T_m)$ and satisfying inequality (2.7).

For $r \geq 2$, system (2.1) and the torus (2.6) satisfy conditions sufficient for us to write down the variational equations corresponding to the torus (2.6) considered as an invariant set of system (2.1).

Since $|u|_{1,\text{Lip}} \leq K|f|_2 \to 0$ as $\epsilon \to 0$, the variational equations for the torus (2.6) for $\epsilon \in [0, \epsilon_0]$ with a sufficiently small ϵ_0 are defined by the functions a, P which are close (with respect to the $C(T_m)$ norm) to the functions a_0, P_0 defining the variational equations (2.5). This suffices for the matrix

$$\hat{S}_1 = \hat{S} + \frac{\partial S}{\partial\phi}(a - a_0) + S(P - P_0) + (P - P_0)^*S$$

constructed by using the matrix $S = S(\phi)$ according to the variational equations for the torus (2.6) to be negative-definite for all $\phi \in T_m$ and any $\epsilon \in [0, \epsilon_0]$, where ϵ_0 is a sufficiently small positive number.

Thus, for $r \geq 2$ all the conditions of Theorem 1 are satisfied and the invariant torus (2.6) of system (2.1) turns out to be exponentially dichotomous for every $\epsilon \in [0, \epsilon_0]$.

Let $r = 1$. We consider the "extended" system of equations:

$$\frac{d\phi}{dt} = a(\phi, \chi(\theta), \epsilon),$$

$$\frac{d\theta}{dt} = P(\phi, \chi(\theta), \epsilon)\chi(\theta) + f(\phi, \epsilon),$$

$$\frac{d\bar{h}}{dt} = P(\phi, \chi(\theta), \epsilon)\bar{h} + f(\phi, \epsilon),$$
(5.29)

where $\chi(\theta) = (\chi_1(\theta_1), \ldots, \chi_1(\phi_n))$ is the function (5.8).

For sufficiently small $\epsilon > 0$ and $\delta_0 > 0$ system of equations (5.29) satisfies the conditions for the existence of an exponentially dichotomous torus and it has such a torus

$$\bar{h} = u_1(\phi, \theta, \epsilon), \quad \phi \in T_m, \quad \theta \in T_n,$$
(5.30)

for all $\epsilon \in [0, \epsilon_0]$. The function u_1 in (5.30) belongs to the space $C(T_m \times T_n)$ and satisfies the inequality

$$\|u_1(\phi, \theta, \epsilon)\| \leq K|f|_1 \leq K(\epsilon),$$
(5.31)

where $K(\epsilon) \to 0$ monotonically as $\epsilon \to 0$.

Since the torus (5.30) is exponentially dichotomous, it follows that all the motions of system (5.29) that are bounded for $t \in \mathbf{R}$ are motions on the torus (5.30).

The system of equations (5.29) has the locally invariant set

$$\theta = \bar{h} = h, \quad \|h\| \leq \epsilon_1, \quad \phi \in T_m,$$
(5.32)

the motions on which are defined by the system of equations (2.1). On this set the invariant torus (5.30) of system (5.29) becomes the locally invariant set of system (2.1) defined by the relations

$$h = u_1(\phi, h, \epsilon), \quad \|h\| < \epsilon_1, \quad \phi \in T_m,$$
(5.33)

for all $\epsilon \in [0, \epsilon_0]$. By virtue of the properties of the torus (5.30), the trajectory of any motion ϕ_t, h_t of system (2.1) that is bounded for $t \in \mathbf{R}$ and for which $\|h_t\| < \epsilon_1$ for all $t \in \mathbf{R}$ lies in the invariant set (5.33) for $t \in \mathbf{R}$.

Since the motion $\phi_t, h_t = u(\phi_t, \epsilon)$ of system (2.1) that starts on the invariant torus (2.6) satisfies the inequality $\|h_t\| = \|u(\phi_t, \epsilon)\| \le K|f|_1 \le K(\epsilon)$ for all $t \in \mathbf{R}$, it follows that for sufficiently small ϵ_0, $\|h_t\| < \epsilon_1$ for all $t \in \mathbf{R}$, $\epsilon \in [0, \epsilon_0]$. This suffices for the torus (2.6) to satisfy the relations (5.33) for all $\epsilon \in [0, \epsilon_0]$. Consequently, $u(\phi, \epsilon) = u_1(\phi, u(\phi, \epsilon), \epsilon)$ for all $\phi \in T_m$, $\epsilon \in [0, \epsilon_0]$ for a sufficiently small $\epsilon_0 > 0$.

We now show that equation (5.33) has no solutions other than $h = u(\phi, \epsilon)$ in the domain $\|h\| < \epsilon_1$, for any $\phi \in T_m$ and $\epsilon \in [0, \epsilon_0]$ provided that ϵ_0 is sufficiently small. Indeed, suppose that for some $\phi = \phi_0' \in T_m$, $\epsilon = \epsilon' \in [0, \epsilon_0]$, equation (5.33) has a solution $h = h_0' = h_0(\phi_0', \epsilon') \ne u(\phi_0', \epsilon')$ for which $\|h_0\| < \epsilon_1$.

Then the motion $\phi = \phi_t(\phi_0', h_0', \epsilon')$, $h = h_t(\phi_0', h_0', \epsilon')$ of system (2.1) for $\epsilon = \epsilon'$ that starts at the point (ϕ_0', h_0') satisfies the identity

$$h_t(\phi_0', h_0', \epsilon') = u_1(\phi_t(\phi_0', h_0', \epsilon'), \ h_t(\phi_0', h_0', \epsilon'), \epsilon') \qquad (5.34)$$

for all those t for which $\|h_t(\phi_0', h_0', \epsilon')\| < \epsilon_1$. It follows from (5.34) that for these t the inequality

$$\|h_t(\phi_0', h_0', \epsilon')\| \le K|f|_1 \le K(\epsilon') \le K(\epsilon_0) < \epsilon_1 \qquad (5.35)$$

holds. It does not allow the motion $\phi = \phi_t(\phi_0', h_0', \epsilon')$, $h = h_t(\phi_0', h_0', \epsilon')$ to leave the domain $\|h\| < \epsilon_1$ for any $t \in \mathbf{R}$. This proves inequality (5.35) for all $t \in \mathbf{R}$. We set

$$z_t = h_t(\phi_0', h_0', \epsilon') - u(\phi_t(\phi_0', h_0', \epsilon'), \epsilon'). \qquad (5.36)$$

The function z_t is defined and absolutely continuous for $t \in \mathbf{R}$ and satisfies the inequality

$$0 < \sup_{t \in \mathbf{R}} \|z_t\| \le 2K(\epsilon_0). \qquad (5.37)$$

It follows from the arguments given in the proof of Theorem 1 of §4.3 for $r = 1$ that z_t is a solution of the system of linear non-homogeneous equations (3.15) with $\phi_t = \phi_t(\phi_0', h_0', \epsilon')$, $\epsilon = \epsilon'$. Since the conditions of Theorem 2 ensure that the system of homogeneous equations corresponding to (3.15) is exponentially dichotomous, it follows that z_t is the unique solution of system (3.15) that is bounded on $\mathbf{R}$ and the following representation holds for it:

$$z_t = \int_{-\infty}^{\infty} G_t(\tau) R(\phi_\tau, u(\phi_\tau, \epsilon), \ z_\tau, \epsilon) d\tau, \qquad (5.38)$$

where $G_t(\tau)$ is a Green's function for the homogeneous system of equations corresponding to system (3.15). The function $G_t(\tau)$ satisfies the following estimate which is the analogue of inequality (5.25):

$$\|G_t(\tau)\| \le K e^{-\gamma|t-\tau|}, \quad t \in \mathbf{R}, \ \tau \in \mathbf{R},$$

where $K > 0$ and $\gamma > 0$ are independent of t, τ and ϵ'.

It follows from formula (5.38) and the above estimate for the function $G_t(\tau)$ that

$$\|z_t\| \le \frac{KK(\epsilon')}{2\gamma} \sup_{t \in \mathbf{R}} \|z_t\|, \qquad (5.39)$$

where $K(\epsilon) \to 0$ and is monotonically decreasing as $\epsilon \to 0$.

For sufficiently small $\epsilon_0 > 0$, inequality (5.39) leads to the estimate

$$\sup_{t \in \mathbf{R}} \|z_t\| \le \tfrac{1}{2} \sup_{t \in \mathbf{R}} \|z_t\|,$$

which contradicts inequality (5.37).

This contradiction shows that only the point (ϕ, h) which belongs to the invariant torus (2.6) of system (2.1) satisfies relations (5.33).

Denote by ρ the number of positive eigenvalues of the matrix $S(\phi)$. Let $\rho < n$. On the locally invariant set (5.32) of system (5.29) the separatrix set M_0^- of invariant torus (5.30) of this system becomes the locally invariant set M^- of system (2.1) defined by the relations

$$C(\phi, h, \epsilon)[h - u_1(\phi, h, \epsilon)] = 0, \quad \|h\| < \epsilon_1, \ \phi \in \mathcal{T}_m \qquad (5.40)$$

for all $\epsilon \in [0, \epsilon_0]$. It has the property that a solution ϕ_t, h_t of system (2.1) for which $(\phi_0, h_0) \in M^-$ satisfies the inequality

$$\|h_t - u_1(\phi_t, h_t, \epsilon)\| \le K e^{\gamma t} \|h_0 - u_1(\phi_0, h_0, \epsilon)\| \qquad (5.41)$$

for all $t \le 0$ as long as $\|h_t\| < \epsilon_1$.

We fix ϕ by setting $\phi = \phi_0$ and consider a solution of equation (5.40) belonging to the domain $\|h\| < \epsilon_1$, for $\phi = \phi_0$. Let $T(\phi_0, h, \epsilon)$ be a matrix that transforms $C(\phi_0, h, \epsilon)$ to Jordan form. Since $C(\phi, h, \epsilon)$ is continuous in ϕ, h for ϕ, h, ϵ in the domain $\phi \in T_m$, $\|h\| < \epsilon_1$, $\epsilon \in [0, \epsilon_0]$ and is a projection with rank ρ, we see that the matrix $T(\phi_0, h, \epsilon)$ can be chosen to be continuous in h when h, ϵ belong to the region $\|h\| < \epsilon_1$, $\epsilon \in [0, \epsilon_0]$. By introducing the variables $x = (x_1, x_2)$ related to h by the equality

$$x = T^{-1}(\phi_0, h, \epsilon)[h - u_1(\phi_0, h, \epsilon)],$$

equation (5.40) takes the form

$$x_1 = 0, \ \phi \in T_m.$$

We define $h = h(c, \epsilon)$ to be a solution of the equation

$$h = u_1(\phi_0, h, \epsilon) + S_2(\phi_0, h, \epsilon)c, \tag{5.42}$$

where $c = (c_1, \ldots, c_{n-\rho})$ is an $(n - \rho)$-dimensional vector in the ball $\|c\| < \delta_1$, and $S_2(\phi, h, \epsilon)$ the block of the matrix $T(\phi, h, \epsilon) = (S_1(\phi, h, \epsilon), \ S_2(\phi, h, \epsilon))$ of appropriate dimension. We choose $\epsilon_0 > 0$ and $\delta_1 > 0$ such that for $\epsilon \in [0, \epsilon_0]$ the inequality

$$\max(2K(\epsilon_0) + M\delta_1, 2KM\delta_1) < \epsilon_1 \tag{5.43}$$

where K is the constant in the estimate (5.41) and

$$M = \sup_{\phi \in T_m, \|h\| < \epsilon_1} \|S_2(\phi, h, \epsilon)\|.$$

Then the right hand side of equation (5.42) defines a continuous operator $h \to h$ mapping the ball $\|h\| < \epsilon_1$ into itself. This suffices for equations (5.42) to have a solution $h = h(c, \epsilon)$ for all $\epsilon \in [0, \epsilon_0]$ and all $\|c\| < \delta_1$. This solution satisfies the inequality $\|h(c, \epsilon)\| < \epsilon_1$, for these ϵ and c. Moreover, since $S_2(\phi, h, \epsilon)c = 0$ only for $c = 0$, it follows that $h(c, \epsilon) \neq u_1(\phi_0, h(c, \epsilon), \epsilon)$ for $c \neq 0$ and consequently, $h(c, \epsilon) \neq u(\phi_0, \epsilon)$ for $c \neq 0$. The solution $\phi = \phi_t(\phi_0, h(c, \epsilon), \epsilon)$, $h = h_t(\phi_0, h(c, \epsilon), \epsilon)$ of system (2.1) that takes the value ϕ_0, $h(c, \epsilon)$ at $t = 0$ satisfies the condition $(\phi_0, h(c, \epsilon)) \in M^-$ and consequently, it satisfies inequality (5.41) for $t \leq 0$ as long as $\|h_t(\phi_0, h(c, \epsilon), \epsilon)\| < \epsilon_1$.

Since $\|h(c,\epsilon) - u_1(\phi_0, h(c,\epsilon), \epsilon)\| < M\delta_1$, it follows from inequality (5.41) that

$$\|h_t(\phi_0, h(c,\epsilon), \epsilon) - u_1(\phi_t(\phi_0, h(c,\epsilon), \epsilon), h_t(\phi_0, h(c,\epsilon), \epsilon), \epsilon)\| \le \frac{\epsilon_1}{2} e^{\gamma t}$$

for $t \le 0$ as long as $\|h_t(\phi_0, h(c,\epsilon), \epsilon)\| < \epsilon_1$. But then $h_t(\phi_0, h(c,\epsilon), \epsilon)$ satisfies the inequality

$$\|h_t(\phi_0, h(c,\epsilon), \epsilon)\| \le K(\epsilon_0) + \frac{\epsilon_1}{2} e^{\gamma t} < \epsilon_1 \qquad (5.44)$$

for these $t \le 0$ and cannot leave the domain $\|h\| < \epsilon_1$ for $t < 0$. Hence it follows that inequality (5.44) holds for all $t \le 0$.

Letting t tend to $-\infty$ in inequality (5.44) we find that the α-limit points of the semi-trajectory of the motion $\phi_t(\phi_0, h(c,\epsilon), \epsilon)$, $h_t(\phi_0, h(c,\epsilon), \epsilon)$ belong to the set (5.33) and consequently, they belong to the invariant torus (2.6) of system (2.1). At the same time the motion $\phi_t(\phi_0, h(c,\epsilon), \epsilon)$, $h_t(\phi_0, h(c,\epsilon), \epsilon)$ itself does not start on invariant torus (2.6) of system (2.1) for $c \ne 0$. Therefore, the motion of system (2.1) that starts on the negative semi-trajectory of the motion $\phi_t(\phi_0, h(c,\epsilon), \epsilon)$, $h_t(\phi_0, h(c,\epsilon), \epsilon)$ lying in an arbitrarily small neighbourhood of torus (2.6), leaves a certain fixed neighbourhood of the latter at some time $t = T > 0$. According to the definition of an unstable invariant set of the dynamic system, the torus (2.6) is unstable.

Let $\rho = n$. The locally invariant set M^+ of system (2.1) is then defined by the relations

$$\|h\| < \epsilon_1, \quad \phi \in T_m$$

for all $\epsilon \in [0, \epsilon_0]$ and has the property that the solution ϕ_t, h_t of system (2.1), for which $\|h_0\| < \epsilon_1$, $\phi_0 \in T_m$, satisfies inequality

$$\|h_t - u_1(\phi_t, h_t, \epsilon)\| \le K e^{-\gamma t} \|h_0 - u_1(\phi_0, h_0, \epsilon)\| \qquad (5.45)$$

for $t \ge 0$, as long as $\|h_t\| < \epsilon_1$.

We fix ϕ by setting $\phi = \phi_0$ and consider a solution $h = h(c,\epsilon)$ of the equation

$$h = u_1(\phi_0, h, \epsilon) + c, \qquad (5.46)$$

where $c = (c_1, \ldots, c_n)$ is an n-dimensional vector in the ball $\|c\| < \delta_1$. If $\epsilon_0 > 0$ and $\delta_1 > 0$ are chosen such that for $\epsilon \in [0, \epsilon_0]$ inequality (5.43) holds with $M = \|E\|$, then the solution $h(c,\epsilon)$ of equation (5.46) exists for all $\epsilon \in [0, \epsilon_0]$,

$\|c\| < \delta_1$, and satisfies inequality $\|h(c, \epsilon)\| < \epsilon_1$ for these ϵ and c. Using
inequality (5.45) and arguing as for the case $\rho < n$, we derive the inequalities

$$\|h_t(\phi_0, h(c, \epsilon), \epsilon) - u_1(\phi_t(\phi_0, h(c, \epsilon), \epsilon),$$

$$h_t(\phi_0, h(c, \epsilon), \epsilon), \epsilon)\| < \frac{\epsilon_1}{2} e^{-\gamma t},$$

$$\|h_t(\phi_0, h(c, \epsilon), \epsilon)\| \le K(\epsilon_0) + \frac{\epsilon_1}{2} e^{-\gamma t} < \epsilon_1 \tag{5.47}$$

for all $t \ge 0$, $\epsilon \in [0, \epsilon_0]$, $\|c\| < \delta_1$.

For a fixed $\epsilon \in [0, \epsilon_0]$ the function $u_1(\phi, h, \epsilon)$ is uniformly continuous in
ϕ, h as it is continuous on the compact set $\phi \in \mathcal{T}_m$, $\|h\| \le \delta_1 < \epsilon_1$. Therefore
we can choose $\delta_2 = \delta_2(\delta_1) > 0$ such that the inequality

$$\|u_1(\phi, h, \epsilon) - u_1(\phi, \bar{h}, \epsilon)\| \le \delta_1/2$$

holds for arbitrary $\phi \in \mathcal{T}_m$ and any h and $\bar{h}$ satisfying the inequalities:

$$\|h\| \le \delta_1, \quad \|\bar{h}\| \le \delta_1, \quad \|h - \bar{h}\| \le \delta_2.$$

We set $\delta_2 < \delta_1/2$ and choose $\epsilon_0 > 0$ such that the inequality

$$\|u(\phi, \epsilon)\| < \delta_1/2, \quad \epsilon \in [0, \epsilon_0]$$

holds for all $\phi \in \mathcal{T}_m$.

For an arbitrary $\phi = \phi_0 \in \mathcal{T}_m$ and any $h = h_0$ in the domain

$$\|h - u(\phi_0, \epsilon)\| < \delta_2 \tag{5.48}$$

we now have the inequality

$$\|h_0 - u(\phi_0, h_0, \epsilon)\| \le \|h_0 - u(\phi, \epsilon)\| + \|u_1(\phi_0, h_0, \epsilon) -$$

$$-u_1(\phi_0, u(\phi_0, \epsilon), \epsilon)\| < \delta_2 + \delta_1/2 < \delta_1,$$

which proves that an arbitrary point (ϕ_0, h_0) in the domain (5.48) coincides
with the point $(\phi_0, h(c, \epsilon))$ when $c = h_0 - u_1(\phi_0, h_0, \epsilon)$ in the ball $\|c\| < \delta_1$.
Thus, for any point (ϕ_0, h_0) satisfying condition (5.48), the solution $\phi_t = \phi_t(\phi_0, h_0, \epsilon)$, $h_t = h_t(\phi_0, h_0, \epsilon)$ of the system of equations (2.1) satisfies the
inequalities (5.47):

$$\|h_t\| < \epsilon_1, \quad \|h_t - u_1(\phi_t, h_t, \epsilon)\| < \frac{\epsilon_1}{2} e^{-\gamma t} \tag{5.49}$$

for all $t \geq 0$.

Since $u(\phi, \epsilon) \in C_{\text{Lip}}(\mathcal{T}_m)$ for $\epsilon \in [0, \epsilon_0]$ and the function $u(\phi, \epsilon)$ satisfies inequality (2.7) for $r = 1$, the set of points (ϕ_0, h_0) satisfying condition (5.48), where ϕ_0 ranges over $\mathcal{T}_m$ contains for some $d = d(\delta_2) > 0$ a d-neighbourhood of the torus $h = u(\phi, \epsilon)$, $\phi \in \mathcal{T}_m$. Thus, inequalities (5.49) hold for solutions ϕ_t, h_t that start at points of this neighbourhood. Letting t tend to $+\infty$ in inequalities (5.49) we see that the ω-limit points of the semi-trajectory of the motion ϕ_t, h_t belong to the invariant torus (2.6) of system (2.1).

None of the α-limit points of the semi-trajectory of the motion ϕ_t, h_t belongs to the torus (2.6). Indeed, if this is not so, then there exists a sequence $t_n \rightarrow -\infty$ as $n \rightarrow \infty$ such that all the points (ϕ_{t_n}, h_{t_n}) of the motion $\phi_t = \phi_t(\phi_0, h_0, \epsilon)$, $h_t = h_t(\phi_0, h_0, \epsilon)$ with fixed initial point (ϕ_0, h_0) in the $d_2(\delta_2)$-neighbourhood of torus (2.6) belong to this neighbourhood for all $n = 1, 2, \ldots$. But then the positive semi-trajectories of the motions $\phi_t = \phi_t(\phi_{t_n}, h_{t_n}, \epsilon)$, $h_t = h_t(\phi_{t_n}, h_{t_n}, \epsilon)$ satisfy inequalities (5.49) for $n = 1, 2, \ldots$.

Since $\phi_t(\phi_{t_n}, h_{t_n}, \epsilon) = \phi_{t+t_n}(\phi_0, h_0, \epsilon)$, $h_t(\phi_{t_n}, h_{t_n}, \epsilon) = h_{t+t_n}(\phi_0, h_0, \epsilon)$ it follows that

$$\|h_{t+t_n}(\phi_0, h_0, \epsilon)\| < \epsilon_1$$

for all $t \geq 0$ and $n = 1, 2, \ldots$. Consequently, the trajectory of the motion $\phi_t(\phi_0, h_0, \epsilon)$, $h_t(\phi_0, h_0, \epsilon)$ satisfies the inequality

$$\|h_t(\phi_0, h_0, \epsilon)\| < \epsilon_1$$

for $t \in [t_n, +\infty)$ $(n = 1, 2 \ldots)$. Thus the motion $\phi_t = \phi_t(\phi_0, h_0, \epsilon)$, $h_t = h_t(\phi_0, h_0, \epsilon)$ of system (2.1) satisfies the inequality $\|h_t\| < \epsilon_1$, for all $t \in \mathbf{R}$. As was proved earlier, such a motion takes place on the invariant torus (2.6). This contradicts the fact that the point (ϕ_0, h_0) does not belong to this torus.

Thus torus (2.6) has a compact neighbourhood that contains neither whole trajectories nor motions having α-limit points on this torus. By Theorem 3 of §2.2, this means that the torus (2.6) considered as an invariant set of system (2.1) is stable. The limit relation necessary for torus (2.6) to be asymptotically stable follows from the fact that all the ω-limit points of the motions ϕ_t, h_t that start in the $d(\delta_2)$-neighbourhood of torus (2.6) belong to this torus. This completes the proof that invariant torus (2.6) of system (2.1) is asymptotically invariant for $\rho = n$. Theorem 2 is proved.

It should be noted that the condition of Theorem 2 can be weakened by replacing the left hand side of inequality (5.24) by a positive number γ determined from the inequality

$$\max_{\|h\|=1} \frac{\langle \hat{S}(\phi)h, h \rangle}{|\langle S(\phi)h, h \rangle|} \leq -2\gamma, \quad \phi \in T_m. \tag{5.50}$$

This follows from the proof of Theorem 2 by taking into account the fact that inequality (5.50) leads to estimate (5.25) with γ equal to the value in (5.50).

However, since $(-\beta)$ is the greatest of the eigenvalues of the matrix $\hat{S}(\phi)/2$, $\phi \in T_m$, and μ is the greatest of the absolute values of the eigenvalues of the matrix $S(\phi)$, $\phi \in T_m$, we see that finding the ratio β/μ is simpler in practice than finding the number γ.

We set α_0 equal to the greatest of the absolute values of the eigenvalues of the matrix $\frac{1}{2}\left[\frac{\partial a_0(\phi)}{\partial \phi} + \left(\frac{\partial a_0(\phi)}{\partial \phi}\right)^*\right]$, $\phi \in T_m$. It is easy to see that inequality (5.24) holds whenever $\beta/\mu > r\alpha_0$ for $r \geq 1$. For example, for system (2.1), the matrix $P_0(\phi)$ of which, when taken in the block form

$$P_0(\phi) = \begin{bmatrix} P_1(\phi) & P_{12}(\phi) \\ P_{21}(\phi) & P_2(\phi) \end{bmatrix},$$

satisfies the condition

$$\hat{S}(\phi) = \begin{bmatrix} P_1(\phi) + P_1^*(\phi) & P_{12}(\phi) + P_{21}^*(\phi) \\ P_{21}(\phi) + P_{12}^*(\phi) & -P_2(\phi) - P_2^*(\phi) \end{bmatrix} \tag{5.51}$$

necessary for it to be a negative-definite matrix, condition (5.24) holds whenever $\beta > r\alpha_0$ for $r \geq 1$, where $(-\beta)$ is the greatest eigenvalue of matrix (5.51) for $\phi \in T_m$.

4.6. An estimate of the smallness of a perturbation and the maximal smoothness of an invariant torus of a non-linear system

The estimate on how small a perturbation must be so that the statements of perturbation theory of an invariant torus for the non-linear system (1.2) remain true is a problem of quantitative analysis of this theory. At present, it is not well studied and is not very effective from the point of view of applications. Therefore we shall give only one result of such analysis, namely, conditions for the preservation of an invariant torus involving an estimate of the smallness of a perturbation admitting practical verification.

We start with the system of equations

$$\frac{d\phi}{dt} = a(\phi, h), \quad \frac{dh}{dt} = P(\phi, h)h + f(\phi), \tag{6.1}$$

where a, P and f are periodic functions in ϕ_ν ($\nu = 1, \ldots, m$) with period 2π, defined in the domain

$$\|h\| \le d, \ \phi \in T_m \tag{6.2}$$

and satisfying Lipschitz conditions with respect to ϕ, h:

$$\|a(\phi, h) - a(\phi', h')\| \le \alpha_0 \|\phi - \phi'\| + \alpha_1 \|h - h'\|,$$

$$\|P(\phi, h) - P(\phi', h')\| \le \beta \|\phi - \phi'\| + \beta_1 \|h - h'\|, \tag{6.3}$$

$$\|f(\phi) - f(\phi')\| \le \lambda \delta_0 \|\phi - \phi'\|,$$

where α_0, α_1, β, β_1 and λ are non-negative constants, $\delta_0 = \max\limits_{\phi \in T_m} \|f(\phi)\|$, ϕ, ϕ', h and h' are arbitrary constants in the region (6.2). Let

$$\max_{\|\eta\|=1} \langle P(\phi, h)\eta, \eta \rangle \le -\beta_0 \tag{6.4}$$

for all ϕ, h in the region (6.2) and some constant $\beta_0 > 0$. Finally, we require that the inequality

$$\beta_0 - \alpha_0 = \gamma_1 > 0 \tag{6.5}$$

hold.

We set

$$\mu_0 = \frac{\beta + \lambda\beta_0}{\beta_1}, \quad \Delta_0 = \left(\mu_0^2 + \frac{\gamma_1}{\alpha_1}\mu_0\right)^{1/2} - \mu_0. \tag{6.6}$$

Theorem 1. *Let conditions (6.3)–(6.5) hold for the system of equations (6.1). If*

$$\delta_0 \le \beta_0 \min\left(d, \frac{\alpha_1}{\beta_1}\left(\frac{\gamma_1}{\alpha_1} - 2\Delta_0\right)\right), \tag{6.7}$$

then the system of equations (6.1) has an invariant torus

$$h = u(\phi), \ \phi \in T_m$$

where the function u belongs to the space $C_{\text{Lip}}(\mathcal{T}_m)$ *and satisfies the inequalities*

$$\|u(\phi)\| \le \delta_0/\beta_0, \quad \|u(\phi) - u(\phi')\| \le \Delta_0\|\phi - \phi'\|$$

for arbitrary $\phi, \phi' \in \mathcal{T}_m$.

Proof. We denote by $C(\Delta)$ the set of functions $F(\phi)$ in $C_{\text{Lip}}(\mathcal{T}_m)$ satisfying the inequalities

$$|F|_0 \le d, \quad \|F(\phi) - F(\phi')\| \le \Delta\|\phi - \phi'\|$$

for any $\phi, \phi' \in \mathcal{T}_m$. By introducing the distance $\rho(F_1, F_2)$ between the elements F_1, F_2 of the set $C(\Delta)$ according to the formula

$$\rho(F_1, F_2) = |F_1 - F_2|_0,$$

we turn $C(\Delta)$ into a complete metric space.

Following the method of integral manifolds [22], [81] we define on $C(\Delta)$ an integral operator S (we shall write $S_\phi(F)$ instead of $S(F)(\phi)$) by setting

$$S_\phi(F) = \int_{-\infty}^0 \Omega_\tau^0(\phi, F) f(\phi_\tau^F(\phi)) d\tau, \tag{6.8}$$

where $\phi_t^F(\phi)$ is a solution of the first equation of system (6.1) for $h = F(\phi)$ and $\Omega_\tau^t(\phi, F)$ is the fundamental matrix of solutions of the system

$$\frac{dh}{dt} = P(\phi_t^F(\phi), F(\phi_t^F(\phi)))h \tag{6.9}$$

that is equal to the identity matrix for $t = \tau$. We now show that the operator S has a fixed point in $C(\Delta)$. Let us make the necessary estimates.

We estimate the difference $\phi_t^F(\phi) - \phi_t^{\overline{F}}(\bar{\phi})$, where $F, \overline{F}$ are any two functions in $C(\Delta)$ and $\phi, \bar{\phi}$ are arbitrary points of $\mathcal{T}_m$. We have:

$$\|\phi_t^F(\phi) - \phi_t^{\overline{F}}(\bar{\phi})\| \le \|\phi - \bar{\phi}\| + \left|\int_0^t \|a(\phi_t^F(\phi), F(\phi_t^F(\phi))) - a(\phi_t^{\overline{F}}(\bar{\phi}),\right.$$

$$\overline{F}(\phi_t^{\overline{F}}(\bar{\phi})))\|dt\Big| \le \|\phi - \bar{\phi}\| + \left|\int_0^t (\alpha_0\|\phi_t^F(\phi) - \phi_t^{\overline{F}}(\bar{\phi})\| + \right.$$

$$\left. + \alpha_1\|F(\phi_t^F(\phi)) - \overline{F}(\phi_t^{\overline{F}}(\bar{\phi}))\|)dt\right| \le$$

$$\le \|\phi - \bar{\phi}\| + \left|\int_0^t (\alpha_0 + \alpha_1\Delta)\|\phi_t^F(\phi) - \phi_t^{\overline{F}}(\bar{\phi})\|dt\right| + \alpha_1|F(\phi) - \overline{F}(\phi)|_0|t|,$$

whence, using the Gronwall-Bellman inequality [43], we obtain

$$\|\phi_t^F(\phi) - \phi_t^{\overline{F}}(\bar{\phi})\| \le e^{(\alpha_0 + \alpha_1 \Delta)|t|}\|\phi - \bar{\phi}\| +$$

$$+\frac{\alpha_1}{\alpha_0 + \alpha_1 \Delta}\left[e^{(\alpha_0 + \alpha_1 \Delta)|t|} - 1\right]|F(\phi) - \overline{F}(\phi)|_0. \qquad (6.10)$$

We estimate the difference $\Omega_\tau^0(\phi, F) - \Omega_\tau^0(\bar{\phi}, \overline{F}) = Z_\tau^0$. To do this we use equations (6.9) to write the system of equations for Z_τ^t:

$$\frac{dZ_\tau^t}{dt} = P(\phi_t^F(\phi), F(\phi_t^F(\phi)))\Omega_\tau^t(\phi, F) - P(\phi_t^{\overline{F}}(\bar{\phi}), \overline{F}(\phi_t^{\overline{F}}(\bar{\phi})))\Omega_\tau^t(\bar{\phi}, \overline{F}) =$$

$$= P(\phi_t^F(\phi), F(\phi_t^F(\phi)))Z_\tau^t + [P(\phi_t^F(\phi), F(\phi_t^F(\phi))) - P(\phi_t^{\overline{F}}(\bar{\phi}), \overline{F}(\phi_t^{\overline{F}}(\bar{\phi})))]\Omega_\tau^t(\bar{\phi}, \overline{F}),$$

from which (since $Z_\tau^\tau = 0$) it follows that

$$Z_\tau^0 = \int_\tau^0 \Omega_s^0(\phi, F)[P(\phi_s^F(\phi), F(\phi_s^F(\phi))) - P(\phi_s^{\overline{F}}(\bar{\phi}), \overline{F}(\phi_s^{\overline{F}}(\bar{\phi})))]\Omega_\tau^s(\bar{\phi}, \overline{F})ds.$$

Inequality (6.4) leads to the estimate

$$\|\Omega_\tau^t(\phi, F)h_0\| \le \exp\left\{-\int_\tau^t \beta_0 ds\right\}\|h_0\| = \exp\{-\beta_0(t - \tau)\}\|h_0\|$$

for $t \ge \tau$, as a result of which we obtain

$$\|Z_\tau^0 h_0\| \le \int_\tau^0 e^{\beta_0 s}[\beta\|\phi_s^F(\phi) - \phi_s^{\overline{F}}(\bar{\phi})\| + \beta_1\|F(\phi_s^F(\phi)) - \overline{F}(\phi_s^{\overline{F}}(\bar{\phi}))\|] \times$$

$$\times e^{-\beta_0(s-\tau)}ds\|h_0\| \le \int_\tau^0 [(\beta + \beta_1 \Delta)\|\phi_s^F(\phi) - \phi_s^{\overline{F}}(\bar{\phi})\| +$$

$$+\beta_1|F(\phi) - \overline{F}(\phi)|_0]dse^{\beta_0 \tau}\|h_0\|$$

for $\tau \le 0$.

The last estimate together with inequality (6.10) shows that

$$\|Z_\tau^0 h_0\| \le \|h_0\|e^{\beta_0 \tau}\left\{(\beta + \beta_1 \Delta)[\|\phi - \bar{\phi}\| + \frac{\alpha_1}{\mu}|F(\phi) - \overline{F}(\phi)|_0]\int_\tau^0 e^{-\mu s}ds + \right.$$

$$\left. +\left(\beta_1 - \frac{\alpha_1(\beta + \beta_1 \Delta)}{\mu}\right)|F(\phi) - \overline{F}(\phi)|_0(-\tau)\right\} =$$

$$= e^{\beta_0 \tau}\left\{\frac{\beta + \beta_1 \Delta}{\mu}[\|\phi - \bar{\phi}\| + \frac{\alpha_1}{\mu}|F(\phi) - \overline{F}(\phi)|_0](e^{-\mu \tau} - 1) + \right.$$

$$+\left(\frac{(\beta+\beta_1\Delta)\alpha_1}{\mu}-\beta_1\right)|F(\phi)-\bar{F}(\phi)|_0\tau\Big\}\|h_0\|=$$

$$=\Big\{\frac{\beta+\beta_1\Delta}{\mu}(e^{(\beta_0-\mu)\tau}-e^{\beta_0\tau})\|\phi-\bar{\phi}\|+$$

$$+\left[\frac{(\beta+\beta_1\Delta)\alpha_1}{\mu^2}(e^{(\beta_0-\mu)\tau}-e^{\beta_0\tau})+\left(\frac{(\beta+\beta_1\Delta)\alpha_1}{\mu}-\beta_1\right)\tau e^{\beta_0\tau}\right]\times$$

$$\times|F(\phi)-\bar{F}(\phi)|_0\Big\}\|h_0\|$$

for $\tau\le 0$ and $\mu=\alpha_0+\alpha_1\Delta$.

We now estimate $\|S_\phi(F)\|$ and $\|S_\phi(F)-S_{\bar{\phi}}(\bar{F})\|$. We have:

$$\|S_\phi(F)\|\le\int_{-\infty}^0\|\Omega_\tau^0(\phi,F)f(\phi_\tau^F(\phi))\|d\tau\le$$

$$\le\int_{-\infty}^0 e^{\beta_0\tau}d\tau|f(\phi)|_0\le\frac{\delta_0}{\beta_0},$$

and similarly,

$$\|S_\phi(F)-S_{\bar{\phi}}(\bar{F})\|\le\int_{-\infty}^0\|Z_\tau^0 f(\phi_\tau^F(\phi))\|d\tau+\int_{-\infty}^0\|\Omega_\tau^0(\bar{\phi},\bar{F})(f(\phi_\tau^F(\phi))-$$

$$-f(\phi_\tau^{\bar{F}}(\bar{\phi})))d\tau\le\Big\{\frac{\beta+\beta_1\Delta}{\mu}\left(\frac{1}{\beta_0-\mu}-\frac{1}{\beta_0}\right)\|\phi-\bar{\phi}\|+$$

$$+\left[\frac{(\beta+\beta_1\Delta)\alpha_1}{\mu^2}\left(\frac{1}{\beta_0-\mu}-\frac{1}{\beta_0}\right)+\left(\frac{(\beta+\beta_1\Delta)\alpha_1}{\mu}-\beta_1\right)\left(-\frac{1}{\beta_0^2}\right)\right]\times$$

$$\times|F(\phi)-\bar{F}(\phi)|_0\Big\}|f(\phi)|_0+\left[\int_{-\infty}^0 e^{\beta_0\tau}\|\phi_\tau^F(\phi)-\phi_\tau^{\bar{F}}(\bar{\phi})\|d\tau\right]\lambda\delta_0,\qquad(6.11)$$

provided that

$$\beta_0>\mu.$$

Since

$$\int_{-\infty}^0 e^{\beta_0\tau}\|\phi_\tau^F(\phi)-\phi_\tau^{\bar{F}}(\bar{\phi})\|d\tau\le\int_{-\infty}^0 e^{\beta_0\tau}[e^{-\mu\tau}\|\phi-\bar{\phi}\|+$$

$$+\frac{\alpha_1}{\mu}(e^{-\mu\tau}-1)|F(\phi)-\bar{F}(\phi)|_0]d\tau=\frac{1}{\beta_0-\mu}\|\phi-\bar{\phi}\|+$$

$$+\frac{\alpha_1}{\mu}\left(\frac{1}{\beta_0-\mu}-\frac{1}{\beta_0}\right)|F(\phi)-\bar{F}(\phi)|_0,\qquad(6.12)$$

we have from (6.11), (6.12) the estimate

$$\|S_\phi(F) - S_{\bar\phi}(\bar F)\| \le \frac{1}{\beta_0 - \mu}\left(\frac{\beta + \beta_1\Delta}{\beta_0}\delta_0 + \lambda\delta_0\right)\|\phi - \bar\phi\| +$$

$$+ \left[\frac{(\beta_0 - \alpha_0)\beta_1 + \beta\alpha_1}{\beta_0^2(\beta_0 - \mu)}\delta_0 + \frac{\alpha_1}{\beta_0(\beta_0 - \mu)}\lambda\delta_0\right]|F(\phi) - \bar F(\phi)|_0 =$$

$$= \frac{1}{\beta_0 - \mu}\left[\frac{\beta + \beta_1\Delta}{\beta_0}\delta_0 + \lambda\delta_0\right]\|\phi - \bar\phi\| + \frac{1}{\beta_0(\beta_0 - \mu)} \times$$

$$\times \left[\frac{(\beta_0 - \alpha_0)\beta_1 + \beta\alpha_1}{\beta_0}\delta_0 + \alpha_1\lambda\delta_0\right]|F(\phi) - \overline{F}(\phi)|_0.$$

When the inequalities

$$\frac{\delta_0}{\beta_0} \le d, \quad \frac{1}{\beta_0 - \mu}\left[\frac{\beta + \beta_1\Delta}{\beta_0}\delta_0 + \lambda\delta_0\right] \le \Delta, \quad \beta_0 > \mu,$$

$$\frac{1}{\beta_0(\beta_0 - \mu)}\left[\frac{(\beta_0 - \alpha_0)\beta_1 + \beta\alpha_1}{\beta_0}\delta_0 + \alpha_1\lambda\delta_0\right] < 1, \tag{6.13}$$

hold, the operator S maps the space $C(\Delta)$ into itself and is a contraction. It follows from the contraction mapping principle that there exists in $C(\Delta)$ a unique fixed point $F(\phi) = u(\phi)$ of the operator S such that the following identity holds:

$$u(\phi) = \int_{-\infty}^0 \Omega_\tau^0(\phi, u) f(\phi_\tau^u(\phi)) d\tau.$$

Here

$$\|u(\phi)\| \le |f(\phi)|_0/\beta_0, \quad \|u(\phi) - u(\phi')\| \le \Delta\|\phi - \phi'\|$$

for any $\phi, \phi' \in T_m$.

We now examine the inequalities (6.13). For this purpose we rewrite them in the form

$$\frac{\delta_0}{\beta_0} \le d, \quad \alpha_1\Delta < \gamma_1, \quad \frac{\delta_0}{\beta_0} \le \frac{\Delta(\gamma_1 - \alpha_1\Delta)}{\beta + \beta_1\Delta + \lambda\beta_0} = g(\Delta),$$

$$\frac{\delta_0}{\beta_0} < \frac{\beta_0(\gamma_1 - \alpha_1\Delta)}{\gamma_1\beta_1 + \beta\alpha_1 + \lambda\beta_0\alpha_1}. \tag{6.14}$$

Since $g(0) = g(\gamma_1/\alpha_1) = 0$, $g(\Delta)$ attains its maximum on the interval $0 < \Delta < \gamma_1/\alpha_1$.

Differentiating $g(\Delta)$, we find that

$$g'(\Delta) \equiv \frac{(\gamma_1 - 2\alpha_1\Delta)(\beta + \lambda\beta_0 + \beta_1\Delta) - \beta_1\Delta(\gamma_1 - \alpha_1\Delta)}{(\beta + \lambda\beta_0 + \beta_1\Delta)^2} = 0$$

if Δ is a solution of the equation

$$\Delta^2 + 2\frac{\beta + \lambda\beta_0}{\beta_1}\Delta - \frac{\gamma_1}{\alpha_1}\frac{\beta + \lambda\beta_0}{\beta_1} = 0.$$

Solving this equation we see that Δ at which the function $g(\Delta)$ takes its maximum is equal to the number Δ_0 defined by equalities (6.6).

Now we have

$$\max_{\Delta} g(\Delta) = g(\Delta_0) = \frac{\Delta_0(\gamma_1 - \alpha_1\Delta_0)}{(\Delta_0 + \mu_0)\beta_1} = \frac{\alpha_1\Delta_0^2}{\beta_1\mu_0} = \frac{\alpha_1}{\beta_1}\left(\frac{\gamma_1}{\alpha_1} - 2\Delta_0\right).$$

We show that

$$\frac{\beta_0(\gamma_1 - \alpha_1\Delta_0)}{\gamma_1\beta_1 + \beta\alpha_1 + \lambda\beta_0\alpha_1} > \frac{\Delta_0(\gamma_1 - \alpha_1\Delta_0)}{(\Delta_0 + \mu_0)\beta_1}. \tag{6.15}$$

To do this we use (6.6) to convert the denominator on the left hand side of inequality (6.15) to the form

$$\gamma_1\beta_1 + \alpha_1(\beta + \lambda\beta_0) = \gamma_1\beta_1 + \alpha_1\beta_1\mu_0 = \beta_1(\gamma_1 + \alpha_1\mu_0) =$$

$$= \frac{\beta_1\alpha_1}{\mu_0}\left(\mu_0^2 + \frac{\gamma_1}{\alpha_1}\mu_0\right) = \frac{\beta_1\alpha_1}{\mu_0}(\Delta_0 + \mu_0)^2$$

and write the inequality

$$\frac{\beta_0\mu_0}{\alpha_1(\Delta_0 + \mu_0)} > \Delta_0, \tag{6.16}$$

equivalent to (6.15). On substituting the expression for Δ_0 into (6.16) and making elementary calculations we obtain the inequality:

$$\frac{\beta_0 - \gamma_1}{\alpha_1} > \mu_0 - \left(\mu_0^2 + \frac{\gamma_1}{\alpha_1}\mu_0\right)^{1/2},$$

which clearly holds for

$$\beta_0 \geq \gamma_1. \tag{6.17}$$

Since $\gamma_1 = \beta_0 - \alpha_0$, inequality (6.17) always holds.

Considering the inequalities (16.14) for $\Delta = \Delta_0$, we see that they hold when

$$\frac{\delta_0}{\beta_0} \leq d, \quad \frac{\delta_0}{\beta_0} = g(\Delta_0) = \frac{\alpha_1}{\beta_1}\left(\frac{\gamma_1}{\alpha_1} - 2\Delta_0\right),$$

that is, when inequalities (6.7) hold. Thus inequalities (6.13) hold if the conditions of Theorem 1 are satisfied.

Consider the torus

$$h = u(\phi) = S_\phi(u), \quad \phi \in T_m. \tag{6.18}$$

We claim that it is an invariant set of system (6.1). To see this, consider the function

$$u(\phi_t^u(\phi)) = \int_{-\infty}^0 \Omega_\tau^0(\phi_t^u(\phi), u) f(\phi_\tau^u(\phi_t^u(\phi))) d\tau,$$

where $\phi_t^u(\phi)$ is the solution of the first equation of system (6.1) for $h = u(\phi)$.

Since

$$\Omega_\tau^0(\phi_t^u(\phi), u) = \Omega_{\tau+t}^t(\phi, u),$$

the equality

$$u(\phi_t^u(\phi)) = \int_{-\infty}^t \Omega_s^t(\phi, u) f(\phi_s^u(\phi)) ds$$

holds, therefore

$$\frac{du(\phi_t^u(\phi))}{dt} = P(\phi_t^u(\phi), u(\phi_t^u(\phi))) u(\phi_t^u(\phi)) + f(\phi_t^u(\phi))$$

for all $t \in \mathbf{R}$. And this implies that the pair of functions $\phi_t = \phi_t^u(\phi)$, $h_t = u(\phi_t^u(\phi))$ is a solution of system (6.1) for $t \in \mathbf{R}$, so that the torus (6.18) is an invariant set of system (6.1). Since the function u belongs to the space $C(\Delta_0)$, inequalities of Theorem 1 hold for the function $u(\phi)$. This completes the proof of Theorem 1.

We note the two limiting cases for inequality (6.7):

$$\delta_0 \le \beta_0 \min\left(d, \frac{\gamma_1}{\beta_1}\right), \quad \alpha_1 = 0,$$

$$\delta_0 \le \beta_0 \min\left(d, \frac{\gamma_1^2}{4\alpha_1(\beta + \lambda\beta_0)}\right), \quad \beta_1 = 0,$$

for which it is significantly simpler to check whether inequality (6.7) holds.

As an example illustrating Theorem 1 we consider the system of three equations

$$\frac{d\phi_1}{dt} = \omega_1 + \sin(\phi_1 - \phi_2)|h|,$$

$$\frac{d\phi_2}{dt} = \omega_2 + \cos(\phi_1 + \phi_2)h, \tag{6.19}$$

$$\frac{dh}{dt} = -2h + \epsilon \sin\phi_1 \cos\phi_2.$$

The conditions of Theorem 1 hold for this system if the positive parameter ϵ satisfies the inequality

$$\epsilon \le 2\min\left(d, \ \frac{(1-d)^2}{2\sqrt{2}}\right) \tag{6.20}$$

for some $d \in (0, 1)$.

The greatest value of ϵ for which inequality (6.20) holds for $d \in (0, 1)$ is given by

$$\epsilon_0 = 2\sqrt{1+\sqrt{2}}(\sqrt{1+\sqrt{2}} - \sqrt{2}) > 0,1 \tag{6.21}$$

and is attained for

$$d = d_0 = \epsilon_0/2. \tag{6.22}$$

If we choose ϵ_0 and d_0 according to equalities (6.21), (6.22), then we obtain the expression

$$\Delta_0 = \frac{1}{\sqrt{2}}(1 - d_0) = \sqrt{1+\sqrt{2}} - 1.$$

Consequently, for all $\epsilon \le \epsilon_0$ system (6.19) has an invariant torus

$$h = u(\phi, \epsilon), \quad \phi = (\phi_1, \phi_2) \in T_2,$$

lying in the region

$$|h| \le d_0 = \epsilon_0/2 \simeq 0.05$$

and satisfying the inequality

$$\|u(\phi, \epsilon) - u(\phi', \epsilon)\| \le \Delta_0 \|\phi - \phi'\| = (\sqrt{1+\sqrt{2}} - 1)\|\phi - \phi'\| \simeq 0.5\|\phi - \phi'\|.$$

The smoothness of an invariant torus of system (6.1) will be called maximal if it is equal to the smoothness of the functions a, P, f on the right hand side of this system. Theorem 1 differs from the other theorems of this chapter in that it ensures that the invariant torus, the existence of which also follows from this theorem, has maximal smoothness. In spite of the smallness of the difference of smoothness between the functions a, P, f and that of an invariant torus in the systems considered before, this difference in the theorems considered before is natural and is explained by an important difference between the conditions of those theorems and those of Theorem 1 of the present section.

Indeed, the conditions of the theorems in which there is a "loss of smoothness" of the invariant torus, the existence of which is affirmed by these theorems, assume that the appropriate argument is carried out for a bounded subset of the space $C^r(\mathcal{T}_m)$, while the conditions of Theorem 1 of the present section assume that we work in a bounded subset of the space $C^r_{\text{Lip}}(\mathcal{T}_m)$. A bounded set of functions in $C^r(\mathcal{T}_m)$ is compact in $C^{r-1}(\mathcal{T}_m)$, whereas a bounded set of functions in $C^r_{\text{Lip}}(\mathcal{T}_m)$ is compact in $C^r(\mathcal{T}_m)$ and any limit function of this set belongs to the space $C^r_{\text{Lip}}(\mathcal{T}_m)$. Since an invariant torus of the systems under consideration is defined by a limit function of bounded sets in spaces $C^r(\mathcal{T}_m)$ and $C^r_{\text{Lip}}(\mathcal{T}_m)$ respectively, it loses smoothness in the first case but not in the second.

The above remarks not only explain but also define the conditions under which it is possible to achieve a maximal smoothness of an invariant torus of the considered systems. Namely, we have the following statement supplementing Lemma 1 of §4.1.

Lemma 1. *Suppose that the functions* a, P *and* f *are defined and periodic in* ϕ_ν *($\nu = 1, \ldots, m$) with period* 2π *in the region* (6.2) *and have continuous partial derivatives with respect to* ϕ, h *up to the order* r *which satisfy a Lipschitz condition with respect to* ϕ, h. *Suppose that for any* $i = 0, 1, \ldots$ *the system of equations* (1.6) *has an invariant torus* (1.5) *belonging to the space* C^r_{Lip} *and satisfying the inequalities*

$$\|u^{i+1}(\phi)\| \le d_1, \quad |u^{i+1}(\phi)|_{r,\text{Lip}} \le \Delta, \quad i = 0, 1, 2, \ldots,$$

where d_1 *and* Δ *are positive constants,* $d_1 < d$.
  If

$$\lim_{i \to \infty} u^i(\phi) = u(\phi)$$

for every $\phi \in \mathcal{T}_m$ *then the limit function* $u(\phi)$ *defines an invariant torus* (1.1) *of system* (6.1) *belonging to the space* C^r_{Lip} *and satisfying the inequalities*

$$\|u(\phi)\| \le d_1, \quad |u(\phi)|_{r,\text{Lip}} \le \Delta.$$

The following theorem is similar to Theorem 1 of §4.3 and Theorem 2 of §4.5 which define the conditions for preservation of an invariant torus of maximal smoothness under small perturbations.

Theorem 2. *Suppose that the functions a, P and f have partial derivatives with respect to ϕ, h up to order l that are continuous in ϕ, h, ϵ and that these partial derivatives satisfy Lipschitz conditions with respect to ϕ, h for all ϕ, h, ϵ in the region (2.2), where the Lipschitz constants $K_l = K_l(\epsilon)$ of the lth derivatives with respect to ϕ, h of the functions $a(\phi, h, \epsilon) - a(\phi, h, 0)$, $P(\phi, h, \epsilon) - P(\phi, h, 0)$ and $f(\phi, \epsilon)$ tend to zero as $\epsilon \to 0$.*

Suppose further that the inequalities (3.10) and (5.24) hold for $r = l + 1$. If $l \geq 1$, then there exists a sufficiently small $\epsilon_0 > 0$ such that for any $\epsilon \in [0, \epsilon_0]$ system (2.1) has an invariant torus (2.6) with function u belonging to the space $C^l_{\text{Lip}}(\mathcal{T}_m)$ and satisfying the inequality:

$$|u|_{l,\text{Lip}} \leq K|f|_{l,\text{Lip}},$$

where K is a positive constant independent of ϵ.

Proof. We "smooth out" the right hand side of system of equations (2.1) by means of the operator T_{NN} defined by the Nash kernel [161] and obtain the system of equations

$$\frac{d\phi}{dt} = a_N(\phi, h, \epsilon), \quad \frac{dh}{dt} = P_N(\phi, h, \epsilon)h + f_N(\phi, \epsilon), \qquad (6.23)$$

where

$$(a_N, P_N, f_N) = (T_{NN}a, T_{NN}P, T_N f)$$

for each $\epsilon \in [0, \epsilon_0]$, N being a large natural number.

It follows from the form of the operator T_{NN} that the functions a_N, P_N and f_N are defined in the region

$$\|h\| \leq d', \ \phi \in \mathcal{T}_m, \ \epsilon \in [0, \epsilon_0], \qquad (6.24)$$

(where $d' = d'(N) \to d$ and is monotonically increasing as $N \to +\infty$), are periodic in ϕ_ν ($\nu = 1, \ldots, m$) with period 2π and have partial derivatives with respect to ϕ, h of arbitrary order that are continuous in ϕ, h, ϵ. Furthermore, the sequences a_N, P_N, f_N together with their derivatives with respect to ϕ, h up to order l are uniformly convergent with respect to ϕ, h, ϵ in the region (6.24) to the functions a, P, f and their respective derivatives. Moreover,

$$f_N(\phi, 0) = 0, \ \phi \in \mathcal{T}_m$$

and the partial derivatives of the functions a_N, P_N and f_N with respect to ϕ, h up to order $l + 1$ are continuous in ϕ, h, ϵ in the region (6.24) and satisfy the inequalities

$$D_{\phi,h}^{l+1} a_N(\phi, h, \epsilon)\| + \|D_{\phi,h}^{l+1} P_N(\phi, h, \epsilon)\| \le CK,$$

$$D_\phi^{l+1} f_N(\phi, \epsilon)\| \le C|f|_{l,\text{Lip}},$$

(6.25)

where C is a constant independent of a, P, f and N, and K is the Lipschitz constant of the derivatives of a and P with respect to ϕ, h up to order l.

These properties of the functions a_N, P_N and f_N are sufficient for us to apply Theorem 1 of §4.3 or Theorem 2 of §4.5 to system (6.23) with $r = l + 1$ and to establish that for all $\epsilon \in [0, \epsilon_0]$ and all $N \ge N_0$ with sufficiently small $\epsilon_0 = \epsilon_0(N_0) > 0$ and sufficiently large N_0, there exists an invariant torus

$$h = u_N(\phi, \epsilon), \quad \phi \in T_m,$$

(6.26)

of system (6.23) with function u_N belonging to the space $C_{\text{Lip}}^l(T_m)$ and satisfying the inequality

$$|u_N(\phi, \epsilon)|_{l,\text{Lip}} \le K_2 |f|_{l,\text{Lip}}$$

(6.27)

where K_2 is a positive constant independent of ϵ and N.

Let $u_0(\phi, \epsilon)$ be the limit of the sequence of functions $u_N(\phi, \epsilon)$ ($N = N_0, N_0 + 1, \ldots$). It follows from inequality (6.27) that

$$|u_0(\phi, \epsilon)|_{l,\text{Lip}} \le K_2 |f|_{l,\text{Lip}}$$

for all $\epsilon \in [0, \epsilon_0]$, while it follows from the identity

$$\frac{\partial u_N(\phi, \epsilon)}{\partial \phi} a_N(\phi, u_N(\phi, \epsilon), \epsilon) = P_N(\phi, u_N(\phi, \epsilon), \epsilon) u_N(\phi, \epsilon) + f_N(\phi, \epsilon)$$

that the torus

$$h = u_0(\phi, \epsilon), \quad \phi \in T_m$$

is an invariant torus of system (2.1) for all $\epsilon \in [0, \epsilon_0]$.

Since an invariant torus of system (2.1) is unique in a neighbourhood of the trivial torus $h = 0$, $\phi \in T_m$, as follows from the exponential dichotomy of the invariant torus (2.6) of system (2.1), we conclude that

$$u_0(\phi, \epsilon) = u(\phi, \epsilon), \quad \phi \in T_m, \ \epsilon \in [0, \epsilon_0].$$

This suffices to prove that the invariant torus (2.6) of system (2.1) satisfies the conditions of Theorem 2, and so completes its proof.

Remark. The theorems on the compactness in $C^{r-1}(\overline{D})$ of any bounded subset of $C^r(\overline{D})$ and the compactness in $C^r(\overline{D})$ of any bounded subset of $C^r_{\text{Lip}}(\overline{D})$, where D is a region in T_m or a bounded region in $\mathbf{R}^m$ are similar to Sobolev's Theorem 2 of §1.2. They are proved by means of the Arzelà-Ascoli criterion [50].

4.7. Galerkin's method for the construction of an invariant torus of a non-linear system of equations and its linear modification

We denote by $L(u_1)u_2$ the quasi-linear operator on the set of functions $u \in C^r(T_m)$ defined by the inequalities

$$|u|_0 < d, \quad |u|_r \leq K,$$

that is given by the expression

$$L(u_1)u_2 = \sum_{\nu=1}^{m} a_\nu(\phi, u_1)\frac{\partial u_2}{\partial \phi_\nu} - P(\phi, u_1)u_2. \tag{7.1}$$

The function $u(\phi)$ defining the smooth invariant torus (1.1) of system (1.2) is a solution of the equation

$$L(u)u = f, \tag{7.2}$$

which belongs to the space $C^1(T_m)$.

Galerkin's method defines the Nth approximation of the solution $u(\phi) \in C^1(T_m)$ of equation (7.2) via the expression

$$W_N(\phi) = \sum_{\|k\| \leq N} W_k^{(N)} e^{i(k,\phi)}, \quad \phi \in T_m, \tag{7.3}$$

the coefficients $W_k^{(N)}$ of which are solutions of the system of non-linear algebraic equations:

$$(L(W_N(\phi))W_N(\phi), e^{i(k,\phi)})_0 = (f(\phi), e^{i(k,\phi)})_0, \quad \|k\| \leq N. \tag{7.4}$$

To define a *linear modification of Galerkin's method* we start with the initial approximation $u_0(\phi) = 0$ for $\phi \in T_m$ and choose a collection of integers N_j $(j = 0, 1, \ldots)$ subject to the condition

$$N_j \geq N_{j-1}, \quad j = 1, 2, \ldots, \tag{7.5}$$

and take the N_jth linear Galerkin's approximation to the solution $u(\phi) \in C^1(\mathcal{T}_m)$ of equation (7.2) to be

$$u_j(\phi) = \sum_{\|k\| \leq N_j} u_k e^{i(k,\phi)}, \qquad (7.6)$$

where the coefficients $u_k = u_k^{(j)}$ are solutions of the system of linear algebraic equations

$$(L(u_{j-1}(\phi))u_j(\phi), e^{i(k,\phi)})_0 = (f, e^{i(k,\phi)})_0, \quad \|k\| \leq N_j. \qquad (7.7)$$

It is clear that the N_jth linear Galerkin approximation to the solution $u(\phi) \in C^1(\mathcal{T}_m)$ of equation (7.2) is the N_jth Galerkin approximation to the solution in $C^1(\mathcal{T}_m)$ of the equation

$$L(u_{j-1}(\phi))u(\phi) = f(\phi). \qquad (7.8)$$

We shall suppose that the $C^r(\mathcal{T}_m)$-norm of the function $f(\phi)$ is sufficiently small. With this assumption the conditions for the existence and convergence of the Galerkin approximations $W_N(\phi)$ and $u_j(\phi)$ are defined by the principal part of the operator $L(u)$, which is

$$L(0) = \sum_{\nu=1}^{m} a_\nu(\phi, 0) \frac{\partial}{\partial \phi_\nu} - P(\phi, 0),$$

and they turn out to be inequalities of the form (14.21) or (14.24) for the coefficients of the operator $L(0)$ in the previous chapter.

The proof of this fact relies on Moser's lemma [89], a particular case of which is given below.

Lemma 1. *Let $a(\phi)$ and $P(\phi)$ belong to the space $C^r(\mathcal{T}_m)$. If inequality (14.21) of Chapter 3 holds, then for an arbitrary $u \in C^{r+1}(\mathcal{T}_m)$ we have the following inequalities*

$$(Lu, u)_0 \geq \gamma \|u\|_0^2,$$
$$(Lu, u)_s \geq \gamma_1 \|u\|_s^2 - \delta \left(1 + \sum_{\nu=1}^{m} \|a_\nu\|_s + \|P\|_s \right)^2, \qquad (7.9)$$

where γ_1 and δ are positive constants which depend only on $C_0 \geq \sum_{\nu=1}^{m} |a_\nu|_2 + |P|_1 + |u|_1$, L is the operator (14.2) of Chapter 3, and $r \geq 2$, $1 \leq s \leq r$.

The proof of Lemma 1 will be given later, but here we use it to substantiate the linear modification of Galerkin's method.

Theorem 1. *Let the functions a and P be defined in region (1.3) and have continuous partial derivatives with respect to ϕ, h up to order r. Suppose that the inequalities (14.21) of Chapter 3 hold for $a(\phi) = a(\phi,0)$, $P(\phi) = P(\phi,0)$, and for $s = 0$ and $s = r$. If*

$$r > m/2 + l, \quad l \geq 2, \tag{7.10}$$

then for any $M > 1$ there exists a sufficiently small $\mu = \mu(M) > 0$ such that if $f \in H^r(T_m)$ and

$$\|f\|_r \leq \mu, \tag{7.11}$$

then the linear Galerkin approximations $u_j(\phi)$ exist for any $j = 1,2,\ldots$ and converge in $H^s(T_m) \cap C^l(T_m)$ for $s < r$ to the function $u(\phi)$ which defines an invariant torus (1.1) of system (1.2) such that

$$\|u - u_j\|_0 \leq C\sigma_j, \quad j = 1,2,\ldots, \tag{7.12}$$

where $C = \text{const}$, $\sigma_j = \sum_{i=j}^{\infty} \sum_{\nu=0}^{i} N_{i-\nu}^{-(r-1)} M^{-\nu}$.

Proof. In equation (7.2) we replace u by ϵu and f by ϵf, taking ϵ to be a small positive parameter. Equation (7.2) then takes the form

$$L(\epsilon u)u = f(\phi). \tag{7.13}$$

Let $w(\phi)$ be a function in $C^r(T_m)$ satisfying the relation

$$|w|_2 \leq 1.$$

Consider the linear operator

$$L(\epsilon w) = \sum_{\nu=1}^{m} a_\nu(\phi, \epsilon w(\phi)) \frac{\partial}{\partial \phi_\nu} - P(\phi, \epsilon w(\phi))$$

as an operator in $C^\infty(T_m)$. Since inequalities (14.21) of Chapter 3 are coarse, they imply similar inequalities for the coefficients of the operator $L(\epsilon w)$ for all $\epsilon \in [0, \epsilon_0]$ for some $\epsilon_0 > 0$. Hence, using Lemma 1 we obtain the estimates

$$(L(\epsilon w)u, u)_0 \geq \gamma\|u\|_0^2, \quad (L(\epsilon w)u, u)_s \geq \gamma\|u\|_s^2 -$$

$$-\delta_0 \left(1 + \sum_{\nu=1}^{m} \|a_\nu(\phi, \epsilon w(\phi))\|_s + \|P(\phi, \epsilon w(\phi))\|_s\right)^2, \quad s = 1,\ldots,r,$$

where γ and δ_0 are positive constants independent of w, u and ϵ. By applying estimate (3.4) of Chapter 1 to the functions $a_\nu(\phi, \epsilon w(\phi))$ and $P(\phi, \epsilon w(\phi))$, the above inequalities become

$$(L(\epsilon w)u, u)_0 \geq \gamma \|u\|_0^2,$$
$$(L(\epsilon w)u, u)_s \geq \gamma \|u\|_s^2 - \delta(1 + \epsilon\|w\|_s)^2, \tag{7.14}$$

where δ does not depend on w, u and ϵ.

Consider the first linear Galerkin approximation $u_1(\phi)$. Its coefficients $u_k^{(1)}$ are determined from the system of equations equivalent to the equation

$$S_{N_1} L(0) u_1(\phi) = S_{N_1} f(\phi), \tag{7.15}$$

where S_N is the truncation operator for the Fourier series of the function $f(\phi) \simeq \sum_k f_k e^{i(k,\phi)}$ defined by the equality $S_N f(\phi) = \sum_{\|k\| \leq N} f_k e^{i(k,\phi)}$. Inequalities (7.14) for the operator $L(0)$ guarantee that equation (7.15) can be solved and that its solution $u_1(\phi)$ has an estimate of the form

$$\|u_1\|_s \leq c\|f\|_s, \quad s = 0, \ldots, r, \tag{7.16}$$

where c is a positive constant independent of u_1 and ϵ.

Let $r > p > m/2 + 2$. By the first Sobolev theorem given in §1.2 and the estimates in §1.3 we obtain from inequality (7.16) the estimate

$$|u_1|_2 \leq cc_1\|f\|_p \leq 2cc_1\|f\|_0^{1-p/r}\|f\|_r^{p/r},$$

where c_1 is a positive constant independent of u_1 and ϵ.

We require that $f(\phi)$ satisfy inequality

$$2cc_1\|f\|_0^{1-p/r}\|f\|_r^{p/r} \leq 1. \tag{7.17}$$

Then $|u_1|_2 \leq 1$ and the operator $L(\epsilon u_1)$ is defined for all $\epsilon \in [0, \epsilon_0]$ and satisfies inequalities (7.14) for $w = u_1$. This guarantees that the approximation $u_2(\phi) = \sum_{\|k\| \leq N_2} u_k^{(2)} e^{i(k,\phi)}$ exists and that it has an estimate of the form

$$\|u_2\|_0 \leq \|f\|_0/\gamma, \quad \gamma\|u_2\|_s^2 \leq \delta(1 + \epsilon\|u_1\|_s)^2 + \|f\|_s\|u_2\|_s,$$

where $s = 1, \ldots, r$.

Solving this inequality we find that

$$\|u_2\|_0 \leq \|f\|_0/\gamma, \quad \|u_2\|_s^2 \leq \delta_1 + \epsilon\|u_1\|_s^2 \tag{7.18}$$

for all $\epsilon \in [0, \epsilon_0]$ if

$$\epsilon_0 \leq \frac{1}{2}\left(1 + \frac{\delta}{\gamma}\right)^{-1}, \quad 2\frac{\delta}{\gamma}\left(1 + \frac{\delta}{\gamma}\right) + \frac{\|f\|_s}{\gamma_2} \leq \delta_1. \tag{7.19}$$

Taking into account the estimate (7.16), the second inequality of (7.18) leads to the inequality

$$\|u_2\|_s^2 \leq \delta_1 + \epsilon c^2\|f\|_s^2 \leq \delta_1(1 + \epsilon), \tag{7.20}$$

if δ_1 satisfies not only inequality (7.19) but also the inequality

$$c^2\|f\|_r^2 \leq \delta_1. \tag{7.21}$$

Inequalities (7.18), (7.20) lead to the estimate

$$|u_2|_2 \leq c_1\|u_2\|_p \leq 2c_1\|u_2\|_0^{1-p/r}\|u_2\|_r^{p/r} \leq$$

$$\leq \frac{2c_1\|f\|_0^{1-p/r}}{\gamma^{1-p/r}}\delta_1^{p/r}(1 + \epsilon)^{p/r} \leq 1,$$

provided that the function $f(\phi)$ satisfies the inequality

$$\frac{2c_1\delta_1^{p/r}(1 + \epsilon_0)^{p/r}}{\gamma^{1-p/r}}\|f\|_0^{1-p/r} \leq 1.$$

Then the operator $L(\epsilon u_2)$ is defined for all $\epsilon \in [0, \epsilon_0]$ and satisfies inequality (7.14) for $w = u_2$. From this it follows that the approximation $u_3(\phi) = \sum_{\|k\| \leq N_3} u_k^{(3)} e^{i(k,\phi)}$ exists and admits an estimate of the form

$$\|u_3\|_0 \leq \|f\|_0/\gamma, \quad \|u_3\|_s^2 \leq \delta_1 + \epsilon\|u_2\|_s^2$$

for all $\epsilon \in [0, \epsilon_0]$. But then

$$\|u_3\|_0 \leq \|f\|_0/\gamma, \quad \|u_3\|_s^2 \leq \delta_1(1 + \epsilon + \epsilon^2),$$

$$\|u_3\|_2 \leq c_1\|u_3\|_p \leq 2c_1\|u_3\|_0^{1-p/r}0\|u_3\|_r^{p/r} \leq$$

$$\leq \frac{2c_1\|f\|_0^{1-p/r}}{\gamma^{1-p/r}}\delta_1^{p/r}(1 + \epsilon + \epsilon^2)^{p/r},$$

which leads to the estimates

$$\|u_3\|_0 \leq \|f\|_0/\gamma, \quad \|u_3\|_s^2 \leq \delta_1(1+\epsilon+\epsilon^2), \quad |u_3|_2 \leq 1,$$

provided that the function $f(\phi)$ satisfies the inequality

$$\frac{2c_1\delta_1^{p/r}(1+\epsilon_0+\epsilon_0^2)^{p/r}}{\gamma^{1-p/r}}\|f\|_0^{1-p/r} \leq 1.$$

It is easy to prove by induction that the approximation $u_j(\phi)$ exists for any $j \geq 3$ and that $u_j(\phi)$ satisfies the inequality

$$\|u_j\|_0 \leq \frac{\|f\|_0}{\gamma}, \quad \|u_j\|_s \leq \delta_1\sum_{\nu=0}^{j-1}\epsilon^{\nu}, \quad |u_j|_2 \leq 1$$

for all $\epsilon \in [0,\epsilon_0]$ if the function $f(\phi)$ satisfies the inequalities

$$\frac{1}{\gamma^{1-p/r}}\left(2c_1\delta_1^{p/r}\left(\sum_{\nu=0}^{j-1}\epsilon_0^{\nu}\right)^{p/r}\right)\|f\|_0^{1-p/r} \leq 1. \tag{7.22}$$

Since $\epsilon_0 < 1/2$, inequalities (7.19), (7.21) and (7.22) hold provided that the function $f(\phi)$ satisfies the inequalities

$$\|f\|_0 \leq \gamma[2^{1+p/r}c_1\delta_1^{p/r}]^{r/(p-r)},$$

$$\|f\|_s \leq \max\left\{\frac{\partial_1^{1/2}}{c}, \left[\delta_1 - \frac{2\delta}{\gamma}\left(1+\frac{\delta}{\gamma}\right)\right]\gamma^2\right\}, \tag{7.23}$$

where $s = 1,\ldots,r$, and δ_1 is a constant chosen so that

$$\frac{2\delta}{\gamma}\left(1+\frac{\delta}{\gamma}\right) < \delta_1. \tag{7.24}$$

When inequalities (7.23), (7.24) hold, the approximations $u_j(\phi)$ are defined for all $\epsilon \in [0,\epsilon_0]$ when $\epsilon_0 > 0$ is sufficiently small, and they satisfy the inequalities

$$\|u_j\|_0 \leq \|f\|_0/\gamma, \quad \|u_j\|_s \leq 2\delta_1, \quad |u_j|_2 \leq 1 \tag{7.25}$$

for all $j = 1, 2, \ldots$ and $s = 1, \ldots, r$.

We now tackle the question of the convergence of the sequence $u_j(\phi)$ $(j = 1, 2, \ldots)$. Since the function $w_{j+1}(\phi) = u_{j+1}(\phi) - u_j(\phi)$ satisfies the equation

$$S_{N_{j+1}}L(\epsilon u_j)w = -S_{N_{j+1}}[(L(\epsilon u_j)-L(\epsilon u_{j-1}))u_j]-(S_{N_{j+1}}-S_{N_j})[L(\epsilon u_{j-1})u_j-f],$$

it satisfies the inequality

$$\gamma\|w_{j+1}\|_0 \leq \| [L(\epsilon u_j) - L(\epsilon u_{j-1})]u_j\|_0 +$$

$$+\|(S_{N_{j+1}} - S_{N_j})[L(\epsilon u_{j-1})u_j - f]\|_0. \qquad (7.26)$$

The first term on the right hand side of inequality (7.26), which we denote by I_1, admits the estimate

$$I_1 \leq \sum_{\nu=1}^{m} \left\|[a_\nu(\phi, \epsilon u_j) - a_\nu(\phi, \epsilon u_{j-1})]\frac{\partial u_j}{\partial \phi_\nu}\right\|_0 +$$

$$+\left\| [P(\phi, \epsilon u_j) - P(\phi, \epsilon u_{j-1})]u_j\right\|_0 \leq c_2|u_j|_1\|\epsilon w_j\|_0.$$

The second term (denoted by I_2) has the estimate

$$I_2 \leq N_j^{-(r-1)}\|L(\epsilon u_{j-1})u_j - f\|_{r-1} \leq c_3(1 + \|\epsilon u_{j-1}\|_{r-1} +$$

$$+\|u_j\|_r + \|f\|_{r-1})N_j^{-(r-1)} \leq c_4(1 + \delta_1)N_j^{-(r-1)},$$

where c_2, c_3 and c_4 are positive constants independent of γ, ϵ and δ_1. The inequalities for I_1 and I_2 lead to the estimate for $\|w_{j+1}\|_0$ in the form

$$\|w_{j+1}\|_0 \leq \frac{\epsilon c_2}{\gamma}\|w_j\|_0 + \frac{c_4(1+\delta_1)}{\gamma}N_j^{-(r-1)}. \qquad (7.27)$$

Let $\alpha = c_4(1 + \delta_1)/\gamma$. Choose ϵ_0 so small that

$$\epsilon_0 c_2/\gamma \leq 1/M. \qquad (7.28)$$

For all $\epsilon \in [0, \epsilon_0]$ we obtain from inequality (7.27) the estimate

$$\|w_{j+1}\|_0 \leq \|w_j\|_0/M + \alpha/N_j^{r-1},$$

which on integration yields

$$\|w_{j+1}\|_0 \leq \alpha \sum_{\nu=0}^{j-1} N_{j-\nu}^{-(r-1)}M^{-\nu} + \frac{\|w_1\|_0}{M^j} \leq \alpha \sum_{\nu=0}^{j} N_{j-\nu}^{-(r-1)}M^{-\nu}.$$

The last inequality proves that the sequence u_j $(j = 1, 2, \ldots)$ is convergent with respect to the $H(\mathcal{T}_m)$ norm whenever the series

$$\sigma = \sum_{j=0}^{+\infty} \sum_{\nu=0}^{j} N_{j-\nu}^{-(r-1)}M^{-\nu}$$

converges. Here the difference of u_j and $u_0 = \lim\limits_{j \to \infty} u_j$ satisfies the estimate

$$\|u_j - u^0\|_0 \le \alpha \sigma_j, \qquad (7.29)$$

where σ_j is the remainder of the series σ defined by the expression given in Theorem 1. Inequalities (7.25) ensure that the sequence $u_j(\phi)$ $(j = 1, 2, \ldots)$ is compact in the space $H^s(T_m)$ for $s < r$. Then

$$\lim\limits_{j \to \infty} \|u_j - u^0\|_s = 0.$$

For $s > m/2 + l$ we have the inclusion $H^s(T_m) \subset C^l(T_m)$ and the estimate

$$|u_j - u^0|_l \le c_5 \|u_j - u^0\|_s \qquad (7.30)$$

with the constant c_5 independent of j. The conditions $r > s > m/2 + l$ hold for $s = p$. Inequality (7.30) leads to the limit relation

$$\lim\limits_{j \to \infty} |u_j - u^0|_l = 0,$$

where the number l is defined by the conditions of Theorem 1.

From inequalities (7.25) we obtain the analogous inequalities for the limit function:

$$\|u^0\|_0 \le \|f\|_0/\gamma, \quad \|u^0\|_s \le 2\delta_1, \quad |u^0|_2 \le 1.$$

Thus the operator $L(\epsilon w)$ is defined for the function $w = u^0$ and the difference $L(\epsilon u^0)u^0 - f$ has the estimate

$$|L(\epsilon u^0)u^0 - f|_0 \le |L(\epsilon u^0)u^0 - L(\epsilon u_j)u_{j+1}|_0 +$$

$$+ |(E - S_{N_{j+1}})[L(\epsilon u_j)u_{j+1} - f]|_0 \le c_6[|u_j - u^0|_0 +$$

$$+ |u_{j+1} - u^0|_1 + N_{j+1}^{-(s-1)}(\|u_j\|_{s-1} + \|u_{j+1}\|_s + \|f\|_s)] \le$$

$$\le c_6(|u_j - u^0|_0 + |u_{j+1} - u^0|_1 + c_7 N_{j+1}^{-(s-1)}\delta_1), \quad c_7 = 4 + \gamma^2$$

for any $j = 1, 2, \ldots$. Consequently, the function $u = u^0(\phi) = u^0(\phi, \epsilon)$ is a solution of equation (7.13) for all the values of ϵ and f considered by us.

We choose $\epsilon_0 > 0$ so small and δ_1 so large that all the conditions in the above argument, in particular, inequalities (7.19), (7.24), (7.28), are satisfied. Then, taking ϵ equal to ϵ_0, we can satisfy inequalities (7.11) for the right hand side of equation (7.13) with $\epsilon = \epsilon_0$ by making inequalities (7.11) hold for the right hand side of equation (7.2) with some small positive $\mu = \mu(M)$. Then the functions $u_j(\phi) = \epsilon_0 u_j(\phi, \epsilon_0) = \epsilon_0 u_j(\phi)$ define linear Galerkin approximations to a solution of equation (7.2) and satisfy all the conditions of Theorem 1 since (initial u) = ϵ (new u).

To complete the proof of Theorem 1 it remains to verify the convergence of the series σ for any collection of integers N_j satisfying inequality (7.5) and arbitrary $M > 1$.

Since the sequence of numbers N_j is strictly monotone, we have

$$N_j \geq j + 1, \quad j = 0, 1, \ldots,$$

and so σ can be majorized by the series

$$\sum_{j=0}^{\infty} \sum_{\nu=0}^{j} \frac{1}{(j+1-\nu)^{r-1} M^{\nu}} = \sigma_0,$$

which converges since

$$\sum_{j=0}^{n} \sum_{\nu=0}^{j} \frac{1}{(j+1-\nu)^{r-1} M^{\nu}} = \sum_{\nu=0}^{n} \sum_{j=\nu}^{n} \frac{1}{(j+1-\nu)^{r-1} M^{\nu}} \leq$$

$$\leq \sum_{\nu=0}^{\infty} \frac{1}{M^{\nu}} \sum_{j=0}^{\infty} \frac{1}{(j+1)^{r-1}} \leq \frac{M}{M-1} \sum_{j=0}^{\infty} \frac{1}{(j+1)^{1+m/2}} = \frac{M}{M-1} c_*$$

for any $n \geq 1$ and an arbitrary $M > 1$. This completes the proof of the theorem.

It should be noted that estimate (7.12), which characterizes the rate of convergence of the approximations $u_j(\phi)$ to the solution $u(\phi)$ of equation (7.2), essentially depends on the choice of the sequence of numbers N_j $(j = 1, 2, \ldots)$ and the number M. By changing them we can control the rate of convergence. In particular, if

$$N_j = N^j, \quad M = N^{r-1}, \quad N \geq 2,$$

then the rate of convergence of u_j to u is given by the inequality

$$\|u - u_j\|_0 \leq 2\alpha \frac{j+1}{N^{(r-1)j}}, \quad j = 1, 2, \ldots.$$

This is a consequence of the following estimate of σ_j for the chosen values of N_j and M:

$$\sigma_j = \sum_{i=j}^{\infty} \sum_{\nu=0}^{i} N^{-\nu(r-1)} N^{(\nu-i)(r-1)} = \sum_{i=j}^{\infty} \frac{i+1}{N^{i(r-1)}} \leq \frac{2(j+1)}{N^{(r-1)j}}.$$

Note also that inequality (7.12) leads to the estimate

$$|u - u_j|_l \leq \bar{c}\sigma_j^{1-p/r}, \quad j = 1, 2, \ldots,$$

which characterizes the rate of convergence of u_j to u with respect to the $C^l(T_m)$ norm. Here $r > p > m/2 + l$ and $\bar{c}$ is a constant independent of j.

Theorem 1 simplifies the problem of the justification of Galerkin's procedure for the construction of the invariant torus (1.1) of system (1.2). The following statement holds [119].

Theorem 2. *Suppose that the conditions of the previous theorem hold. Then there exists a sufficiently small $\mu > 0$ such that if $f \in H^r(T_m)$ and inequality (7.11) holds, then the Galerkin approximations $W_N(\phi)$ exist for any $N = 1, 2, \ldots$ and converge in $H^s(T_m) \cap C^l(T_m)$ for $s < r$ to a function $u(\phi)$ which defines an invariant torus (1.1) of system of equations (1.2); furthermore,*

$$\|u - W_N\|_0 \leq CN^{-(r-1)}, \tag{7.31}$$

where $C = \text{const}.$

Proof. Consider equation (7.13). The Galerkin procedure defines the Nth approximation $W_N(\phi)$ to its solution $u(\phi) \in C^1(T_m)$ as the trigonometric polynomial (7.3) satisfying the equation

$$S_N L(\epsilon W_N) W_N(\phi) = S_N f(\phi). \tag{7.32}$$

To solve equation (7.32) we use a linear modification of Galerkin's procedure, according to which we define the approximation $W_N^j(\phi)$ to the solution $W_N(\phi)$ of equation (7.32), starting with the initial approximation $W_N^0 = 0$, by the polynomial

$$W_N^j(\phi) = \sum_{\|k\| \leq N} W_{k,N}^{(j)} e^{i(k,\phi)},$$

satisfying the equation

$$S_N L(\epsilon W_N^{j-1}) W_N^j(\phi) = S_N f(\phi), \quad j = 1, 2, \dots. \tag{7.33}$$

Equation (7.33) has the form of the equation considered in the proof of the previous theorem. According to this theorem, the approximations $W_N^j(\phi)$ exist for any $j = 1, 2, \dots$ and any fixed $N = 1, 2, \dots$. They satisfy inequalities similar to (7.25):

$$\|W_N^j\|_0 \le \|f\|_0/\gamma, \quad \|W_N^j\|_s \le 2\delta_1, \quad |W_N^j|_2 \le 1 \tag{7.34}$$

for all $j = 1, 2, \dots, s = 1, \dots, r$, $\epsilon \in [0, \epsilon_0]$ when $\epsilon_0 > 0$ and $\|f\|_0$ are sufficiently small. An estimate of the difference $V_N^{j+1}(\phi) = W_N^{j+1}(\phi) - W_N^j(\phi)$ satisfying the equation

$$S_N L(\epsilon W_N^j) V(\phi) = S_N [L(\epsilon W_N^{j-1}) - L(\epsilon W_N^j)] W_N^j(\phi),$$

shows that

$$\|V_N^{j+1}\|_0 \le \bar{c}\epsilon \|V_N^j\|_0 \le \tfrac{1}{2} \|V_N^j\|_0, \quad j = 1, 2, \dots, \tag{7.35}$$

for all $\epsilon \in [0, \epsilon_0]$ and sufficiently small $\epsilon_0 > 0$.

Inequalities (7.35) imply that the sequence $W_N^j(\phi)$ ($j = 1, 2, \dots$) converges in the $H^0(\mathcal{T}_m)$ norm as $j \to \infty$. Then inequalities (7.31) ensure that the sequence $W_N^j(\phi)$ ($j = 1, 2, \dots$) converges in the $H^s(\mathcal{T}_m) \cap C^l(\mathcal{T}_m)$ norm as $j \to \infty$ when $s < r$. On passing to the limit we see that the limit function $W_N(\phi) = \lim\limits_{j \to \infty} W_N^j(\phi)$ is a solution of equation (7.32) satisfying the inequalities similar to (7.34):

$$\|W_N\|_0 \le \|f\|_0/\gamma, \quad \|W_N\|_s \le 2\delta_1, \quad |W_N|_2 \le 1 \tag{7.36}$$

for $s = 1, \dots, r$ and $\epsilon \in [0, \epsilon_0]$ when $\epsilon_0 > 0$ and $\|f\|_0$ are sufficiently small.

Estimating the difference $W_{N+1}^j(\phi) - W_N^j(\phi)$ with respect to the $H(\mathcal{T}_m)$ norm, we obtain

$$\gamma \|W_{N+1}^j - W_N^j\|_0 \le \bar{c}_1 \epsilon \|W_{N+1}^{j-1} - W_N^{j-1}\|_0 + \bar{c}_2 \|(E - S_N)[L(\epsilon W_N^{j-1}) W_N^j - f]\|_0 \le$$

$$\le \bar{c}_1 \epsilon \|W_{N+1}^{j-1} - W_N^{j-1}\|_0 + \bar{c}_3 N^{-(r-1)},$$

where the constants $\bar{c}_1$ and $\bar{c}_3$ do not depend on j and ϵ. The last inequality shows that

$$\|W_{N+1} - W_N\|_0 \le \bar{c}_4 N^{-(r-1)}, \tag{7.37}$$

where $\bar{c}_4$ does not depend on N and $\epsilon \in [0, \epsilon_0]$.

It follows from inequality (7.37) that the sequence $W_N(\phi)$ $(N = 1, 2, \ldots)$ converges to the function $u^0(\phi) = \lim\limits_{N \to \infty} W_N(\phi)$ as $N \to \infty$. Inequalities (7.36) guarantee that this sequence converges to the function $u^0(\phi)$ with respect to the $H^s(T_m) \cap C^l(T_m)$ norm for $s < r$. As can be seen by passing to the limit in identity (7.32) with $W_N = W_N(\phi)$, the limit function is a solution of equation (7.2). The estimate (7.31) for the difference $u^0(\phi) - W_N(\phi)$ follows from inequalities (7.37). We now complete the proof of Theorem 2 by setting $u(\phi) = u^0(\phi) = u^0(\phi, \epsilon)$ and taking into account the relationship between the solutions of equations (7.13) and (7.2).

As for the case of Theorem 1, inequality (7.31) implies the following estimate for the difference $u(\phi) - W_N(\phi)$ with respect to the $C^l(T_m)$ norm:

$$|u - W_N|_l \leq \bar{C} N^{-(r-1)(1-p/r)}, \quad N = 1, 2, \ldots,$$

where $r > p > m/2 + l$ and $\bar{C}$ is a constant independent of N.

4.8. Proof of Moser's Lemma

In effect, we have already started the proof of Moser's lemma in the version given in §3.15. To complete this proof it remains to determine more precisely how the parameters γ and δ in the inequality for the product $(Lu, u)_s$ given by (14.6) in Chapter 3 depend on the coefficients of the operator

$$L = \sum_{\nu=1}^{m} a_\nu(\phi)\frac{\partial}{\partial\phi_\nu} - P(\phi).$$

It follows from inequalities (15.1) and (15.3) in Chapter 3 that for $s = 1$

$$(Lu, u)_1 \geq \gamma\|u\|_1^2 - \delta\|u\|_0^2, \tag{8.1}$$

where γ and δ are positive constants which depend only on $\sum_{\nu=1}^{m} |a_\nu|_1 + |P|_1$. The following estimate clearly follows from estimate (8.1):

$$(Lu, u)_1 \geq \gamma\|u\|_1^2 - \delta\Big(\sum_{\nu=1}^{m} \|a_\nu\|_1 + \|P\|_1 + \|u\|_0\Big)\|u\|_1. \tag{8.2}$$

Consider the product $(Lu, u)_s$ for $s \geq 2$. Write it in the form

$$(Lu, u)_s = (L_2^s u, u)_{s-1} + \Phi_s(u),$$

where L_2^s is an elliptic operator of the form

$$L_2^s = -\sum_{\nu,j=1}^{m} \left(b_0 E_{\nu j} + s\frac{\partial a_\nu}{\partial \phi_j} E \right) \frac{\partial^2}{\partial\phi_\nu\partial\phi_j} = \sum_{\nu,j=1}^{m} A_{\nu j}(\phi) \frac{\partial^2}{\partial\phi_\nu\partial\phi_j} = L_2,$$

where $b_0 = -\frac{1}{2}(P + P^* + \sum_{\nu=1}^{m} \frac{\partial a_\nu}{\partial\phi_\nu} E)$, $E_{\nu j}$ is the identity matrix E for $\nu = j$ and the zero matrix for $\nu \neq j$, and $\Phi_s(u)$ is the functional equal to the difference $(Lu, u)_s - (L_2^s u, u)_{s-1}$.

Let $s - 1 = 2p$. Using standard transformations we find that

$$(L_2 u, u)_{2p} = (K^p L_2 u, K^p u)_0 = (L_2 K^p u, K^p u)_0 +$$

$$+((K^p L_2 - L_2 K^p)u, K^p u)_0 \geq \gamma_1 C \|K^p u\|_1^2 - \delta' \|K^p u\|_0^2 +$$

$$+((K^p L_2 - L_2 K^p)u, K^p u)_0, \quad K = 1 - \Delta,$$

where δ' is a positive constant which depends on $\sum_{\nu=1}^{m} |a_\nu|_2 + |P|_1$, and C is a positive constant independent of u and the coefficients of L_2.

The operator

$$\Psi_p(u) = (K^p L_2 - L_2 K^p)u$$

contains terms of the form $D^{s-i} A_{\nu j} \cdot D^i u$ $(i = 0, \ldots, s-1)$ which admit the estimate

$$\|D^{s-i} A_{\nu j} D^i u\|_0 \leq C\left(\sum_{\nu,j=1}^{m} |A_{\nu j}|_0 + |u|_0 \right) \left(\sum_{\nu,j=1}^{m} \|A_{\nu j}\|_s + \|u\|_s \right) \leq$$

$$\leq C'\left(\sum_{\nu=1}^{m} |a_\nu|_1 + |P|_0 + |u|_0 \right) \left(\sum_{\nu,j=1}^{m} \|a_\nu\|_{s+1} + \|P\|_s + \|u\|_s \right),$$

from which it follows that

$$\|\Psi_p(u)\|_0 \leq C'' C_0 \left(\sum_{\nu,j=1}^{m} \|a_\nu\|_{2p+1} + \|P\|_{2p} + \|u\|_{2p} \right),$$

where $C_0 \geq \sum_{\nu=1}^{m} |a_\nu|_2 + |P|_1 + |u|_0$.

From this we obtain the following inequality

$$(L_2 u, u)_{s-1} \geq \gamma_1' \|u\|_s^2 - \delta_1 \left(\sum_{\nu=1}^{m} \|a_\nu\|_s + \|P\|_{s-1} + \|u\|_{s-1} \right) \|u\|_s, \qquad (8.3)$$

where γ_1' and δ_1 are positive constants that depend only on C_0.

Let $s - 1 = 2p + 1$. Then

$$(L_2 u, u)_{2p+1} = (K^p L_2 u, K^p u)_1 = (L_2 K^p u, K^p u)_1 + (\Psi_p(u), K^p u)_1 \geq$$

$$\geq (L_2 K^p u, K^p u)_1 - \|\Psi_p(u)\|_0 \|K^p u\|_2 \geq (L_2 K^p u, K^p u)_1 -$$

$$- \delta'' \left(\sum_{\nu=1}^m \|a_\nu\|_{2p+1} + \|P\|_{2p} + \|u\|_{2p} \right) \|u\|_{2p+2},$$

and by using Gårding's inequality we obtain

$$(L_2 u, u)_{2p+1} \geq \gamma_2 \|K^p u\|_2^2 - \delta' \|K^p u\|_0^2 -$$

$$- \delta'' \left(\sum_{\nu=1}^m \|a_\nu\|_{2p+1} + \|P\|_{2p} + \|u\|_{2p} \right) \|u\|_{2p+2} \geq$$

$$\geq \gamma_2 \|u\|_{2p+2}^2 - \delta_2 \left(\sum_{\nu=1}^m \|a_\nu\|_{2p+1} + \|P\|_{2p} + \|u\|_{2p} \right) \|u\|_{2p+2},$$

where the positive constants γ_2 and δ_2 depend only on C_0.

Thus estimate (8.3) holds for both even and odd $s \geq 2$.

Let us estimate the functional $\Phi_s(u)$. Suppose that for $s = 2, \ldots, p$ the following estimate holds

$$|\Phi_s(u)| \leq c \left[\sum_{\substack{i+j \leq s \\ j \neq s}} \left(\sum_\nu \|D^i a_\nu D^j u\|_0 + \|D^i P D^j u\|_0 \right) + \right.$$

$$\left. + \sum_{2+i+j \leq s} \sum_\nu \|D^i (D^2 a_\nu) D^j (Du)\|_0 \right] \|u\|_s, \qquad (8.4)$$

where $c = $ const. We now prove that a similar estimate holds for $s = p + 1$.

It follows from the definition of the functional $\Phi_s(u)$ that for $s = p + 1$ we have

$$|\Phi_{p+1}(u)| \leq \left[\sum_\nu \left\| a_\nu \frac{\partial u}{\partial \phi_\nu} \right\|_{p-1} + \|Pu\|_{p-1} + \sum_\nu \left(\left\| \Delta a_\nu \frac{\partial u}{\partial \phi_\nu} \right\|_{p-1} + \left\| \frac{\partial P}{\partial \phi_\nu} u \right\|_p + \right. \right.$$

$$\left. \left. + \left\| \frac{\partial b_0}{\partial \phi_\nu} \Delta u \right\|_{p-2} \right) + \sum_{\nu, i, j} \left\| \frac{\partial^2 a_\nu}{\partial \phi_i \partial \phi_j} \frac{\partial^2 u}{\partial \phi_\nu \partial \phi_i} \right\|_{p-2} \right] \|u\|_{p+1} + \sum_\nu \left| \Phi_p \left(\frac{\partial u}{\partial \phi_\nu} \right) \right| \leq$$

$$\leq \sum_\nu \left| \Phi_p \left(\frac{\partial u}{\partial \phi_\nu} \right) \right| + c_1 \left[\sum_{\substack{i+j \leq p \\ j \geq 1}} \sum_\nu \|D^i a_\nu D^j u\|_0 + \sum_{i+j \leq p-1} \|D^i P D^j u\|_0 + \right.$$

$$+ \sum_{i+j\le p-1}\sum_{\nu}\|D^i(D^2 a_\nu)D^j(Du)\|_0 + \sum_{\substack{i+j\le p+1 \\ i\ge 1}}\|D^i P D^j u\|_0 +$$

$$+ \sum_{\substack{i+j\le p+1 \\ i\ge 1, j\ge 2}}\|D^i P D^j u\|_0 + \sum_{\substack{i+j\le p-1 \\ j\ge 1}}\sum_{\nu}\|D^i(D^2 a_\nu)D^j(Du)\|_0\Big]\|u\|_{p+1},$$

whence it follows that

$$|\Phi_{p+1}(u)| \le \sum_{\nu}\Big|\Phi_p\Big(\frac{\partial u}{\partial \phi_\nu}\Big)\Big| + c_2\Big[\sum_{\substack{i+j\le p+1 \\ j\ne p+1}}\Big(\sum_{\nu}\|D^i a_\nu D^j u\|_0 +$$

$$+\|D^i P D^j u\|_0\Big) + \sum_{i+j\le p-1}\sum_{\nu}\|D^i(D^2 a_\nu)D^j(Du)\|_0\Big]\|u\|_{p+1}. \qquad (8.5)$$

Using inequality (8.4) for $\Phi_p(\partial u/\partial \phi_\nu)$, we find that

$$\Big|\Phi_p\Big(\frac{\partial u}{\partial \phi_\nu}\Big)\Big| \le c\Big[\sum_{\substack{i+j\le p \\ j\ne p}}\Big(\sum_{\nu}\|D^i a_\nu D^j(Du)\|_0 + \|D^i P D^j(Du)\|_0\Big) +$$

$$+ \sum_{i+j\le p-2}\sum_{\nu}\|D^i(D^2 a_\nu)D^j(D^2 u)\|_0\Big]\|u\|_{p+1} \le$$

$$\le c\Big[\sum_{\substack{i+j\le p+1 \\ j\ne p+1}}\Big(\sum_{\nu}\|D^i a_\nu D^j u\|_0 + \|D^i P D^j u\|_0\Big) +$$

$$+ \sum_{i+j\le p-1}\sum_{\nu}\|D^i(D^2 a_\nu)D^j(Du)\|_0\Big]\|u\|_{p+1}. \qquad (8.6)$$

Estimate (8.4) for the functional $\Phi_{p+1}(u)$ follows from inequalities (8.5) and (8.6).

It follows from the calculations in §3.15 for $s = 2$ that

$$\Phi_2(u) = \Big(-\tfrac{1}{2}\Big(P + P^* + \sum_\nu \frac{\partial a_\nu}{\partial \phi_\nu}E\Big)u - \sum_\nu(\Delta a_\nu)\frac{\partial u}{\partial \phi_\nu} +$$

$$+2\sum_j \frac{\partial P}{\partial \phi_j}\frac{\partial u}{\partial \phi_j} + (\Delta P)u, u\Big)_1,$$

and thus inequality (8.4) holds for $\Phi_2(u)$:

$$|\Phi_2(u)| \le c\Big[\sum_{\substack{i+j\le 2 \\ j\ne 2}}\Big(\sum_\nu\|D^i a_\nu D^j u\|_0 + \|D^i P D^j u\|_0\Big) + \sum_\nu\|D^2 a_\nu Du\|_0\Big]\|u\|_2,$$

where $c = \text{const}$. We see by induction that estimate (8.4) holds for all $s \geq 2$.

From the inequalities

$$\sum_{\substack{i+j\leq s \\ j\neq s}} [\|D^i a_\nu D^j u\|_0 + \|D^i P D^j u\|_0] \leq \sum_{i+j\leq s-1} [\|D^i a_\nu D^j u\| +$$

$$+\|D^i P D^j u\|_0] + \sum_{i+j=s-2} [\|D^i(D\alpha_\nu)D^j(Du)\|_0 +$$

$$+\|D^i D^j(\Delta u)\|_0] + \leq C_1[(|a_\nu|_0 + |P|_0 + |u|_0)(\|a_\nu\|_{s-1} +$$

$$+\|P\|_{s-1} + \|u\|_{s-1}) + (|a_\nu|_1 + |P|_1 + |u|_1)(\|a_\nu\|_{s-1} + \|P\|_{s-1} + \|u\|_{s-1})],$$

$$\sum_{i+j\leq s-2} \|D^i(D^2 a_\nu)D^j(Du)\|_0 \leq C_2(|a_\nu|_2 + |u|_1)(\|a_\nu\|_s + \|u\|_{s-1}),$$

we obtain for $\Phi_s(u)$ the estimate

$$|\Phi_s(u)| \leq C_* \Big(\sum_{\nu=1}^m |a_\nu|_2 + |P|_1 + |u|_1\Big)\Big(\sum_{\nu=1}^m \|a_\nu\|_s + \|P\|_s + \|u\|_{s-1}\Big)\|u\|_s,$$

where C_* does not depend on a, P or u.

The last estimate together with estimate (8.3) proves that

$$(Lu, u)_s \geq \bar\gamma\|u\|_s^2 - \bar\delta\Big(\sum_{\nu=1}^m \|a_\nu\|_s + \|P\|_s + \|u\|_{s-1}\Big)\|u\|_s,$$

where $\bar\gamma$ and $\bar\delta$ are positive constants which depend only on C_0.

Now if we take into consideration the fact that for any $\epsilon > 0$,

$$\Big(\sum_{\nu=1}^m \|a_\nu\|_s + \|P\|_s + \|u\|_{s-1}\Big) \leq$$

$$\leq \Big(\sum_{\nu=1}^m \|a_\nu\|_s + \|P\|_s\Big)^2/(2\epsilon) + \|u\|_0^2/(2\epsilon^{2s-1}) + 2\epsilon\|u\|_s^2,$$

we finally obtain for $(Lu, u)_s$ estimate (7.9) of the lemma.

4.9. Invariant tori of systems of differential equations with rapidly and slowly changing variables

We shall apply the results of the perturbation theory of an invariant torus of a system of differential equations to systems of equations that contain slow and rapid variables. We consider several types of such equations.

$1°$. *Equations with slowly changing phase*

By this we mean a system of equations of the form

$$\frac{d\phi}{dt} = \epsilon a(\phi, x, \epsilon), \quad \frac{dx}{dt} = X(\phi, x, \epsilon), \tag{9.1}$$

the right hand side of which is defined in the domain

$$x \in D, \quad \phi \in T_m, \quad \epsilon \in [0, \epsilon_0] \tag{9.2}$$

and is periodic in ϕ_ν ($\nu = 1, \ldots, m$) with period 2π. Here D is a domain in n-dimensional Euclidean space E^n, T_m is an m-dimensional torus and ϵ is a small positive parameter.

We shall assume that for $\epsilon = 0$ system (9.1) has an m-parameter family of equilibrium points $x = u(\phi)$:

$$X(\phi, u(\phi), 0) = 0, \quad \phi \in T_m, \tag{9.3}$$

which cover the torus

$$x = u(\phi), \quad \phi \in T_m, \quad u \in C(T_m). \tag{9.4}$$

By applying the perturbation theory of an invariant torus to system (9.1) we can obtain the following result on the invariant tori of this system of equations.

Theorem 1. *Let the functions a and X have partial derivatives with respect to ϕ, x up to order r which are continuous with respect to ϕ, x, ϵ in the domain (9.2). Suppose that torus (6.4) consists of equilibrium points of system (9.1) when $\epsilon = 0$, that a neighbourhood of it lies in the domain D, and that $u \in C^r(T_m)$.*

If $r \geq 2$ and the matrix $P(\phi) = \partial X(\phi, u(\phi), 0)/\partial x$ admits the representation

$$P(\phi) = T(\phi)\operatorname{diag}\{-D_1(\phi), D_2(\phi)\}T^{-1}(\phi), \tag{9.5}$$

where $T(\phi) \in C^{r-1}(T_m)$ and the eigenvalues of the matrices $D_1(\phi)$, $D_2(\phi)$ have positive real parts for all $\phi \in T_m$, then there exists a sufficiently small $\epsilon_0 > 0$ such that for any $\epsilon \in [0, \epsilon_0]$ system (9.1) has an invariant torus

$$x = u(\phi, \epsilon), \quad \phi \in T_m, \quad u \in C^{r-2}_{\text{Lip}}(T_m),$$

satisfying the condition

$$\lim_{\epsilon \to 0} |u(\phi, \epsilon) - u(\phi)|_{r-2, \mathrm{Lip}} = 0, \tag{9.6}$$

and this torus is exponentially stable when $P(\phi) = -T(\phi)D_1(\phi)T^{-1}(\phi)$, *and is unstable when* $P(\phi) \neq -T(\phi)D_1(\phi)T^{-1}(\phi)$. *When* $r \geq 3$, *this torus is exponentially dichotomous.*

Proof. Under the above assumptions on the variables ϕ and $x - u(\phi) = h$, system (9.1) takes the form of system (2.1) in which the functions a, f and the matrix P defined in the domain (2.2) for sufficiently small $d > 0$ and $\epsilon_0 > 0$ have partial derivatives with respect to ϕ, h up to order $r-1$ which are continuous in ϕ, h, ϵ. Here the system of variational equations corresponding to the invariant torus (9.4) of the generating system of equations has the form (2.5) with $a_0 = 0$ and $P_0(\phi) = P(\phi) = \partial X(\phi, u(\phi), 0)/\partial x$.

In this case the fundamental matrix $\Omega_\tau^t(\phi)$ of the system of variational equations is equal to the matrix $e^{P(\phi)(t-\tau)}$. We set

$$C(\phi) = T(\phi)I_1 T^{-1}(\phi),$$

where $I_1 = \mathrm{diag}\{E_1, 0\}$, E_1 is the identity matrix of dimension equal to that of the matrix $D_1(\phi)$. The matrix $C(\phi)$ is a projection and belongs to the space $C^{r-1}(T_m) \subset C^1(T_m)$. Moreover, under the assumptions concerning the matrix $P(\phi)$ we have:

$$\|e^{P(\phi)(t-\tau)}C(\phi)\| = \|T(\phi)\mathrm{diag}\{e^{-D_1(\phi)(t-\tau)}, 0\}T^{-1}(\phi)\| \leq K_1 e^{-\gamma_1(t-\tau)}$$

for all $t \geq \tau$ and $\tau \in \mathbf{R}$,

$$\|e^{P(\phi)(t-\tau)}(C(\phi) - E)\| = \|T(\phi)\mathrm{diag}\{0, e^{D_2(\phi)(t-\tau)}\}T^{-1}(\phi)\| \leq K_2 e^{\gamma_2(t-\tau)}$$

for all $t \leq \tau$ and $\tau \in \mathbf{R}$, where K_1, K_2, γ_1 and γ_2 are positive constants independent of $\phi \in T_m$. Thus the trivial torus of the system of variational equations corresponding to the invariant torus (9.4) of system (9.1) for $\epsilon = 0$ is exponentially dichotomous (stable if $P(\phi) = -T(\phi)D_1(\phi)T^{-1}(\phi)$ and unstable otherwise).

The derivative of the matrix $e^{P(\phi)t}$ with respect to the variable ϕ_ν ($\nu = 1, \ldots, m$) satisfies the identity

$$\frac{\partial e^{P(\phi)t}}{\partial \phi_\nu} = \int_0^t e^{P(\phi)(t-\tau)} \frac{\partial P(\phi)}{\partial \phi_\nu} e^{P(\phi)\tau} d\tau \tag{9.7}$$

for all $t \in \mathbf{R}$ and $\phi \in \mathcal{T}_m$. If we replace the matrix $P(\phi)$ by the matrix $-D_1(\phi)$ in (9.7) we find that

$$\left\| \frac{\partial e^{-D_1(\phi)t}}{\partial \phi_\nu} \right\| \le K_1^2 \int_0^t e^{-\gamma_1(t-\tau)} \left\| \frac{\partial D_1(\phi)}{\partial \phi_\nu} \right\| e^{-\gamma_1 \tau} d\tau \le$$

$$\le K_1^2 \left| \frac{\partial D_1(\phi)}{\partial \phi_\nu} \right|_0 t e^{-\gamma_1 t}, \quad t \ge 0. \tag{9.8}$$

Similarly, we find that

$$\left\| \frac{\partial e^{D_2(\phi)t}}{\partial \phi_\nu} \right\| \le K_2^2 \left| \frac{\partial D_2(\phi)}{\partial \phi_\nu} \right|_0 |t| e^{\gamma_2 t}, \quad t \le 0. \tag{9.9}$$

Inequalities (9.8) and (9.9) imply that the matrix $S(\phi)$ defined by relations (9.13), (9.14) of Chapter 3 is continuously differentiable with respect to ϕ_ν ($\nu = 1, \ldots, m$). Since for this $S(\phi)$ the matrix $\hat{S}(\phi) = S(\phi)P(\phi) + P^*(\phi)S(\phi)$ is negative-definite for all $\phi \in \mathcal{T}_m$, all the conditions of Theorem 1 of §4.3 or Theorem 2 of §4.5 hold for system (9.1) written in terms of the variables ϕ, $x - u(\phi) = h$. The assertions of these theorems prove Theorem 1.

It should be noted that the statements of Theorem 1 remain true when the matrix $P(\phi)$ has the representation (9.5) with the matrix $T(\phi)$ whose period with respect to ϕ_ν ($\nu = 1, \ldots, m$) is a multiple of 2π. Indeed, if the period of the matrix $T(\phi)$ with respect to ϕ_ν ($\nu = 1, \ldots, m$) is equal to $2\pi p$, where p is an integer, $p \ge 2$, then the substitution $\phi = p\psi$ transforms this system of equations to a system of equations of the same kind, but now satisfying all the conditions of Theorem 1.

Then the system of equations (9.1) has an invariant surface

$$x = u(\phi, \epsilon), \quad \phi \in \mathbf{R}^m, \tag{9.10}$$

where the function $u(\phi, \epsilon)$ has period $2\pi p$ with respect to ϕ_ν ($\nu = 1, \ldots, m$). Since the right hand side of system (9.1) is periodic in ϕ_ν ($\nu = 1, \ldots, m$) with period 2π, it follows that the surfaces given by the equations

$$x = u(\phi + 2k\pi, \epsilon), \quad \phi \in \mathcal{T}_m, \tag{9.11}$$

where $k = (k_1, \ldots, k_m)$ is an integer vector for which $|k| = \sum_{\nu=1}^m |k_\nu| < p$, are invariant surfaces of this system.

All the surfaces (9.11) lie in a δ-neighbourhood of surface (9.10) since

$$\|u(\phi, \epsilon) - u(\phi + 2k\pi, \epsilon)\| \leq$$

$$\leq \|u(\phi, \epsilon) - u(\phi)\| + \|u(\phi + 2k\pi, \epsilon) - u(\phi + 2k\pi)\| \leq \eta(\epsilon),$$

where $\eta(\epsilon) \to 0$ as $\epsilon \to 0$. As follows from the proofs of Theorem 1 of §4.3 and Theorem 2 of §4.5, the invariant set (9.10) of system (9.1) has the stability or instability property that there are no whole trajectories of system (9.1) in some δ-neighbourhood of it which does not depend on ϵ. For sufficiently small $\epsilon_0 > 0$ we have the inequality $\eta(\epsilon) < \delta$ for all $\epsilon \in [0, \epsilon_0]$, and this leads to the identities

$$u(\phi, \epsilon) = u(\phi + 2k\pi, \epsilon), \quad \phi \in \mathbf{R}^m \qquad (9.12)$$

for all integer $k = (k_1, \ldots, k_m)$ and all $\epsilon \in [0, \epsilon_0]$. Identities (9.12) show that $u(\phi, \epsilon) \in C(\mathcal{T}_m)$ for all $\epsilon \in [0, \epsilon_0]$, as required.

It follows from this remark and Sibuja's Theorem given in §1.9 that even without the requirement expressed by relation (9.5) the statements of Theorem 1 remain true for $m = 1$. In particular, this leads to the following statement for systems of the form

$$\frac{dx}{dt} = X(\tau, x), \qquad (9.13)$$

where $\tau = \epsilon t$.

Theorem 2. *Let the right hand side of the system of equations* (9.13), *defined in the domain*

$$x \in D, \quad \tau \in \mathbf{R}, \quad \epsilon \in [0, \epsilon_0],$$

be periodic in τ with period 2π and have derivatives up to order r which are continuous in τ, x. Suppose that the equation

$$X(\tau, x) = 0$$

has a solution

$$x = u(\tau), \quad \tau \in [0, 2\pi]$$

that is periodic in τ with period 2π, some neighbourhood of which lies in the domain D, the function $u(\tau)$ having continuous derivatives with respect to τ up to order r.

If $r \geq 2$ and the real parts of the eigenvalues of the matrix $P(\tau) = \partial X(\tau, u(\tau))/\partial x$ are non-zero for all $\tau \in [0, 2\pi]$, then there exists a sufficiently small $\epsilon_0 > 0$ such that for any $\epsilon \in (0, \epsilon_0]$ system (9.13) has a solution

$$x = x(\tau, \epsilon), \quad \tau = \epsilon t$$

which is periodic in $\tau = \epsilon t$ with period 2π and satisfies the condition

$$\lim_{\epsilon \to 0} |x(\tau, \epsilon) - u(\tau)|_{r-1} = 0. \tag{9.14}$$

Furthermore, this solution is asymptotically stable if the real parts of the eigenvalues of the matrix $P(\tau)$ are negative and is unstable otherwise.

To prove Theorem 2 we further note that $x(\tau, \epsilon)$ is a solution of the equation obtained from (9.13) by substituting $\epsilon t = \tau$ for t. Thus, the function $x(\tau, \epsilon)$ is $r + 1$ continuously differentiable with respect to τ for any $\epsilon \in (0, \epsilon_0]$. This enables us to replace in relation (9.6) the norm $|\cdot|_{r-2, \text{Lip}}$ by $|\cdot|_{r-1}$ for the difference $x(\tau, \epsilon) - u(\tau)$, and this completes the proof of Theorem 2.

To illustrate the statements of Theorem 2, we consider the system defined by Van der Pol's equation

$$\frac{d^2 x}{dt^2} + x = \alpha(\tau)(1 - \beta(\tau)x^2)\frac{dx}{dt} + f(\tau), \tag{9.15}$$

where $\alpha(\tau), \beta(\tau), f(\tau)$ are twice continuously differentiable functions which are periodic in τ with period 2π, and $\tau = \epsilon t$ is slowly changing time. By introducing the variables $x, y = dx/dt$, equation (9.15) takes the form of system (9.13) which consists of two equations. The function $u(\tau)$ is defined by the expression

$$x = f(\tau), \quad y = 0, \quad \tau = [0, 2\pi],$$

and the matrix $P(\tau)$ by the equation

$$P(\tau) = \begin{bmatrix} 0 & 1 \\ -1 & \alpha(\tau)(1 - \beta(\tau)f^2(\tau)) \end{bmatrix}.$$

The eigenvalues of the matrix $P(\tau)$ have negative real parts if

$$\alpha(\tau)(1 - \beta(\tau)f^2(\tau)) < 0, \quad \tau \in [0, 2\pi], \tag{9.16}$$

and positive real parts if

$$\alpha(\tau)(1 - \beta(\tau)f^2(\tau)) > 0, \quad \tau \in [0, 2\pi]. \tag{9.17}$$

Consequently, if inequalities (9.16) hold, then for all sufficiently small $\epsilon > 0$ Van der Pol's equation has a solution

$$x = x(\tau, \epsilon), \quad \tau = \epsilon t \tag{9.18}$$

which is asymptotically stable periodic in $\tau = \epsilon t$ with period 2π, while if inequalities (9.17) hold, then it has an unstable solution (9.18), and this solution satisfies the condition

$$\lim_{\epsilon \to 0} |u(\tau, \epsilon) - f(\tau)|_1 = 0$$

in both cases.

$2°$. *Periodic systems of equations with rapidly and slowly changing time*

By this we mean a system

$$\frac{dx}{dt} = X(t, \tau, x), \tag{9.19}$$

the right hand side of which is defined in the domain

$$x \in D, \quad t \in \mathbf{R}, \quad \tau \in \mathbf{R} \tag{9.20}$$

and periodic with respect to t, $\tau = \epsilon t$ with period 2π.

We shall suppose that the system of equations (9.19) considered as a system of equations that depends on the parameter τ has a solution

$$x = u(t, \tau) \tag{9.21}$$

for all $\tau \in [0, 2\pi]$ that is periodic in t and τ with period 2π. We write down the variational equation for solution (9.21) of system (9.19) for $\tau = \text{const}$:

$$\frac{dy}{dt} = \frac{\partial X(t, \tau, u(t, \tau))}{\partial x} y, \tag{9.22}$$

and denote by $\Omega_0^t(\tau)$ the fundamental matrix of its solutions which is equal to the identity matrix when $t = 0$.

Theorem 3. *Let the right hand side of system (9.19) be r times continuously differentiable with respect to t, τ, x in the domain (9.20) and periodic in t and τ with period 2π. Suppose that when $\tau = \text{const}$ some neighbourhood of the solution (9.21) of system (9.19) lies in the domain D for all $(t, \tau) \in \mathbf{R} \times \mathbf{R}$ and that $u \in C^r(\mathcal{T}_2)$.*

If $r \geq 2$ and the eigenvalues of the matrix $\Omega_0^{2\pi}(\tau)$ lie on curves that do not intersect the unit circle and do not encircle the origin of the complex plane when $\tau \in [0, 2\pi]$, then there exists a sufficiently small $\epsilon_0 > 0$ such that for any $\epsilon \in [0, \epsilon_0]$ system (9.19) has a solution

$$x = x(t, \tau, \epsilon), \quad \tau = \epsilon t$$

that is periodic in t and $\tau = \epsilon t$ with period 2π and satisfies the condition

$$\lim_{\epsilon \to 0} |x(t, \tau, \epsilon) - u(t, \tau)|_{r-1} = 0;$$

furthermore this solution is asymptotically stable if the eigenvalues of the matrix $\Omega_0^{2\pi}(\tau)$ lie inside the unit circle with centre at the origin and is unstable if there are eigenvalues lying outside this circle.

Proof. Using the coordinates $\phi = (\theta, \tau)$, $x - u(\theta, \tau) = h$, we rewrite system (2.1) in the form of system (9.19), where $a = (1, \epsilon)$, $f = -\partial u(\theta, \tau)/\partial \tau$ and P is a function defined in the domain (2.2) for sufficiently small $d > 0$ and $\epsilon_0 > 0$ and having partial derivatives with respect to ϕ, h up to order $r - 1$ which are continuous in ϕ, h, ϵ.

The system of variational equations corresponding to the invariant torus

$$x = u(\phi), \quad \phi \in T_2,$$

of system (9.19) for $\epsilon = 0$, written in the coordinates ϕ, $x - u(\phi) = h$ has the form (2.5) with $a_0 = (1, 0)$ and $P_0(\phi) = \partial X(\phi, u(\phi))/\partial x$. Consequently, it is equivalent to the system

$$\frac{dh}{dt} = \frac{\partial X(t + \theta, \tau, u(t + \theta, \tau))}{\partial x} h \tag{9.23}$$

where θ and τ are parameters. The fundamental matrix $\Omega_0^t(\theta, \tau)$ of solutions of system (9.23) is clearly related to the fundamental matrix $\Omega_0^t(\tau)$ of solutions of system (9.22) by the formula

$$\Omega_0^t(\theta, \tau) = \Omega_\theta^{t+\theta}(\tau)$$

for all $(t, \theta, \tau) \in \mathbf{R}^3$.

The right hand side of system (9.22) is periodic in t with period 2π. Therefore we can apply to it the Floquet-Lyapunov reducibility theorem, according to which,

$$\Omega_0^t(\tau) = \Phi(t, \tau) e^{H(\tau)t},$$

where $\Phi(t, \tau)$ is a real periodic matrix with respect to t with period 4π, $4\pi H(\tau)$ is the real branch of the logarithm of the matrix $[\Omega_0^{2\pi}(\tau)]^2$:

$$H(\tau) = \frac{1}{4\pi} \ln[\Omega_0^{2\pi}(\tau)]^2 = \frac{1}{4\pi} \ln \Omega_0^{4\pi}(\tau).$$

Since $\Omega_0^{2\pi}(\tau) \in C^{r-1}(T_1)$, we see that under the above assumptions it follows from the lemma on the logarithm given in §1.9 that $H(\tau)$ is also a matrix in $C^{r-1}(T_1)$. Moreover, according to Sibuja's theorem also given in §1.9, the matrix $H(\tau)$ can be represented as

$$H(\tau) = T(\tau)\mathrm{diag}\{-D_1(\tau), D_2(\tau)\}T^{-1}(\tau),$$

where $T(\tau)$ is a real matrix that is periodic in τ with period 4π and has continuous derivatives withe respect to τ up to order $r-1$, and $D_1(\tau)$ and $D_2(\tau)$ are matrices, the real parts of the eigenvalues of which are positive for all $\tau \in [0, 4\pi]$.

The matrix $e^{H(\tau)t}$ can be represented as

$$e^{H(\tau)t} = T(\tau)\mathrm{diag}\{e^{-D_1(\tau)t}, e^{D_2(\tau)t}\}T^{-1}(\tau),$$

where for all $\tau \in [0, 4\pi]$

$$\|e^{-D_1(\tau)t}\| \le K_1 e^{-\gamma_1 t}, \quad t \ge 0,$$

$$\|e^{D_2(\tau)t}\| \le K_2 e^{\gamma_2 t}, \quad t \le 0,$$

and K_1, K_2, γ_1 and γ_2 are positive constants independent of τ. If we set $\phi = \tau$, then the derivatives of the matrices $e^{-D_1(\tau)t}$ and $e^{D_2(\tau)t}$ with respect to τ satisfy inequalities (9.8) and (9.9).

We set

$$C(\theta, \tau) = \Phi(\theta, \tau)T(\tau)I_1 T^{-1}(\tau)\Phi^{-1}(\theta, \tau),$$

where $I_1 = \mathrm{diag}\{E_1, 0\}$ and E_1 is the identity matrix of the same dimension as $D_1(\tau)$. The matrix $C(\theta, \tau)$ is a projection which is periodic in θ, τ with period 4π and has continuous partial derivatives up to order $r-1$ with respect to θ, τ for all $(\theta, \tau) \in \mathbf{R}^2$.

Since

$$\Omega_0^t(\theta, \tau)C(\theta, \tau) = \Omega_0^{t+\theta}(\tau)[\Omega_0^\theta(\tau)]^{-1}C(\theta, \tau) =$$

$$= \Phi(t + \theta, \tau)e^{H(\tau)(t+\theta)}e^{-H(\tau)\theta}\Phi^{-1}(\theta, \tau)C(\theta, \tau) =$$

$$= \Phi(t + \theta, \tau)T(\tau)\mathrm{diag}\{e^{-D_1(\tau)t}, 0\}T^{-1}(\tau)\Phi^{-1}(\theta, \tau),$$

it follows from the estimates of the matrix $e^{-D_1(\tau)t}$ and its derivatives with respect to τ that the matrix $S_1(\theta,\tau)$, defined in terms of $\Omega_0^t(\theta,\tau)C(\theta,\tau)$ by relation (9.14) in Chapter 3, is continuously differentiable with respect to θ,τ and periodic in to θ,τ with period 4π.

Similar reasoning shows that the matrix $S_2(\theta,\tau)$, defined by relation (9.14) in Chapter 3 in terms of $\Omega_0^t(\theta,\tau)C_1(\theta,\tau)$, where $C_1(\theta,\tau) = E - C(\theta,\tau)$, is continuously differentiable with respect to θ,τ and differentiable and periodic with respect to θ,τ with period 4π. Then the matrix

$$\hat{S}(\theta,\tau) = \frac{\partial S(\theta,\tau)}{\partial \theta} + S(\theta,\tau)P_0(\theta,\tau) + P_0^*(\theta,\tau)S(\theta,\tau)$$

corresponding to $S(\theta,\tau) = S_1(\theta,\tau) - S_2(\theta,\tau)$ is negative-definite for all $(\theta,\tau) \in [0,4\pi] \times [0,4\pi]$.

This is sufficient for system (9.19), written in terms of the variables $\phi = (\theta,\tau)$, $x - u(\theta,\tau) = h$, to satisfy all the conditions of Theorem 1 of §4.3 or Theorem 2 of §4.5. We complete the proof of Theorem 3 by taking into account the reasoning given after the proof of Theorem 1 and the assertions of those theorems.

As an example illustrating Theorem 3 we consider the system given by the equation

$$\frac{d^2x}{dt^2} + 2\lambda(\tau)\frac{dx}{dt} + n^2(\tau)x = F(\tau)\cos\omega t, \tag{9.24}$$

where $\lambda(\tau)$, $n(\tau)$ and $F(\tau)$ are periodic functions in $\tau = \epsilon t$ with period 2π which are twice continuously differentiable and are such that $\lambda(\tau) \neq 0$, $n(\tau) \neq 0$ for all $\tau \in [0,2\pi]$. In terms of the variables $x,y = dx/dt$ we can write equation (9.24) in the form of system (9.19), which consists of two equations.

The function $u(t,\tau)$ defined by the relations

$$x = u(t,\tau) = \frac{F(\tau)\cos(\omega t - \alpha(\tau))}{\sqrt{[n^2(\tau) - \omega^2]^2 + 4\omega^2\lambda^2(\tau)}},$$

$$\alpha(\tau) = \tan^{-1}\frac{2\omega\lambda(\tau)}{n^2(\tau) - \omega^2}, \quad \tau \in [0,2\pi], \tag{9.25}$$

is a periodic solution of equation (9.24) for constant τ. The variational equation corresponding to the torus (9.25) is the same as equation (9.24) when $F \equiv 0$; consequently the quantities

$$\lambda_{1,2} = -\lambda(\tau) \pm \sqrt{\lambda^2(\tau) - n^2(\tau)}, \quad \tau \in [0,2\pi],$$

are the eigenvalues of the matrix $H(\tau)$.

When $\lambda(\tau) < 0$ or $\lambda(\tau) > 0$ for all $\tau \in [0, 2\pi]$ the conditions of Theorem 3 hold; consequently for all sufficiently small $\epsilon > 0$ equation (9.24) has a solution

$$x = x(t, \tau, \epsilon)$$

that is periodic in t and $\tau = \epsilon t$ with period satisfying the condition

$$\lim_{\epsilon \to 0} |x(t, \tau, \epsilon) - u(t, \tau)|_1 = 0.$$

This solution is asymptotically stable for $\lambda(\tau) < 0$ and unstable for $\lambda(\tau) > 0$.

Bibliography

1. P.S. Aleksandrov, *Combinatorial topology*, Gostekhizdat, Moscow, 1947 (in Russian).
2. A.A. Andronov, *Collected works*, Izdat. Akad. Nauk. Moscow, 1956 (in Russian).
3. A.A. Andronov and A.A. Witt, 'On quasi-periodic motions'. In: *Collected works of A.A. Andronov*, Izdat. Akad. Nauk SSSR, Moscow, 1956 (in Russian).
4. —, — and S.E. Haikin, *Oscillation theory*, Fizmatgiz, Moscow, 1959 (in Russian).
5. V.I. Arnol'd, 'Small denominators I: On the mappings of a circle onto itself', *Izv. Akad. Nauk SSSR Ser. Mat.* **25**:1 (1961), 21–86.
6. —, 'Small denominators. A proof of A.N. Kolmogorov's theorem on conservation of conditionally periodic motions under a small change of the Hamiltonian', *Uspekhi Mat. Nauk* **18**:5 (1963), 13–40.
7. —, 'Small denominators and the stability problem in classical and celestial mechanics', *Uspekhi Mat. Nauk* **18**:6 (1963), 91–192.
8. —, *Mathematical methods in classical mechanics*, Nauka, Moscow, 1974 (in Russian).
9. —, *Supplementary chapters to the theory of ordinary differential equations*, Nauka, Moscow, 1978 (in Russian).
10. N.I. Akhiezer, *Lectures on approximation theory*, Nauka, Moscow, 1965 (in Russian).
11. A.S. Bakai and Yu.P. Stepanovskii, *Adiabatic invariants*, Naukova Dumka, Kiev, 1981 (in Russian).
12. Ya.S. Baris, 'On the integral manifold of a non-regularly perturbed differential system, *Ukrain. Mat. Zh.* **20** (1968), 439–448.

13. — and V.I. Fodchuk, 'Study of bounded solutions of non-linear non-regu-
    larly perturbed systems using the integral manifold method', *Ukrain. Mat.
    Zh.* **22** (1970), 3–11.

14. L. Bers, F. John and M. Schechter, *Partial differential equations*, Wiley,
    New York, 1964.

15. Yu.N. Bibikov, 'On the existence of invariant tori in a neighbourhood of
    an equilibrium position of a system of differential equations', *Dokl. Akad.
    Nauk SSSR* **185** (1969), 9–13.

16. — and V.A. Pliss, 'On the existence of invariant tori in a neighbourhood
    of the zero solution of a system of ordinary differential equations', *Differ-
    entsial'nye Uravneniya* **3** (1967), 1864–1881.

17. G.D. Birkhoff, *Dynamical systems*, American Mathematical Society, New
    York, 1927.

18. Yu.S. Bogdanov, 'On the transformation of a variable matrix to canonical
    form', *Dokl. Akad. Nauk BSSR* **7**:3 (1963), 152–154.

19. N.N. Bogolyubov, *On some statistical methods in mathematical physics*,
    Izdat. Akad. Nauk Ukrain. SSR, Kiev, 1945 (in Russian).

20. —,'On quasi-periodic solutions in problems of non-linear mechanics, In:
    Proc. First Summer Math. School, Naukova Dumka, Kiev, 1964, pp. 11–
    102 (in Russian).

21. —, *Collected works, Vol.* 1, Naukova Dumka, Kiev, 1969 (in Russian).

22. — and Yu.A. Mitropolskii, 'The method of integral manifolds in non-linear
    mechanics', In: *Proc. Internat. Conf. on Non-linear Oscillations* Part I,
    Analytic methods, Inst. Mat. Akad. Nauk Ukrain. SSR, Kiev, 1963, pp. 93–
    154 (in Russian).

23. — and —, *Asymptotic methods in the theory of non-linear oscillations*,
    Nauka, Moscow, 1974 (in Russian).

24. —, — and A.M. Samoilenko, *The accelerated convergence method in non-
    linear mechanics* , Naukova Dumka, Kiev, 1969 (in Russian).

25. P.G. Bol, *Collected works*, Akad. Nauk Latv. SSR, Riga, 1961 (in Russian).

26. H. Bohr, *Almost periodic functions*, Gostekhizdat, Moscow-Leningrad, 1934
    (in Russian).

27. I.Yu. Bronshtein, *Extensions of minimal transformation groups*, Shtiintsa,
    Kishinev, 1975 (in Russian).

28. —, 'Weak regularity and Green's functions of a linear extension of dynamical systems', *Differentsial'nye Uravneniya* **19** (1983), 2031–2038.

29. — and V.P. Burdaev, 'Invariant manifolds of weakly linear extensions of dynamical systems', Preprint *Akad. Nauk Moldavsk. SSR*, Kishinev, 1983.

30. — and V.F. Chernii, 'Linear extensions satisfying the exponential dichotomy condition', *Izv. Akad. Nauk Moldavsk. SSR* **3** (1976), 12–16.

31. N.V. Butenin, *Elements of the theory of non-linear oscillations*, Sudpromgiz., Leningrad, 1962 (in Russian).

32. B.F. Bylov, 'On the structure of solutions of a system of linear differential equations with almost-periodic coefficients', *Mat. Sb.* **66** (1965), 215–229.

33. —, 'On the reduction of a system of differential equations to diagonal form', *Mat. Sb.* **66** (1965), 338–344.

34. —, R.E. Vinograd, V.Ya. Lin' and O.V. Lokutsievskii, 'On the topological reasons for the abnormal behaviour of certain almost periodic systems, In: *Problems of asymptotic theory of non-linear oscillations*, Naukova Dumka, Kiev, 1977, pp. 54–61 (in Russian).

35. —, —, — and —, 'On topological obstacles to block diagonalization of exponentially split almost-periodic systems', Preprint No. 58, *Inst. Prikl. Mat. Akad. Nauk SSSR*, Moscow, 1977.

36. V. Vazov, *Asymptotic expansions of solutions of ordinary differential equations*, Mir, Moscow, 1968 (in Russian).

37. B. Van der Pol, *Non-linear theory of electric oscillations*, Svyaz'tekhizdat, Moscow, 1935 (in Russian).

38. V.M. Volosov and V.I. Morgunov, *The averaging method in the theory of non-linear oscillatory systems*, Moscow University Press, Moscow, 1971 (in Russian).

39. F.R. Gantmakher, *Theory of matrices*, Nauka, Moscow, 1967 (in Russian).

40. V.L. Golets, 'On the perturbation problem for a stable invariant torus of a dynamical system', *Ukrain. Mat. Zh.* **23** (1971), 130–137.

41. Eu.A. Grebenikov and Yu.A. Ryabov, *Constructive methods of analysis of non-linear systems*, Nauka, Moscow, 1979 (in Russian).

42. Yu.L. Daletskii and M.G. Krein, *Stability of solutions of differential equations in Banach space*, Nauka, Moscow, 1970 (in Russian).

43. B.P. Demidovich, *Lectures on the mathematical theory of stability*, Nauka, Moscow, 1967 (in Russian).

44. P.P. Zabreiko and S.O. Strygina, 'Cesari's equalities and Galerkin's procedure for finding periodic solutions of ordinary differential equations', *Dokl.
Akad. Nauk Ukrain. SSR Ser. A* **7** (1970), 583–586.

45. K.V. Zadiraka, 'On the integral manifold of a system of differential equations containing a small parameter', *Dokl. Akad. Nauk SSSR* **115** (1957),
646–649.

46. V.I. Zubov, *Lyapunov's methods and their applications*, Leningrad State
Univ. Press., Leningrad, 1957 (in Russian).

47. —, *Stability of motion*, Vysshaya Shkola, Moscow, 1973 (in Russian).

48. A.N. Kolmogorov, 'On the preservation of conditionally periodic motions
under a small change of the Hamiltonian', *Dokl. Akad. Nauk SSSR* **98**
(1954), 527–530.

49. —, 'General theory of dynamical systems and classical mechanics', In:
International Math. Congress in Amsterdam, Fizmatgiz, Moscow, 1961,
187–208 (in Russian).

50. — and S.V. Fomin, *Elements of the theory of functions and functional
analysis*, Nauka, Moscow, 1968 (in Russian).

51. M.A. Krasnosel'skii, 'On the theory of periodic solutions of non-autonomous
differential equations', *Uspekhi Mat. Nauk* **21**:3 (1966), 53–74.

52. —, *The shift operator along trajectories of differential equations*, Nauka,
Moscow, 1966 (in Russian).

53. —, V.Sh. Burd and Yu.S. Kolesov, *Non-linear almost-periodic oscillations*,
Nauka, Moscow (1970) (in Russian).

54. A.N. Krylov, *On some differential equations of mathematical physics that
have applications in technical problems*, Izdat. Akad. Nauk SSSR, Moscow,
1950 (in Russian).

55. — and N.N. Bogolyubov, *Application of methods of non-linear mechanics
to the theory of stationary oscillations*, Izdat. Akad. Nauk Ukrain. SSR,
Kiev, 1934 (in Russian).

56. — and —, *New methods in non-linear mechanics*, ONTI, Moscow-Leningrad, 1934 (in Russian).

57. — and —, *Introduction to non-linear mechanics*, Izdat. Akad. Nauk Ukrain.
SSR, Kiev, 1937 (in Russian).

58. V.L. Kulik, 'Quadratic forms and dichotomy of solutions of linear differential equations', *Ukrain. Mat. Zh.* **34** (1982), 43–49.

59. —, 'Converse of a theorem on the decomposability of linear extensions of dynamical systems on a torus', *Ukrain. Mat. Zh.* **35** (1983), 67–72.

60. A.G. Kurosh, *A course in higher algebra*, Nauka, Moscow, 1968 (in Russian).

61. Y. Kurtsveil', 'Invariant manifolds of differential systems', *Differentsial'nye Uravneniya* **4** (1968), 785–797.

62. O.A. Ladyzhenskaya, *Boundary-value problems in mathematical physics*, Nauka, Moscow, 1973 (in Russian).

63. I.A. Lappo-Danilevskii, *Application of matrix functions to the theory of linear systems of ordinary differential equations*, Gostekhizdat, Moscow, 1957 (in Russian).

64. B.M. Levitan, *Almost-periodic functions*, Gostekhizdat, Moscow, 1953 (in Russian).

65. — and V.V. Zhikov, *Almost-periodic functions and differential equations*, Moscow State Univ. Press, Moscow, 1978 (in Russian).

66. D.K. Lika and Yu.A. Ryabov, *The methods of iteration and majorizing Lyapunov's equations in the theory of non-linear oscillations*, Shtiintsa, Kishinev, 1974 (in Russian).

67. L.A. Lyusternik and V.I. Sobolev, *Elements of functional analysis*, Nauka, Moscow, 1965 (in Russian).

68. A.M. Lyapunov, *The general problem on stability of motion*, Gostekhizdat, Moscow, 1950 (in Russian).

69. —, *Collected works*, *Vol.* 2, Izdat. Akad. Nauk SSSR, Moscow, 1956 (in Russian).

70. I.G. Malkin, *The methods of Lyapunov and Poincaré in the theory of non-linear oscillations*, Gostekhizdat, Moscow, 1949, (in Russian).

71. —, *Some problems of the theory of non-linear oscillations*, Gostekhizdat, Moscow (1956) (in Russian).

72. —, *The theory of stability of motion* , Nauka, Moscow, 1966 (in Russian).

73. L.I. Mandel'shtam, *Lectures on oscillation theory*, Nauka, Moscow, 1972 (in Russian).

74. D.I. Martynyuk and A.M. Samoilenko, 'Invariant tori of systems with after-effect', *Godishnik Vissh. Uchebn. Zaved. Prilozhna Mat.* **11** (1976), 47–53.

75. — and N.S. Tsiganovskii. 'Invariant manifolds of a system with delay under the action of impulses', *Differentsial'nye Uravneniya* **15** (1979), 1783–1795.

76. Yu.A. Mitropol'skii, *Non-stationary processes in non-linear oscillatory systems*, Izdat. Akad. Nauk Ukrain. SSR, Kiev, 1955 (in Russian).

77. —, 'On the construction of the general solution of non-linear differential equations using an "accelerated" convergence method', *Ukrain. Mat. Zh.* **16** (1964), 475–501.

78. —, 'An accelerated convergence method in problems of non-linear mechanics', *Funkcialaj Ekvacioj* **9** (1966), 27–42.

79. —, 'On the construction of solutions and the reducibility of differential equations with quasi-periodic coefficients', In: *Third USSR Congress on Theoretical and Applied Mechanics*, Nauka, Moscow, 1968, pp. 212–213 (in Russian).

80. —, *The averaging method in non-linear mechanics*, Naukova Dumka, Kiev, 1971 (in Russian).

81. — and O.B. Lykova, *Integral manifolds in non-linear mechanics*, Nauka, Moscow, 1973 (in Russian).

82. — and A.M. Samoilenko, 'On the structure of trajectories on toroidal manifolds', *Dokl. Akad. Nauk Ukrain. SSR, Ser. A* **8** (1964), 984–985.

83. — and —, 'On the method of accelerated convergence in non-linear mechanics', In: *Methods of integral manifolds in non-linear differential equations*, Inst. Mat. Akad. Nauk Ukrain. SSR, Kiev, 1973, 5–31.

84. — and —, 'Some problems in the theory of multi-frequency oscillations', In: *VII Internationale Konferenz über Nichtlineare Schwingungen* 1, Academic-Verlag, Berlin, 1977, No. 4, pp. 107–116.

85. — and —, 'Some problems in the theory of multi-frequency oscillations', Preprint No. 14, *Inst. Akad. Nauk Ukrain. SSR*, Kiev, 1977.

86. —, — and V.L. Kulik, 'The study of linear differential equations using quadratic forms', Preprint No. 10, *Inst. Mat. Akad. Nauk Ukrain. SSR*, Kiev, 1982.

87. —, — and D.I. Martynyuk, *Systems of evolutionary equations with periodic and conditionally periodic coefficients*, Naukova Dumka, Kiev, 1984 (in Russian).

88. —, — and K.B. Tsidylo, 'On invariant toroidal manifolds of non-linear systems with delay', In: *Differential equations with deviating arguments*, Naukova Dumka, Kiev, 1977, 207–214 (in Russian).

89. J. Moser, 'A rapidly convergent iteration method and non-linear differen-

tial equations', *Uspekhi Mat. Nauk* **23**:4 (1968), 179–238.

90. —, 'On the decomposition of conditionally periodic motions into convergent power series', *Uspekhi Mat. Nauk* **24**:2 (1969), 165–211.

91. —, *Lectures on Hamiltonian systems*, Mir, Moscow, 1973 (in Russian).

92. Yu.I. Neimark, *The method of point mappings in the theory of non-linear oscillations*, Nauka, Moscow, 1972 (in Russian).

93. V.V. Nemytskii, 'Oscillations in autonomous systems', In: *Fifth Summer Math. School*, Naukova Dumka, Kiev, 1968, 436–472 (in Russian).

94. — and V.V. Stepanov, *Qualitative theory of differential equations*, OGIZ GITTL, Moscow-Leningrad, 1947 (in Russian).

95. G.S. Osipenko, 'Perturbations of dynamical systems near invariant manifolds' I, *Differentsial'nye Uravneniya* **15** (1979), 1967–1979; II **16** (1980), 620–628.

96. N.A. Perestyuk, 'Invariant sets of one class of discontinuous dynamical systems', *Ukrain. Mat. Zh.* **1** (1984), 63–69.

97. A.I. Perov, *Variational methods in the theory of non-linear oscillations*, Voronezh Univ. Press, Voronezh, 1981 (in Russian).

98. V.A. Pliss, 'The reduction principle in the theory of stability of motion', *Izv. Akad. Nauk SSSR Ser. Mat.* **28** (1964), 1297–1324.

99. —, *Non-local problems in oscillation theory*, Nauka, Moscow-Leningrad, 1964 (in Russian).

100. H. Poincaré, *On curves defined by differential equations*, OGIZ GITTL, Moscow-Leningrad, 1947 (in Russian).

101. —, *Collected works*, I, II, Nauka, Moscow, 1971, 1972 (in Russian).

102. C. Pugh, 'A closure lemma', *Matematika* **12** (1968), 80–135 (in Russian).

103. A.M. Samoilenko, 'On the structure of trajectories on a torus', *Ukrain. Mat. Zh.* **16** (1964), 769–782.

104. —, 'On the reducibility of a system of ordinary differential equations in a neighbourhood of a smooth toroidal manifold', *Izv. Akad. Nauk SSSR Ser. Mat.* **30** (1966), 1047–1072.

105. —, 'On the reducibility of systems of linear differential equations with quasi-periodic coefficients', *Ukrain. Mat. Zh.* **20** (1968), 279–281.

106. —, 'On the preservation of an invariant torus under perturbations', *Izv. Akad. Nauk SSSR Ser. Mat.* **34** (1970), 1219–1240.

107. —, 'On perturbation theory of invariant manifolds of dynamical systems',

In: *Proc. Fifth Internat. Congress on Non-linear Oscillations* I, Analytical Methods, Inst. Mat. Akad. Nauk Ukrain. SSR, Kiev, 1970, pp. 495–499 (in Russian).

108. —, 'Study of dynamical systems by using functions of constant sign', *Ukrain. Mat. Zh.* **24** (1972), 374–384.

109. —, 'On the exponential stability of an invariant torus of a dynamical system', *Differentsial'nye Uravneniya* **11** (1975), 820–834.

110. —, 'Quasi-periodic solutions of systems of linear algebraic equations with quasi-periodic coefficients', In: *Analytical methods of study of solutions of non-linear differential oscillations*, Inst. Mat. Akad. Nauk Ukrain. SSR, Kiev, 1975, 5–26 (in Russian).

111. —, 'Invariant toroidal manifolds of systems with slowly changing variables', In: *Problems of the asymptotic theory of non-linear oscillations*, Naukova Dumka, Kiev, 1977, 181–191 (in Russian).

112. —, 'Green's function for a linear extension of a dynamical system on a torus, its uniqueness conditions and properties following from these conditions', *Ukrain. Mat. Zh.* **32** (1980), 791–797.

113. —, 'Necessary conditions for the existence of invariant tori of linear extensions of dynamical systems on a torus', *Differentsial'nye Uravneniya* **16** (1980), 1427–1437.

114. —, 'Separatrix manifolds and decomposability of linear extensions of dynamical systems on a torus', *Ukrain. Mat. Zh.* **33** (1981), 31–38.

115. — and A.V. Dvorak, 'Splitting of a dynamical system in a neighbourhood of a conditionally stable invariant manifold', *Ukrain. Mat. Zh.* **29** (1977), 557–562.

116. — and V.L. Kulik, 'On the problem of the existence of a Green's function for the invariant torus problem', *Ukrain. Mat. Zh.* **27** (1975), 348–359.

117. — and —, 'Exponential dichotomy of an invariant torus of dynamical systems', *Differentsial'nye Uravneniya* **15** (1979), 1434–1444.

118. — and —, 'On decomposability of linearized systems of differential equations', *Ukrain. Mat. Zh.* **34** (1982), 587–597.

119. — and I.O. Parasyuk, 'On Galerkin's procedure in the perturbation theory of invariant tori', *Dokl. Akad Nauk Ukrain. SSR, Ser. A*, 1977, No. 2, 112–115.

120. — and N.A. Perestyuk, 'On the existence of invariant toroidal sets of

systems with instantaneous change', In: *Integral manifold method in non-linear differential equations*, Inst. Mat. Akad. Nauk Ukrain. SSR, Kiev, 1973, 262–273.

121. — and I.V. Polesya, 'Birth of invariant manifolds in a neighbourhood of a position of equilibrium', *Differentsial'nye Uravneniya* **11** (1975), 1409–1416.

122. — and N.J. Ronto, *Numerical-analytic methods of studying periodic solutions*, Vishcha Shkola, Kiev, 1976 (in Russian).

123. — and V.I. Tkachenko, 'Decomposability of linear extensions of dynamical systems on a two- or three-dimensional torus', *Dokl. Akad Nauk Ukrain. SSR, Ser. A*, 1983, No. 2, 22–25.

124. —, V.A. Mal'kov and S.I. Trofimchuk, 'Splitting of a linear extension of a dynamical system on a torus', *Dokl. Akad Nauk Ukrain. SSR, Ser. A*, 1982, No. 11, 19–22.

125. —, D.I. Martynyuk and N.A. Perestyuk, 'Existence of invariant tori of systems of difference equations', *Differentsial'nye Uravneniya* **10** (1973), 1904–1910.

126. —, Yu.V. Teplinskii and N.S. Tsyganovskii, *On invariant tori of countable systems of differential equations*, Preprint Inst. Mat. Akad. Nauk Ukrain. SSR, Kiev, 1983.

127. S.L. Sobolev, *Some applications of functional analysis in mathematical physics*, Leningrad State Univ. Press, Leningrad, 1950 (in Russian).

128. —, *Equations of mathematical physics*, Nauka, Moscow, 1966 (in Russian).

129. S. Sternberg, *Lectures on differential geometry*, Prentice-Hall, Englewood Cliffs, N.J., 1964.

130. V.I. Tkachenko, 'On block diagonalization of almost periodic systems', *Dokl. Akad Nauk Ukrain. SSR, Ser. A*, 1983, No. 6, 18–20.

131. S.I. Trofimchuk, 'Necessary conditions for the existence of an invariant manifold of a linear extension of a dynamical system on a compact manifold', *Ukrain. Mat. Zh.* **36** (1984), 390–393.

132. D.U. Umbetzhanov, *Almost multiperiodic solutions of partial differential equations*, Nauka, Alma-Ata, 1979 (in Russian).

133. A.N. Filatov, *Asymptotic methods in the theory of differential and integro-differential equations*, FAN, Tashkent, 1974 (in Russian).

134. V.Kh. Kharasakhal, *Almost periodic solutions of ordinary differential equa-*

tions, Nauka, Alma-Ata, 1970 (in Russian).

135. F. Hartman, *Ordinary differential equations*, Wiley, New York, 1964.

136. T. Hayasi, *Forced oscillations in non-linear systems*, Inost. Lit., Moscow, 1967 (in Russian).

137. —, *Non-linear oscillations in physical systems*, Mir, Moscow, 1968 (in Russian).

138. J.K. Hale, *Oscillations in non-linear systems*, Mir, Moscow, 1966 (in Russian).

139. A.Ya. Khinchin, *Continuous fractions*, Gostekhizdat, Moscow-Leningrad, 1949 (in Russian).

140. A.Ya. Khokhryakov, 'On the periodic boundary-value problem for the differential equation $y^{(n)} + f(t, y, y', \ldots, y^{(n-1)}) = 0$, *Differentsial'nye Uravneniya* 3 (1967), 1643–1655.

141. L. Cesari, *Asymptotic behaviour and stability problems in ordinary differential equations*, Springer-Verlag, Berlin, 1963.

142. A. Besicovitch, *Almost periodic functions*, Cambridge Univ. Press, Cambridge, 1932.

143. H. Bohr, *Collected mathematical works. II. Almost periodic functions*, Copenhagen, 1952.

144. L. Cesari, 'Functional analysis and periodic solutions of non-linear differential equations', In: *Contributions to differential equations*, New York, 1963, pp. 148–187.

145. S.P. Diliberto, 'An application of periodic surfaces', In: *Contributions to the theory of non-linear oscillations*, Princeton Univ. Press, Princeton, N.J., 1956, pp. 257–261.

146. —, 'Perturbation theorems for periodic surfaces I, II, *Rend. Circolo Mat. Palermo* 9 (1960), 265–299; 10 (1961), 111–161.

147. —, 'Application of periodic surfaces to dynamical systems', In: *Proc. Fifth Internat. Conf. on Non-linear Oscillations* Vol. 1: Analytical methods of the theory of non-linear oscillations, Inst. Mat. Akad. Nauk Ukrain. SSR, Kiev, 1970, 257–264.

148. N. Fenichel, 'Persistence and smoothness of invariant manifolds for flows', *Indiana Univ. Math. J.* 21 (1971), 193–226.

149. K.O. Friedrichs, 'Symmetric positive linear differential equations', *Comm. Pure Appl. Math.* 11 (1958), 333–418.

150. —, 'Symmetric hyperbolic linear differential equations', *Comm. Pure Appl. Math.* **7** (1954), 345–392.

151. J.K. Hale, 'Integral manifolds of perturbed differential systems', *Ann. of Math.* **73** (1961), 496–531.

152. M.W. Hirsch, C.C. Pugh and M. Shub, *Invariant manifolds*, Lecture Notes in Math. **583**, Springer, Berlin, 1977.

153. J. Kyner, 'Invariant manifolds', *Rend. Circolo Mat. Palermo* **9** (1961), 98–110.

154. I. Kupka, 'Stabilité des variétés invariantes d'un champ de vecteurs pour les petites perturbations', *C.R. Acad. Sci., Paris*, **258** (1964), 4197–4200.

155. Lord Rayleigh, 'On maintained vibrations', *Philosophical magazine* **15** (1883), 229.

156. J. Moser, 'A new technique for the construction of solutions of non-linear differential equations', *Proc. Nat. Acad. Sci.* **47** (1961), 1824–2831.

157. —, 'On invariant curves of area-preserving mappings of an annulus', *Nachr. Akad. Wiss. Gottingen, Math.-Phys.* **11a** (1962), 1–20.

158. —, 'On invariant surfaces and almost periodic solutions for ordinary differential equations', *Notices Amer. Math. Soc.* **12** (1965), 124.

159. —, 'A rapidly convergent iteration method and non-linear differential equations', *Ann. Scuola Norm Sup. Pisa* **20**:3 (1966), 499–535.

160. —, 'Convergent series expansions for quasi-periodic motions', *Math. Ann.* **169** (1967), 136–176.

161. J. Nash, 'The imbedding problem for Riemannian manifolds', *Ann. of Math.* **63** (1956), 20–63.

162. L. Nirenberg, 'On elliptic partial differential equations', *Ann. Scuola Norm. Sup. Pisa* **13**:11 (1959), 115–162.

163. R.J. Sacker, ' A new approach to perturbation theory of invariant surfaces', *Comm. Pure Appl. Math.* **18** (1965), 717–732.

164. —, 'A perturbation theorem for invariant manifolds and Hölder continuity', *J. Math. Mech* **18** (1969), 705–761.

165. — and G.R. Sell, 'Existence of dichotomies and invariant splittings for linear differential systems', *J. Differential Eqations* **27** (1978), 106–137.

166. — and —, 'A spectral theory for linear differential systems', *J. Differential Eqations* **27** (1978), 320–358.

167. G.R. Sell, 'The Floquet problem for almost periodic linear differential equations', *Lecture Notes in Math.* **415**, Springer, Berlin, 1974, 239–251.

168. F.R. Sell, 'Bifurcation of higher dimensional tori', *Arch. Rational Mech. Anal.* **69** (1979), 199–230.

169. Y. Sibuja, 'Some global properties of matrices of functions of one variable', *Math. Ann.* **161** (1965), 67–77.

170. M. Urabe, 'Existence theorems of quasi-periodic solutions to non-linear differential systems', *Funkcialaj. Ekvacioj.* **15** (1972), 75–100.

171. —, 'On a modified Galerkin's procedure for non-linear quasi-periodic differential systems', In: *Équations différentielles et fonctionelles non-linéaires*, Paris, 1973, 223–258.

172. H. Weyl, 'Über die Gleichvereilung von Zahlen Mod Eins', *Math. Ann.* **77** (1916), 313–352.

Author Index

Index of Notation

Subject Index